Contaminants in the Water Environment

Contaminants in the Water Environment

Guest Editors

Abdul Qadeer
Xia Jiang
Kelly Kirsten

MDPI

Basel • Beijing • Wuhan • Barcelona • Belgrade • Novi Sad • Cluj • Manchester

Guest Editors

Abdul Qadeer
National Engineering
Laboratory for Lake Pollution
Control and Ecological
Restoration
Chinese Research Academy
of Environmental Science
Beijing
China

Xia Jiang
National Engineering
Laboratory for Lake Pollution
Control and Ecological
Restoration
Chinese Research Academy
of Environmental Science
Beijing
China

Kelly Kirsten
The School of the
Environment
Coventry University
Coventry
UK

Editorial Office
MDPI AG
Grosspeteranlage 5
4052 Basel, Switzerland

This is a reprint of the Special Issue, published open access by the journal *Water* (ISSN 2073-4441), freely accessible at: https://www.mdpi.com/journal/water/special_issues/4671N8F525.

For citation purposes, cite each article independently as indicated on the article page online and as indicated below:

Lastname, A.A.; Lastname, B.B. Article Title. *Journal Name* **Year**, *Volume Number*, Page Range.

ISBN 978-3-7258-4879-9 (Hbk)
ISBN 978-3-7258-4880-5 (PDF)
https://doi.org/10.3390/books978-3-7258-4880-5

Contents

About the Editors

Abdul Qadeer

Abdul Qadeer is currently working as an Associate Professor at the Chinese Research Academy of Environmental Sciences in Beijing, China. He obtained his PhD in Environmental Science from East China Normal University, Shanghai, and gained his postdoctoral experience at Indiana University, USA. His research focuses on environmental chemistry and the fate and transport of pollutants in aquatic, atmospheric, and terrestrial systems, with a particular emphasis on both legacy and emerging contaminants. He applies advanced screening techniques using GC-MS/MS and LC-MS/MS alongside modeling approaches to assess contaminant dynamics and ecological risks. Abdul Qadeer has published over 60 peer-reviewed articles in high-impact journals and actively serves as a Guest Editor and reviewer for several international journals. His scholarly contributions span analytical method development, environmental monitoring, and modeling-based decision support for pollution control.

Xia Jiang

Xia Jiang is a faculty member and professor at the Chinese Research Academy of Environmental Sciences, Beijing, and a leading expert in lake sciences, particularly focusing on the impacts of anthropogenic activities on lake health. She has authored several books and book chapters and has published over 100 peer-reviewed research articles.

Kelly Kirsten

Kelly Kirsten is a lecturer at Coventry University and aims to expand her research into geospatial modeling for environmental monitoring, aligning her work with global conservation and sustainability efforts. Dr. Kirsten is a paleoenvironmental scientist specializing in reconstructing past climates, hydrological changes, and ecosystem responses using diatoms as a proxy. Her work specifically focuses on sediment core analyses from coastal systems, lakes, and paleolakes across South Africa, significantly contributing to understanding Holocene climate fluctuations, the development of and natural changes in aquatic environments, and the influence of anthropogenic forces over various timescales. Dr. Kirsten has contributed to interdisciplinary research on human–environment interactions and collaborates internationally with institutions such as the University of York, Nanjing University, the University of Cape Town, and the University of KwaZulu-Natal. Her research aligns with international initiatives such as the UN Sustainable Development Goals and the PAGES Planetary Boundaries working group and aims to enhance interdisciplinary collaborations, bridging palaeoclimatology, sustainability science, and geospatial analytics to address contemporary environmental challenges.

Preface

The Reprint *Contaminants in the Water Environment* presents a curated selection of original research articles and reviews that collectively address the occurrence, behavior, and impacts of various environmental contaminants in aquatic systems. The primary aim of this Reprint is to provide a scientific platform that reflects the latest developments in detecting, modeling, and managing pollutants such as heavy metals, nutrients, emerging contaminants, microplastics, and pathogens in water environments.

The scope of this Reprint includes surface water, groundwater, wastewater, and integrated water systems, incorporating interdisciplinary approaches from environmental chemistry, hydrology, engineering, and public health. The motivation for compiling this publication stems from the urgent global need to understand contaminant dynamics in light of increasing anthropogenic pressures, climate variability, and the demand for sustainable water management practices.

This Reprint is intended for researchers, professionals, policymakers, and graduate students who are engaged in environmental science, water quality monitoring, pollution mitigation, and risk assessment. By bringing together studies from diverse geographical regions and methodological frameworks, we hope that this volume will support informed decision-making and inspire further research in safeguarding water resources worldwide.

Abdul Qadeer, Xia Jiang, and Kelly Kirsten
Guest Editors

water

MDPI

Editorial

Contaminants in the Water Environment: Significance from the Perspective of the Global Environment and Health

Xia Jiang [1], Kelly L. Kirsten [2,3] and Abdul Qadeer [1,4,5,*]

1 State Key Laboratory of Environmental Criteria and Risk Assessment, National Engineering Laboratory for Lake Pollution Control and Ecological Restoration, State Environmental Protection Key Laboratory for Lake Pollution Control, Chinese Research Academy of Environmental Science, Beijing 100012, China; jiangxia@craes.org.cn
2 The School of the Environment, Coventry University, Coventry CV1 2TU, UK
3 Human Evolution Research Institute, University of Cape Town, Cape Town 7700, South Africa
4 Institute of Soil & Environmental Sciences, University of Agriculture, Faisalabad 37000, Pakistan
5 O'Neil School of Public and Environmental Affairs, Indiana University, Bloomington, IN 47405, USA
* Correspondence: dr.aq.geographer@gmail.com or aqadeer@iu.edu

1. Introduction

Water is essential for all life, yet it faces increasing threats from contamination due to various human activities and natural processes. Pollutants enter the water environment through industrial discharge, agricultural runoff, household waste, and natural processes, leading to the widespread pollution of freshwater and marine ecosystems. These contaminants can be categorized as chemical (e.g., heavy metals, persistent organic pollutants (POPs), pharmaceuticals, and emerging contaminants such as per- and polyfluoroalkyl substances (PFASs)); biological (e.g., microbial pathogens); and physical (e.g., microplastics, sediments) [1–4]. The presence of these substances in water systems poses significant risks to both environmental and human health.

Water contamination is a global crisis with severe consequences for biodiversity, food security, and public health on an unprecedented scale. Contaminants disrupt aquatic ecosystems, bioaccumulate in the food chain, and contribute to severe health conditions, including cancer, endocrine disruption, and developmental disorders [5–9]. As water contamination transcends geographical boundaries through atmospheric deposition and hydrological cycles, addressing this issue requires a global perspective that integrates scientific research, policy implementation, and technological innovation.

This Special Issue on *Contaminants in the Water Environment* (https://www.mdpi.com/journal/water/special_issues/4671N8F525: accessed on 22 April 2025) presents a diverse range of studies examining the sources, distribution, and impacts of various pollutants in aquatic ecosystems. Research on nutrient dynamics highlights how human activities and climate factors influence the nitrogen and phosphorus levels in urban river basins. Heavy metal contamination is explored through an assessment of the iron pollution in Egypt's groundwater and an elemental analysis in Utah Lake using ICP-OES. Microplastic pollution is investigated in Central Europe's largest shallow lake, while industrial contributions to water pollution are evaluated in Indonesia's Cikembar sub-watershed. This Issue also addresses the impacts of climate and land use on water quality seen in Colorado's Fountain Creek watershed and Indonesia's Cirata Reservoir. Analytical advancements are featured in pathogen surveillance studies comparing methods for the detection of SARS-CoV-2 in hospital wastewater. Additionally, toxicological studies in this Issue assess the combined effects of glyphosate and copper on aquatic organisms and the presence of heavy metals in fish from the Yangtze River, raising concerns in terms of food safety. These contributions collectively provide valuable insights into the complex interactions of contaminants in the water environment, underscoring the urgent need for improved monitoring, regulation, and sustainable water management practices.

2. Sources and Types of Contaminants

Water contaminants can be categorized based on their sources and chemical nature. The major sources include the following:

(a) Industrial discharge: Factories release heavy metals (e.g., lead, mercury, cadmium), solvents, and POPs, contaminating nearby water bodies;
(b) Agricultural runoff: Fertilizers, pesticides, and herbicides introduce nitrates, phosphates, and toxic compounds into aquatic systems, leading to eutrophication and ecosystem imbalances;
(c) Urban wastewater: Household products, pharmaceuticals, and personal care products enter sewage systems, ultimately reaching natural water sources;
(d) Atmospheric deposition: Airborne contaminants, including polycyclic aromatic hydrocarbons (PAHs) and mercury, settle into water bodies through precipitation, illustrating the global environmental transport of pollutants;
(e) Natural sources: Environmental processes such as volcanic eruptions, wildfires, dust storms, the natural weathering of heavy metals from soils and rocks, and the decomposition of organic matter contribute to contamination.

3. The Global Impact of Water Contamination

Water contamination has far-reaching implications, potentially destabilizing global ecological balance and socioeconomic stability. Its key consequences include (i) bioaccumulation and biomagnification, where toxic substances accumulate in aquatic organisms and intensify through the food chain, endangering higher trophic levels [10,11]; (ii) eutrophication, where excessive nutrient runoff from agriculture promotes algal blooms, depleting oxygen levels and causing large-scale fish mortality [12,13]; and (iii) habitat degradation, where chemical pollutants reduce biodiversity by negatively impacting aquatic flora and fauna [14–17]. Additionally, water contamination poses substantial socioeconomic challenges. Polluted water reduces agricultural productivity, increases healthcare costs, and burdens governments and communities financially.

4. The Health Risks Associated with Water Contaminants

Waterborne contaminants, including both legacy and emerging pollutants, contribute to a wide range of health risks, ranging from acute poisoning to chronic diseases, often resulting in fatalities and significant economic losses. For instance, heavy metals such as lead, mercury, and arsenic can cause neurotoxicity, kidney damage, and developmental disorders [18–22]. Pesticides and herbicides can cause endocrine disruption, reproductive toxicity, and an increased risk of cancer [23–25]. Pharmaceuticals can contribute to antibiotic resistance and hormonal imbalances in both humans and aquatic organisms [26,27]. Microplastics and nanoparticles can be associated with inflammatory responses, oxidative stress, and potential carcinogenic effects [28–30].

Water contamination intensifies global water scarcity and poses a significant threat to food security by reducing the availability of clean water for drinking and irrigation, directly impacting agriculture and global food supply chains. Exposure to waterborne contaminants leads to severe health consequences, resulting in millions of deaths worldwide. According to the World Health Organization, in 2022, at least 1.7 billion people relied on unsafe or contaminated drinking water, with an estimated 1 million deaths annually due to diarrhea caused by unsafe water [31]. Similarly, other contaminants, such as microplastics and antimicrobial-resistant pathogens, also contribute to mortality on a global scale. Figure 1 below illustrates the global distribution of deaths associated with unsafe drinking water, highlighting the disparities in water quality worldwide. Furthermore, the economic burden associated with healthcare costs, water treatment, and ecosystem restoration places significant financial strain on governments and communities worldwide.

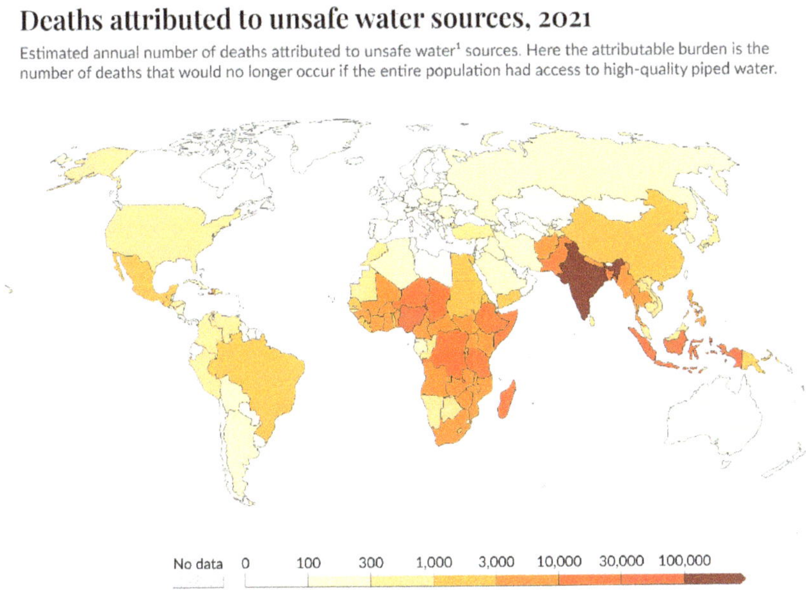

Figure 1. This figure illustrates the global distribution of deaths associated with unsafe drinking water sources, providing a broad indication of this critical issue across all continents. The map highlights regions where unsafe water poses significant health risks, emphasizing the disparities in water quality and access worldwide. The data, sourced from the Institute for Health Metrics and Evaluation (IHME) as part of the Global Burden of Disease study (2024), were reproduced under the Creative Commons license from *Our World in Data* (2024) [32]. This visual representation underscores the urgent need for improved water management policies and interventions to mitigate the adverse health impacts of contaminated drinking water.

5. Advances in Detection and Monitoring

The ability to detect and quantify contaminants is essential for assessing environmental and health risks while developing effective mitigation strategies. Recent advancements in analytical techniques have significantly improved detection capabilities. Gas Chromatography-Mass Spectrometry (GC-MS) and Liquid Chromatography-Mass Spectrometry (LC-MS) allow for the identification of trace contaminants, while high-resolution mass spectrometry, such as Liquid Chromatography-High-Resolution Mass Spectrometry (LC-HRMS) (e.g., Orbitrap and Quadrupole Time-of-Flight [QTOF] and Gas Chromatography-High-Resolution Mass Spectrometry (GC-HRMS), enhances suspect screening for a broader range of pollutants [33–36]. Additionally, Inductively Coupled Plasma Mass Spectrometry (ICP-MS) provides precise analyses of heavy metal concentrations [37,38]. Artificial intelligence and machine learning can contribute further by enabling predictive modeling of contaminant transport and risk assessments, as well as automating the detection and classification of pollutants through big data approaches [39–42]. Meanwhile, biosensors and smart technologies can revolutionize real-time water quality monitoring, with portable sensors allowing for rapid field-based assessments [43–45]. These innovations collectively enhance the accuracy, efficiency, and timeliness of responses to water contamination, strengthening environmental and public health protection.

6. Conclusions and Call to Action

Water contamination is an escalating global challenge with profound environmental, economic, and health implications. While scientific advancements have improved contaminant detection and regulation, significant gaps remain, particularly as concerns emerging contaminants in the water bodies, policy enforcement, and public awareness. Addressing these challenges requires a collaborative approach involving researchers, policymakers, industries, and communities. Strengthening the global regulations, promoting sustainable water management, and leveraging AI-driven solutions can pave the way for a cleaner and healthier water environment.

To protect future generations, it is important to prioritize research on water quality, advocate for stricter environmental policies, and foster a culture of responsible water use. Taking decisive action today can ensure access to clean and safe water for all. We hope that the articles published in this Special Issue will engage readers and contribute to our scientific understanding of water contamination significantly, advancing global scientific discourse further.

Author Contributions: A.Q. and X.J.: Conceptualization; investigation; writing, original draft preparation. K.L.K.: Review and editing; investigation. All authors have read and agreed to the published version of the manuscript.

Acknowledgments: We are grateful to CRAES for providing the resources and the workplace necessary to complete this task.

Conflicts of Interest: The authors declare no conflict of interest.

References

1. Kim, S.; Kim, Y.; Moon, H.B. Contamination and historical trends of legacy and emerging plasticizers in sediment from highly industrialized bays of Korea. *Sci. Total Environ.* **2021**, *765*, 142751. [CrossRef]
2. Shen, A.; Lee, S.; Ra, K.; Suk, D.; Moon, H.B. Historical trends of perfluoroalkyl substances (PFASs) in dated sediments from semi-enclosed bays of Korea. *Mar. Pollut. Bull.* **2018**, *128*, 287–294. [CrossRef] [PubMed]
3. Zhao, L.; Hu, G.; Yan, Y.; Yu, R.; Cui, J.; Wang, X.; Yan, Y. Source apportionment of heavy metals in urban road dust in a continental city of eastern China: Using Pb and Sr isotopes combined with multivariate statistical analysis. *Atmos. Environ.* **2019**, *201*, 201–211. [CrossRef]
4. Liu, M.; Qadeer, A.; Anis, M.; Ajmal, Z.; Bekele, T.G.; Wang, S.; Jiang, X.; Leung, K.M.Y. Occurrence of per- and polyfluoroalkyl substances (PFAS): A special reference to their monitoring, distribution, and environmental fate. In *Emerging Aquatic Contaminants, One Health Framework for Risk Assessment and Remediation in the Post COVID-19 Anthropocene*; Elsevier: Amsterdam, The Netherlands, 2023; pp. 173–202. [CrossRef]
5. Bekele, T.G.; Zhao, H.; Wang, Q.; Chen, J. Bioaccumulation and Trophic Transfer of Emerging Organophosphate Flame Retardants in the Marine Food Webs of Laizhou Bay, North China. *Environ. Sci. Technol.* **2019**, *53*, 13417–13426. [CrossRef] [PubMed]
6. Qadeer, A.; Kirsten, K.L.; Ajmal, Z.; Jiang, X.; Zhao, X. Alternative Plasticizers As Emerging Global Environmental and Health Threat: Another Regrettable Substitution? *Environ. Sci. Technol.* **2022**, *56*, 1482–1488. [CrossRef]
7. Qadeer, A.; Mubeen, S.; Liu, M.; Bekele, T.G.; Ohoro, C.R.; Adeniji, A.O.; Alraih, A.M.; Ajmal, Z.; Alshammari, A.S.; Al-Hadeethi, Y.; et al. Global environmental and toxicological impacts of polybrominated diphenyl ethers versus organophosphate esters: A comparative analysis and regrettable substitution dilemma. *J. Hazard. Mater.* **2024**, *466*, 133543. [CrossRef]
8. Yang, J.; Qadeer, A.; Liu, M.; Zhu, J.; Huang, Y.; Du, W.; Wei, X. Occurrence, source, and partition of PAHs, PCBs, and OCPs in the multiphase system of an urban lake, Shanghai. *Appl. Geochem.* **2019**, *106*, 17–25. [CrossRef]
9. Jiang, Q.; Hanari, N.; Miyake, Y.; Okazawa, T.; Lau, R.K.F.; Chen, K.; Wyrzykowska, B.; Ka, M.; Yamashita, N.; Lam, P.K.S. Health risk assessment for polychlorinated biphenyls, polychlorinated dibenzo-p-dioxins and dibenzofurans, and polychlorinated naphthalenes in seafood from Guangzhou and Zhoushan, China. *Environ. Pollut.* **2007**, *148*, 31–39. [CrossRef]

10. Qadeer, A.; Liu, M.; Yang, J.; Liu, X.; Khan, S.; Huang, Y.; Mamun, H.A.; Gao, D.; Yang, Y. Trophodynamics and parabolic behaviors of polycyclic aromatic hydrocarbons in an urbanized lake food web, Shanghai. *Ecotoxicol. Environ. Saf.* **2019**, *178*, 17–24. [CrossRef]
11. Zhang, Y.; Luo, X.J.; Wu, J.P.; Liu, J.; Wang, J.; Chen, S.J.; Mai, B.X. Contaminant pattern and bioaccumulation of legacy and emerging organhalogen pollutants in the aquatic biota from an e-waste recycling region in South China. *Environ. Toxicol. Chem.* **2010**, *29*, 852–859. [CrossRef]
12. Cheng, X.; Li, X. 20-year variations of nutrients (N and P) and their impacts on algal growth in Lake Dianshan, China. *J. Lake Sci.* **2008**, *20*, 409–419.
13. Le, C.; Zha, Y.; Li, Y.; Sun, D.; Lu, H.; Yin, B. Eutrophication of Lake Waters in China: Cost, Causes, and Control. *Environ. Manag.* **2010**, *45*, 662–668. [CrossRef]
14. Dudgeon, D.; Arthington, A.H.; Gessner, M.O.; Kawabata, Z.I.; Knowler, D.J.; Lévêque, C.; Naiman, R.J.; Prieur-Richard, A.H.; Soto, D.; Stiassny, M.L.J.; et al. Freshwater biodiversity: Importance; threats, status and conservation challenges. *Biol. Rev.* **2006**, *81*, 163–182. [CrossRef]
15. Hu, Z.; Wang, S.; Wu, H.; Chen, Q.; Ruan, R.; Chen, L.; Liu, Q. Temporal and spatial variation of fish assemblages in Dianshan Lake, Shanghai, China. *Chin. J. Oceanol. Limnol.* **2014**, *32*, 799–809. [CrossRef]
16. Baroudi, F.; Al Alam, J.; Fajloun, Z.; Millet, M. Snail as sentinel organism for monitoring the environmental pollution; a review. *Ecol. Indic.* **2020**, *113*, 106240. [CrossRef]
17. Muqi, X.; Jiang, Z.; Yuyao, H.; Yurong, G.; Shen, Z.; Yijian, T.; Chengqing, Y.; Zijian, W. The ecological degradation and restoration of baiyangdian lake, China. *J. Freshw. Ecol.* **1998**, *13*, 433–446. [CrossRef]
18. Qadeer, A.; Saqib, Z.A.; Ajmal, Z.; Xing, C.; Khalil, S.K.; Usman, M.; Huang, Y.; Bashir, S.; Ahmad, Z.; Ahmed, S.; et al. Concentrations, pollution indices and health risk assessment of heavy metals in road dust from two urbanized cities of Pakistan: Comparing two sampling methods for heavy metals concentration. *Sustain. Cities Soc.* **2020**, *53*, 101959. [CrossRef]
19. Lim, H.S.; Lee, J.S.; Chon, H.T.; Sager, M. Heavy metal contamination and health risk assessment in the vicinity of the abandoned Songcheon Au-Ag mine in Korea. *J. Geochem. Explor.* **2008**, *96*, 223–230. [CrossRef]
20. Wei, X.; Gao, B.; Wang, P.; Zhou, H.; Lu, J. Pollution characteristics and health risk assessment of heavy metals in street dusts from different functional areas in Beijing, China. *Ecotoxicol. Environ. Saf.* **2015**, *112*, 186–192. [CrossRef]
21. Chen, X.; Liu, M.; Ma, J.; Liu, X.; Liu, D.; Chen, Y.; Li, Y.; Qadeer, A. Health risk assessment of soil heavy metals in housing units built on brownfields in a city in China. *J. Soils Sediments* **2017**, *17*, 1741–1750. [CrossRef]
22. Jan, A.T.; Azam, M.; Siddiqui, K.; Ali, A.; Choi, I.; Haq, Q.M.R. Heavy metals and human health: Mechanistic insight into toxicity and counter defense system of antioxidants. *Int. J. Mol. Sci.* **2015**, *16*, 29592–29630. [CrossRef]
23. Hu, G.; Dai, J.; Mai, B.; Luo, X.; Cao, H.; Wang, J.; Li, F.; Xu, M. Concentrations and Accumulation Features of Organochlorine Pesticides in the Baiyangdian Lake Freshwater Food Web of North China. *Arch. Environ. Contam. Toxicol.* **2010**, *58*, 700–710. [CrossRef]
24. Zawiyah, S.; Man, Y.B.C.; Nazimah, S.A.H.; Chin, C.K.; Tsukamoto, I.; Hamanyza, A.H.; Norhaizan, I. Determination of organochlorine and pyrethroid pesticides in fruit and vegetables using SAX/PSA clean-up column. *Food Chem.* **2007**, *102*, 98–103. [CrossRef]
25. Gilden, R.C.; Huffling, K.; Sattler, B. Pesticides and Health Risks. *J. Obstet. Gynecol. Neonatal Nurs.* **2010**, *39*, 103–110. [CrossRef] [PubMed]
26. Ji, X.; Shen, Q.; Liu, F.; Ma, J.; Xu, G.; Wang, Y.; Wu, M. Antibiotic resistance gene abundances associated with antibiotics and heavy metals in animal manures and agricultural soils adjacent to feedlots in Shanghai; China. *J. Hazard. Mater.* **2012**, *235–236*, 178–185. [CrossRef] [PubMed]
27. Kumar, A.; Chang, B.; Xagoraraki, I. Human Health Risk Assessment of Pharmaceuticals in Water: Issues and Challenges Ahead. *Int. J. Environ. Res. Public Health* **2010**, *7*, 3929–3953. [CrossRef]
28. Qadeer, A.; Ajmal, Z.; Usman, M.; Zhao, X.; Chang, S. Agricultural plastic mulching as a potential key source of microplastic pollution in the terrestrial ecosystem and consequences. *Resour. Conserv. Recycl.* **2021**, *175*, 105855. [CrossRef]
29. Sun, A.; Wang, W.X. Human Exposure to Microplastics and Its Associated Health Risks. *Environ. Health* **2023**, *1*, 139–149. [CrossRef] [PubMed]
30. Vethaak, A.D.; Legler, J. Microplastics and human health: Knowledge gaps should be addressed to ascertain the health risks of microplastics. *Science* **2021**, *371*, 672–674. [CrossRef]
31. WHO. Drinking-Water. 2023. Available online: https://www.who.int/news-room/fact-sheets/detail/drinking-water (accessed on 27 March 2025).
32. Deaths Attributed to Unsafe Water Sources, 2021, (n.d.). Available online: https://ourworldindata.org/grapher/deaths-due-to-unsafe-water-sources?time=latest (accessed on 27 March 2025).
33. Fang, J.; Zhao, H.; Zhang, Y.; Lu, M.; Cai, Z. Atmospheric pressure chemical ionization in gas chromatography-mass spectrometry for the analysis of persistent organic pollutants. *Trends Environ. Anal. Chem.* **2020**, *25*, e00076. [CrossRef]

34. da Costa Filho, B.M.; Duarte, A.C.; Santos, T.A.P.R. Environmental monitoring approaches for the detection of organic contaminants in marine environments: A critical review. *Trends Environ. Anal. Chem.* **2022**, *33*, e00154. [CrossRef]
35. Guo, W.; Pan, B.; Sakkiah, S.; Yavas, G.; Ge, W.; Zou, W.; Tong, W.; Hong, H. Persistent Organic Pollutants in Food: Contamination Sources, Health Effects and Detection Methods. *Int. J. Environ. Res. Public Health* **2019**, *16*, 4361. [CrossRef] [PubMed]
36. Caroleo, F.; Magna, G.; Naitana, M.L.; Di Zazzo, L.; Martini, R.; Pizzoli, F.; Muduganti, M.; Lvova, L.; Mandoj, F.; Nardis, S.; et al. Advances in Optical Sensors for Persistent Organic Pollutant Environmental Monitoring. *Sensors* **2022**, *22*, 2649. [CrossRef] [PubMed]
37. Van Acker, T.; Theiner, S.; Bolea-Fernandez, E.; Vanhaecke, F.; Koellensperger, G. Inductively coupled plasma mass spectrometry. *Nat. Rev. Methods Prim.* **2023**, *3*, 52. [CrossRef]
38. Malik, L.A.; Bashir, A.; Qureashi, A.; Pandith, A.H. Detection and removal of heavy metal ions: A review. *Environ. Chem. Lett.* **2019**, *17*, 1495–1521. [CrossRef]
39. Ibn-Mohammed, T.; Mustapha, K.B.; Abdulkareem, M.; Fuensanta, A.U.; Pecunia, V.; Dancer, C.E.J. Toward artificial intelligence and machine learning-enabled frameworks for improved predictions of lifecycle environmental impacts of functional materials and devices. *MRS Commun.* **2023**, *13*, 795–811. [CrossRef]
40. Brown, N.; Ertl, P.; Lewis, R.; Luksch, T.; Reker, D.; Schneider, N. Artificial intelligence in chemistry and drug design. *J. Comput. Aided Mol. Des.* **2020**, *34*, 709–715. [CrossRef]
41. Basile, A.O.; Yahi, A.; Tatonetti, N.P. Artificial Intelligence for Drug Toxicity and Safety. *Trends Pharmacol. Sci.* **2019**, *40*, 624–635. [CrossRef]
42. Jeong, J.; Choi, J. Artificial Intelligence-Based Toxicity Prediction of Environmental Chemicals: Future Directions for Chemical Management Applications. *Environ. Sci. Technol.* **2022**, *56*, 7532–7543. [CrossRef]
43. Hui, Y.; Huang, Z.; Alahi, M.E.E.; Nag, A.; Feng, S.; Mukhopadhyay, S.C. Recent Advancements in Electrochemical Biosensors for Monitoring the Water Quality. *Biosensors* **2022**, *12*, 551. [CrossRef]
44. Yaroshenko, I.; Kirsanov, D.; Marjanovic, M.; Lieberzeit, P.A.; Korostynska, O.; Mason, A.; Frau, I.; Legin, A. Real-Time Water Quality Monitoring with Chemical Sensors. *Sensors* **2020**, *20*, 3432. [CrossRef] [PubMed]
45. Krishnan, S.; Giwa, A. Advances in real-time water quality monitoring using triboelectric nanosensors. *J. Mater. Chem. A* **2025**, *13*, 11134–11158. [CrossRef]

water

MDPI

Article

Regulated Inductively Coupled Plasma–Optical Emission Spectrometry Detectible Elements in Utah Lake: Characterization and Discussion

Rachel A. Valek [1], Kaylee B. Tanner [1], Jacob B. Taggart [1], Rebecca L. Ryan [1], Anna C. Cardall [2], Lauren M. Woodland [1], Maddeline J. Oxborrow [1], Gustavious P. Williams [1,*], A. Woodruff Miller [1] and Robert B. Sowby [1]

1 Department of Civil and Construction Engineering, Brigham Young University, Provo, UT 84602, USA; ravalek8756@gmail.com (R.A.V.); kbt36@byu.edu (K.B.T.); rebecca.l.butters@gmail.com (R.L.R.); lolo25@byu.edu (L.M.W.); mt678@byu.edu (M.J.O.); wood_miller@byu.edu (A.W.M.); rsowby@byu.edu (R.B.S.)

2 Department of Chemical Engineering, Brigham Young University, Provo, UT 84602, USA; cardalla@byu.edu

* Correspondence: gus.williams@byu.edu; Tel.: +1-(801)-422-7810

Abstract: During the 2021 (n = 15) and 2022 (n = 13) summers, we measured the total and dissolved (<0.45 μm) concentration of 25 elements in Utah Lake using Inductively Coupled Plasma–Optical Emission Spectrometry (ICP-OES) with detection limits in the order of a few parts-per-billion (ppb). This resulted in 1400 measurements, which is a unique dataset in terms of sensitivity and temporal resolution. Regulated elements are not commonly measured at the ppb-level; thus, these data provide insight into both the behavior and existence of these elements in an aquatic environment and have implications for both the management and regulation of the lake. Utah regulates twelve of these elements. While ICP-OES has ppb-level sensitivity, it is not the approved regulatory analysis method for these elements. All regulations are for dissolved concentrations, except aluminum (Al) and phosphorus (P), which are for total recovery. We found total Al above the allowable concentration, but dissolved concentrations were well below allowable concentrations. We attribute high total concentrations to suspended clays. This suggests that regulatory methods should be reviewed for lakes with a high suspended-solid content. Dissolved copper (Cu) concentrations were below regulatory levels in 2021, but some samples were above regulatory levels in 2022. This could be related to the use of Cu-based algaecide treatments, or from other sources. Lead (Pb) data were inconclusive; dissolved Pb concentrations were well below the acute (1 h average) limit, but the chronic concentration limit (4 h average) was below the ICP-OES minimal detection limit. Arsenic (As) concentrations exhibited a seasonal trend that we attribute to groundwater inflows—they were below regulatory levels for aquatic environments but around the levels for drinking water. This ppb-level study with high temporal resolution provides insight into regulated elements in Utah Lake previously not available due to the high sensitivity of the method and measurements of both total and dissolved concentrations.

Keywords: Utah Lake; inductively coupled plasma–optical emission spectrometry; water quality; micronutrient; lake management; trace elements; state regulations

1. Introduction

1.1. Study Overview

Utah Lake (UL) is approximately 40 km by 21 km, with a surface area of 390 km^2 at maximum fill, and is in north-central Utah in the United States. UL waters are characteristically turbid and unstratified because of wave action and bioturbation and are near the solubility limit of calcite, with calcite precipitation commonly occurring [1].

There is a significant body of published research on UL. Early published UL research from 1931 [2] addresses algae, with additional work on phytoplankton communities published from the 1960s to the present [3–12]. Numerous studies on nutrients have been published [13–20], with studies on the local geology that contains formations with elevated phosphorus (P) [20–22]. Early remote-sensing studies from the 1970s used UL in some of the first papers from NASA's Landsat satellite [23,24]. Other early remote-sensing studies from the 1980s demonstrated the heat capacity mapping mission (HCMM) [25] and AVHRR sensor [26] on UL. Recent studies using remote-sensing data address long-term trends and variability in algal blooms [27–37].

With the exception of P, there has been little published work on regulated elements in UL water [38]. A 2014 study focused on compliance with EPA standards and identifying pollution sources for arsenic (As) and other heavy metals in Utah Lake and its tributaries [39]. Despite the lack of studies on regulated elements in UL, their concentrations are a concern as excessive or deficient amounts of these elements can have damaging effects on aquatic life [40–42] and affect biogeochemical processes [43]. The impacts of trace elements in aquatic systems have been studied worldwide [44,45], demonstrating that in most ecosystems, nutrients are the limiting factor in algal growth, though sometimes other trace elements limit growth. Bayer, et al. [46] studied Lake Hayes in New Zealand in regard to the importance of macro- and micronutrients. They reported on nitrogen (N), P, copper (Cu), iron (Fe), boron (B), molybdenum (Mo), zinc (Zn), and silicon (Si) and found that the limiting factors for phytoplankton growth in Lake Hayes were N, Zn, and B. P. Dengg, et al. [47] evaluated the trace elements of manganese (Mn), cobalt (Co), Fe, Zn, and Mo in the waters of the Taupo Volcanic Zone (TVZ) in New Zealand and found that Fe was a colimiting micronutrient for cyanobacteria. Studies in southeastern Australia showed that trace micronutrients are "an important regulator of the severity of cyanobacterial blooms" [48,49]. This research shows that understanding trace micronutrients is required to effectively manage algal blooms, as these trace elements may control ecological processes. Our study starts to provide these data for UL.

We provide data on regulated elements that can be detected using Inductively Coupled Plasma–Optical Emission Spectrometry (ICP-OES). We selected ICP-OES for this study for two reasons: (1) it measures 25 elements per sample, providing a comprehensive dataset, and (2) it has very low parts-per-billion (ppb) detection limits for these elements. These regulated detectible elements (DEs) include heavy metal and nonmetal elements in UL. Over two summers, we measured both dissolved and suspended concentrations of DEs in UL approximately weekly, with 15 samples in 2021 and 13 in 2022. We analyzed both total and dissolved concentrations of 25 elements for each sample. This resulted in 1400 data points over the critical warm growing season.

For this manuscript, we evaluated the 12 regulated elements that are detected by ICP-OES. For five of these elements, which are well under regulatory levels of concern, we provide only a minimal discussion. We provide a more detailed discussion of the seven elements more likely to be of concern, either from toxic excess or deficiency. We did not follow regulatory methods for the elements presented in this analysis as we used a standard ICP-OES method—ICP-OES is not generally stated for most standard methods as it is complex and expensive, and for most elements, the low detection levels are not required. Also, ICP-OES measures elemental concentrations, while many regulations are based on ionic or molecular forms. Our analysis of this large range of elemental concentrations (25) at the very low ppb detection level for both total and dissolved concentrations provides insight into UL conditions and potential elements of concern, at both toxic excess and toxic deficiency levels. It characterizes if the elements are in the dissolved state, and thus more biologically available, or associated with suspended solids. This analysis is not meant to determine if UL waters are out of compliance with Utah regulatory standards, but rather to provide important information about the condition of the lake and facilitate future research.

1.2. ICP-OES Detectable Elements in Aquatic Systems

Trace metals and other elements are natural and present in all ecosystems, and water concentrations are influenced by geology, land use, erosion, ecology, and geochemistry. Ecological processes also affect concentrations as aquatic biota ingest and absorb metals. These organisms are affected in both beneficial and adverse ways by the concentrations of these elements. These trace elements typically have no visible indicators in surface waters, although at toxic or deficient levels, they can have visual impacts on the ecology [50]. The toxicity of trace elements is a particular concern for UL as it hosts a large migratory bird population along with the endemic and threatened June Sucker fish species. All metals are toxic at certain thresholds [51] and impair the survival, reproduction, and behavior of aquatic life.

Some common toxic DEs in lakes and reservoirs, including metals, found at concentrations that can cause detrimental impacts include arsenic (As), cadmium (Cd), chromium (Cr), lead (Pb), nickel (Ni), selenium (Se), copper (Cu), and zinc (Zn) [50]. Even at non-toxic concentrations, As, Cd, and Pb, can affect the mobility, feeding, and navigation behaviors of invertebrates [52–54]. These effects cascade through aquatic ecosystems [55,56]. In larger organisms such as fish, Pb, Cd, Ni, and Cr can impact growth rates, biological processes, and reproductive health [57,58]. The impacts of aluminum (Al), Cd, Cu, Zn, Pb, Ni, As, and Se on birds include behavioral impairments and reduced reproductive success [59–61].

Some metals are toxic at excess levels and also harmful when deficient. Zn, Cu, and Ni negatively impact biological activity if they are present in excess levels as well as if they are missing or below toxic deficit concentrations [40–42,55,56,62]. Some elements, such as boron (B), iron (Fe), manganese (Mn), and molybdenum (Mo), are beneficial at certain concentrations and act as micronutrients that are essential for aquatic life. Micronutrients have been found to be limiting or colimiting factors for algal growth [46,63–66]. In 2008, Downs, et al. [42] conducted a review of micronutrients in 56 freshwater lakes and found that "the proportion of the lakes analyzed in which micronutrient limitation was found was 76% for molybdenum; 74% for iron; 67% for boron, 67% for cobalt, and 20% for copper." Similar to the effects of toxicity, nutrient deficiencies for algae and phytoplankton affect the entire biota of lake systems, including invertebrates, migratory birds, fish, and other organisms [55,56].

The US Environmental Protection Agency (EPA) states that sources and activities "such as mines, smelters, firing ranges, municipal wastewater treatment outfalls, industrial point sources, urban runoff, landfills, and junkyards are potential sources for heavy metals and other harmful elements that can cause impairment" [50]. Other sources include the combustion of fossil fuels, phosphate fertilizers, metallo-pesticides, and road salts [51,67]. All of these sources are currently or have historically been present within the UL watershed. The Geneva Steel plant operated from 1944 to 2002 on the northeastern shore of UL. This plant produced steel using "coal-derived coke," with effluent drains near the lake that served as a substantial source of metals to UL [38,68,69]. Anthropogenic nutrient sources include seven wastewater treatment plants (WWTPs) that discharge into the lake, while the southern end of UL is surrounded by agricultural land, and the western side has active gravel mines. The population in the last 40 years has nearly tripled, increasing from ~220,000 in 1980 to ~640,000 in 2020 [70].

In addition to the potentially toxic DEs, ICP-OES measures several other elements that have important functions or impacts on aquatic ecosystems, including barium (Ba), calcium (Ca), potassium (K), sodium (Na), sulfur (S), B, silicon (Si) and P.

B and S are naturally present in aquatic environments and serve as essential nutrients. However, human activities can lead to elevated concentrations, potentially reaching toxic levels [71–73]. Toxic S concentrations are typically rare in surface waters unless there is a direct source of pollution. B is also influenced by human activities, and the range between toxic deficit and excess B concentrations is relatively small compared to other nutrients [74].

Ba occurs naturally and is also released by industrial processes primarily in the form of airborne particles that can eventually settle in nearby surface waters. UL is particularly susceptible to capturing atmospherically deposited dust due to its location and large surface area [16,17,19,20]. Ba usually exists in a precipitated form in aqueous systems, with alkaline environments, such as UL, further limiting its solubility [75].

K and P are essential nutrients for aquatic primary production. Since plants require less K than P, and K is not usually a limiting nutrient. Neither element becomes toxic at the concentrations typically found in surface waters, although high concentrations of P can lead to eutrophication, which negatively impacts water quality. Natural P and K [76] sources include geologic formations after weathering. In the UL watershed, the Delle Phosphatic and Meade Peak geologic formations contribute large amounts of P to the lake [1,18].

Na is a key component of salinity and can severely degrade freshwater systems if present in high concentrations. It enters surface waters through natural weathering, but also from road salts [77].

The impacts of Se are not well understood in freshwater systems due to the complexity of its geochemical interactions and cycling. However, it is toxic at very low concentrations and can degrade aquatic systems when it enters surface waters through anthropogenic activity [78]. While Se is necessary for aquatic biota, there is a very narrow concentration window in which Se turns from beneficial to toxic [79].

These discussions of the origins and impacts of DEs on the UL ecosystem are not comprehensive. We included this review to demonstrate the potential impacts and functions of these DEs in the UL ecosystem and illustrate the importance of understanding their ambient concentrations and behaviors.

1.3. Regulated DE Water Quality Standards Applicable to UL

We analyzed 25 elements detectable by ICP-OES, 12 of which are regulated by water quality standards promulgated by the Division of Water Quality of the Utah Department of Environmental Quality in compliance with the Clean Water Act [80–82]. To contextualize our discussion of water column concentrations, we outline the relevant standards.

The publication "Standards of Quality for Waters of the State" [80–82] categorizes waterbodies according to their use and significance and prescribes different standards for each category. UL is classified as follows:

- Class 2A: "Protected for frequent primary contact recreation where there is a high likelihood of ingestion of water or a high degree of bodily contact with the water" (swimming, kayaking, diving, water skiing, etc.);
- Class 2B: "Protected for infrequent primary contact recreation. Also protected for secondary contact recreation where there is a low likelihood of ingestion of water or a low degree of bodily contact with water" (e.g., boating, wading, etc.) [82];
- Class 3B: "Protected for warm-water species of game fish and other warm water aquatic life, including the necessary aquatic organisms in their food chain";
- Class 3D: "Protected for waterfowl, shore birds, and other water-oriented wildlife not included in Classes 3A, 3B, or 3C, including the necessary aquatic organisms in their food chain";
- Class 4: "Protected for agricultural uses including irrigation of crops and stock watering" [80–82].

For the Class 3B and 3D standards, the regulations are for dissolved elements in units of μg/L. For Class 4 standards, regulations are for dissolved elements in units of mg/L. We converted the Class 3B and 3D standards to units of mg/L to match our data units. Table 1 lists the most stringent acute (1 h average) and chronic (4-day average) standards for the 12 regulated DE analytes [80].

The standards for Al, Cd, Cr, Cu, Ni, Pb, and Zn depend on water hardness. Typically, the adjustment factor for these "hardness dependencies" would be determined using the hardness level detected in each sample; however, our samples were not tested for hardness, so we used a representative average hardness level for the lake. Using data from the Utah Ambient Water Quality Management System (AWQMS) collected from 1978 to 2015, we found the mean hardness, based on $CaCO_3$, of UL water to be 197.4 mg/L, with a range from 123 to 291 mg/L. We used the mean value to calculate a reference concentration for elements with hardness dependencies for discussions. For Cd, Cr, Pb, and Zn, the conversion is a natural log equation based on the hardness of the water with a specified conversion factor [80]. Cr standards are for hexavalent chromium—a toxic ion. We measured elemental Cr, which provides an upper bound, but does not measure hexavalent Cr. Again, we want to stress that our analytical methods do not follow required regulatory methods. Instead, we determined elemental concentrations using ICP-OES, which has sensitivity in the few parts-per-billion range, depending on the element, and uses either filtration for dissolved concentrations or digestion for total concentration.

Table 1. Most stringent applicable standard for each regulated element in UL.

DE	Acute Standard (mg/L)	Chronic Standard (mg/L)	Designated Use
Aluminum (Al) *	0.75	0.75	3B, 3D
Arsenic (As) **	0.10	0.10	4
Boron (B) **	0.75	0.75	4
Cadmium (Cd)	0.0018	0.00072	3B, 3D
Chromium (Cr) (Hexavalent) ***	0.016	0.011	3B, 3D
Copper (Cu)	0.013	0.009	3B, 3D
Iron (Fe) **	1	1	3B, 3D
Nickel (Ni)	0.468	0.052	3B,3D
Phosphorus (P) *	0.025	0.025	3B
Lead (Pb)	0.065	0.0025	3B, 3D
Selenium (Se)	0.0184	0.0046	3B, 3D
Zinc (Zn)	0.12	0.12	3B, 3D

Notes: * Concentration based on total recovery criteria. ** Measured as maximum not acute and chronic. *** We measured elemental Cr, not hexavalent.

The standards for Al depend on pH and hardness. Since the pH of UL water is nearly always greater than 7.0 and the hardness, based on $CaCO_3$, is greater than 50 mg/L, we use the acute Al criterion of 750 µg/L (expressed as total recoverable Al) for our discussion. The Utah Department of Water Quality (UDWQ) is currently in the process of adopting the EPA's recommended Al criteria, which depends on dissolved organic carbon (DOC), hardness, and pH. We were told in conversations with representatives of the UDWQ that the current value of 750 µg/L is acceptable as a reference point. Al and P are the only regulated elements with a standard for total recoverable concentration—all other elements require the analysis of dissolved concentrations. As discussed below, we performed a complete digestion of unfiltered water samples for our total concentration measurements.

Total elemental Al concentrations can be misleading, as clay minerals, which make up a significant fraction of suspended solids, have high Al concentrations and are captured by a total digestion approach. However, the Al found in clay minerals is not bioavailable or reactive and is of less concern than dissolved Al. For this study, we analyzed filtered and unfiltered samples to show the impact of suspended solids on elemental concentrations. We performed a complete digestion of the unfiltered samples, which includes the suspended clay minerals in the sample. The filtered samples only measured dissolved elements. For Al concentrations, we did not attempt to follow the UDWQ analysis methods but assumed that "total recoverable Al" required unfiltered lake water to be analyzed after aggressive acid digestion, which would include the nonreactive Al bound up in clay particles. The regulatory methods under consideration use less-aggressive digestion.

1.4. Study Goals and Objectives

This study presents research based on a unique high-sensitivity (ppb-level) weekly dataset collected over two years, measuring both dissolved and total concentrations of 25 elements. This dataset includes 1400 measurements that characterize the weekly variation and temporal behavior of these elements over the study period. These data and our analysis provide insights into lake geochemistry and potential anthropogenic impacts that are not found in the literature, for either Utah Lake or other water bodies, because of the number of elements analyzed for both total and dissolved concentrations, the high sensitivity of the analytical method, and the temporal resolution of the dataset—all of these factors are unique in studies of aquatic environments (to our knowledge). Regulated elements are not commonly measured with detection limits in the order of a few ppb, except for lead (Pb).

This dataset helps characterize both the behavior and presence of these elements in an aquatic environment. For example, the analysis of what percentage of the 1400 samples were above detection limits for the various elements could only be conducted on this one-of-a-kind dataset and helps in understanding both the variability and behaviors of these 25 elements in both total and dissolved phases. These measurements start to characterize how these elements are present in the water column—whether they are dissolved and readily bioavailable or associated with suspended solids. We provide an in-depth analysis of the regulated elements that have concentrations near regulatory limits or have other interesting behaviors. The goal is not to evaluate regulatory compliance, as our methods are not the approved analysis methods, but to provide data for managers and regulators that can be used to design sampling schemes, evaluate management practices, and review regulatory analysis methods. We show interesting seasonal variations along with differences between total and dissolved concentrations that can help inform decisions.

Our goal in this manuscript is to present and describe a unique, ppb-level dataset collected weekly in Utah Lake. These data, consisting of 1400 measurements, provide managers and regulators with information that can be used to develop management strategies that encourage beneficial use and mitigate potential harmful conditions. Our research goals present our characterization of this dataset to include insights such as which elements are commonly present in Utah Lake, in what phase, and their seasonal variation. For regulated elements, we compare our data to regulatory limits and discuss the implications. Unique aspects of this research include the analysis of the percent of measurements below the ppb-level detection limits and how often this occurred, and data for elements with concentrations at or near regulatory limits that characterize their behavior and variability. One distinctive insight includes a discussion of current aluminum (Al) regulations, which are based on total recovery standards but can easily be exceeded if there are suspended clay-minerals that contain Al but are not bioavailable, and another is the presence of Cu above regulatory limits and potential Cu sources. Both provide managers with information to help them make decisions on the monitoring and management of Utah Lake. While these findings and data are specific to Utah Lake, our data collection and analysis methods provide a case study for other researchers. This approach can be used in other water bodies to better understand complex geochemistry with relatively limited study requirements.

2. Materials and Methods

2.1. Field Study

We collected UL water samples approximately once per week over two sampling seasons: June–October 2021 and May–August 2022. We collected lake background samples roughly 400 m offshore (Figure 1).

At the beginning of the 2021 and 2022 sampling seasons, the depth of the water column was 2.5 m at the sample location, which is a typical depth for the lake at that distance from shore. By the end of the season, the depth was 1.5 m as the lake levels decreased due to evaporation and outflow through the Jordan River. The seasonal pattern of drawdown and refill is also typical for the lake. We analyzed the samples using ICP-OES analysis on a Thermo-Scientific iCAP7400 (ThermoFisher Scientific, Waltham, MA, USA) for both dissolved (filtered on a 0.45 μm filter) and total (unfiltered and digested) samples.

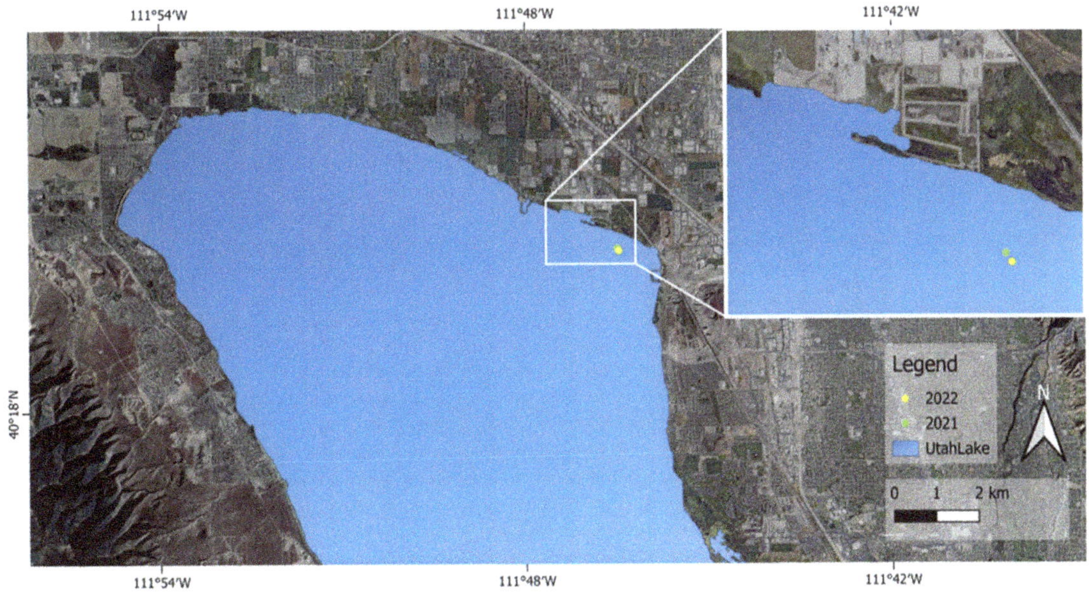

Figure 1. Study locations: 0.4 km (0.25 mi) offshore, northwest of Lindon marina.

This is a unique and comprehensive longitudinal dataset for UL, with near-weekly samples collected on 28 days over two summer sampling campaigns. While spatially limited to the northeastern corner of the lake, we assume these data characterize general conditions in UL, though it is an area likely to be impacted by human activities.

The purpose of our study is to begin to characterize and understand the presence and behaviors of ICP-OES regulated DEs in UL. The sample locations are near the largest WWTP outfall and the former site of the Geneva Steel Plant, and they consequently represent an area more likely to have higher DE concentrations than other areas of the lake.

2.2. Sample Collection

Over the two sampling seasons, weather permitting, we collected water samples approximately weekly from June to October in 2021 (n = 15) and from May to August in 2022 (n = 13) near the northeast corner of Utah Lake (Figure 1). We followed sampling procedures from the "Standard Operating Procedure for Lake Water Sampling and Data Collection" from the UDEQ [37]. We used 1 L dip samplers that were triple-rinsed in lake water at the sample location prior to collecting the sample. After the triple-rinse, we inverted the dip sampler and submerged it to approximately elbow depth. We then inverted the sample cup to fill it with water. This reduced potential contamination from surface particles. We used the collected water to triple-rinse a new, pre-labeled, opaque 250 mL plastic sample bottle, then filled the bottles, leaving no headspace, and immediately placed the sample on ice in a cooler. Samples were either analyzed or frozen within a few hours of collection.

Coincident with water samples, we used YSI ProDSS (YSI, Yellow Springs, OH, USA) water quality sondes (probes) to collect data on chlorophyll-a and phycocyanin. For these measurements, we fully submerged the probes and logged the sensor data at a depth of ~30 cm.

2.3. Analysis Methods

2.3.1. Laboratory Analysis

We split each water sample for ICP-OES analysis for dissolved (particles < 0.45 μm) and total concentrations. For dissolved concentrations, we filtered each sample using a new 0.45 μm membrane filter. For total concentrations, we performed microwave-assisted acid digestion using the EPA 3015A method. Table 2 summarizes the methods used for the laboratory analysis of the water samples.

Table 2. Lab analytes and methods.

Analyte	Model/Method	Equipment
DE Total (digested)	EPA 3015A	Thermo Scientific™ 7400 ICP-OES
DE Dissolved (filtered)	0.45 μ filter	Thermo Scientific™ 7400 ICP-OES (ThermoFisher Scientific, Waltham, MA, USA)

Table 3 lists the minimum detection limits (MDLs) for our ICP-OES method. MDLs vary by element but are approximately in the range of a few μg/L or at the part-per-billion (ppb) level. Se has the highest MDL of 7.36 ppb, while Ca has the lowest MDL at 0.02 ppb or 20.0 ppt. We selected ICP-OES for sample analysis because it is highly sensitive and allows us to easily analyze multiple trace elements in the UL water column.

Table 3. ICP-OES minimum detection limits (MDLs).

Analyte	Detection Limit (μg/L)	Analyte	Detection Limit (μg/L)
Aluminum (Al)	1.51	Molybdenum (Mo)	1.11
Arsenic (As)	4.74	Sodium (Na)	1.80
Boron (B)	1.26	Nickel (Ni)	2.29
Barium (Ba)	0.17	Phosphorus (P)	5.66
Calcium (Ca)	0.02	Lead (Pb)	4.50
Cadmium (Cd)	0.19	Sulfur (S)	2.22
Cobalt (Co)	1.16	Selenium (Se)	7.36
Chromium (Cr)	0.85	Silicon (Si)	7.20
Copper (Cu)	2.36	Strontium (Sr)	0.04
Iron (Fe)	0.80	Titanium (Ti)	0.58
Potassium (K)	5.10	Vanadium (V)	0.80
Magnesium (Mg)	0.04	Zinc (Zn)	0.60
Manganese (Mn)	0.21		

Additionally, water samples were tested for volatile suspended solids and total suspended solids. These results were correlated with ICP-OES measurements to further contextualize if total DE concentrations are associated with suspended sediments.

2.3.2. Filtration

Dissolved particles are often defined as smaller than 0.45 μm, while suspended particles are larger than 0.45 μm [50,83]. To separate the dissolved from the suspended samples, we used Nyaflo® Membrane Disc Filters (Cytiva, Marlborough, MA, USA) with 0.45 μm pores. For this study, we used the terms dissolved and filtered interchangeably, and the terms total and unfiltered interchangeably. Total or unfiltered concentrations include both the suspended and dissolved fractions.

We found through our quality assurance analyses that trace elements on the filters contaminated our samples for As, B, K, Mg, Na, S, and Zn. We used the same brand of 0.45 μm filters for all samples, though due to the number of samples we analyzed, using the same filter batch was not possible. Overall, we found that the level of contamination in the filtered samples was small. In this study, we assumed that all samples were equally impacted by filter contaminants, which allowed us to make comparisons and observe trends based on the sample results, which would still be accurate, even if the specific concentrations for these elements amounts are not.

2.4. Data Cleaning

Prior to analysis, we replaced measurements with concentrations below the MDL (Table 3) with a value of half of the MDL to preserve the detection of the element for analysis. The final dataset contained 28 samples collected over the 2021 and 2022 sampling periods (June through October and May through August, respectively), for which we analyzed 50 DE analytes, comprising 25 filtered and 25 unfiltered elements. The 2021 sampling season has a sampling gap from late July through late August due to equipment challenges but resumed in September and October. The 2022 sampling period was more consistent during the sampling season but was shorter, ending in August.

3. Results and Discussion: Regulation Levels

3.1. MDL Analysis

We analyzed how often a detection for an element was below the MDL. This analysis showed how often a sample contained the element at or below detection limits. Figure 2 shows the percentage of DE measurements that were above the MDL for both unfiltered and filtered samples by element. If every sample for both the unfiltered and filtered samples was above the MDL, we excluded that DE from the graph for simplification. Forty percent of the DEs—Ba, Ca, K, Mg, Mo, Na, S, Si, Sr, and V—had all measurements above the MDL and are not included in the graph.

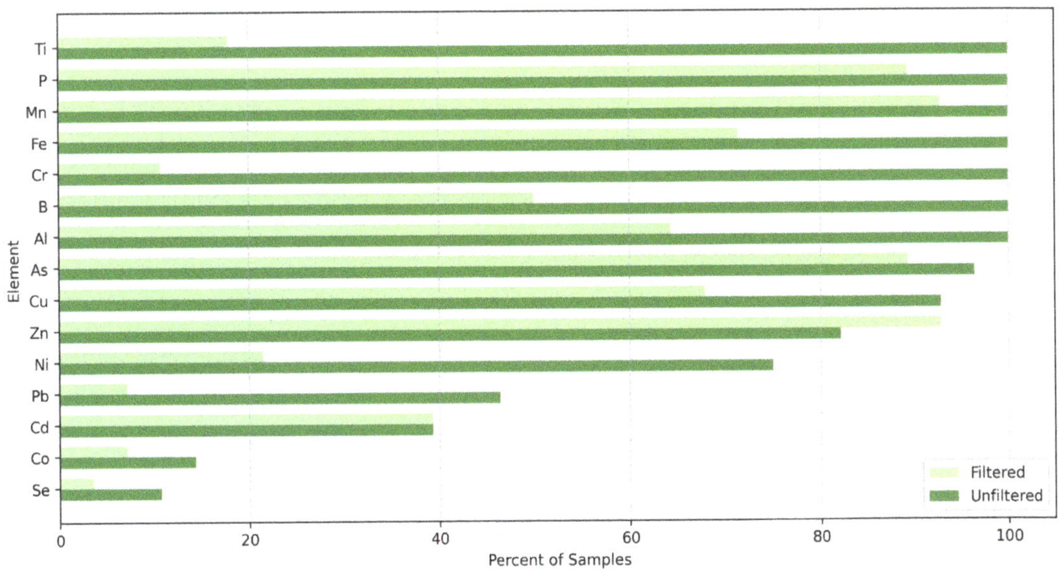

Figure 2. Percentage of samples for each ICP-OES analyte with measured concentrations above the MDL. Analytes with 100% detection in both filtered and unfiltered samples are excluded from the plot.

Co and Se concentrations fluctuate around the MDL in UL, with less than 20% of filtered or unfiltered samples having Co and Se concentrations above the MDL, and the remainder had concentrations only slightly above the MDL. These elements are present in UL, but in trace amounts.

Ti, Cr, Ni, and Pb were above the MDLs in less than 20% of filtered samples but were over the MDLs in unfiltered samples. These elements are mainly associated with suspended solids, with very low dissolved concentrations. Ti and Cr were above the MDLs in 100% of the unfiltered samples but only 20% of the filtered samples. From this, we can infer that these metals are mainly contained in the suspended solids and sediments of UL and are not present as dissolved particles.

Unfiltered Ni and Pb were above the MDLs in about 75% and 45% of cases, respectively. These elements similarly show that they are prevalent within the water column more in the unfiltered than the filtered state.

One metal, Zn, had more filtered samples over the MDL than unfiltered samples. This was caused by contamination from the filters, which consequently only affected the dissolved samples. We assume that although the dissolved Zn values are not accurate, the changes and trends in concentration are still relatively correct. As Zn is regulated and a vital micronutrient, we included the filtered results in this study. The unfiltered results are not contaminated because they did not come into contact with the filter and are therefore representative of our samples.

3.2. Specific Element Selection

We highlight eight ICP-OES elements—Cu, Zn, Ni, Al, Pb, P, Ba, and As—that were close to or over the most stringent applicable standards or exhibited interesting characteristics. The other five regulated DEs—B, Cd, Cr, Fe, and Se—are well below regulatory limits and we do not discuss or present them in this study.

We analyzed the distributions as box-and-whisker plots grouped by year (2021, $n = 15$; 2022, $n = 13$) and phase (filtered and unfiltered). In these plots, the line in the middle of the box is the median concentration, the box ends are the 25th and 75th percentiles, and the whiskers represent 1.5 times the interquartile range (1.5×IQR). Outliers, or values outside 1.5×IQR, are shown as dots.

We included lines on each boxplot indicating the most stringent acute (1 h average) and chronic (4-day average) standards (Table 1). All regulations are for dissolved (filtered) concentrations except Al and P, which are based on the total (unfiltered) concentration.

We did not follow tate-approved methods of analysis. Consequently, our findings cannot be used to determine or indicate the impairment status of UL; we include the regulatory criteria only in order to provide context.

In the following figures, the dark-blue boxplots represent unfiltered data, and the light-blue boxplots represent filtered data. The dark-green dotted horizontal line represents the most stringent chronic standard, and the light-green line represents the most stringent acute standard. In a few cases where the acute and chronic regulations are the same, only the dark-green line is visible.

3.3. Copper (Cu)

Cu (Figure 3) has harmful effects at both excess and deficient concentrations. It is a micronutrient vital for a healthy ecosystem. For the unfiltered Cu concentrations, the median concentration for both years is above or close to the regulatory criterion, with concentrations more varied in 2021 ($n = 15$ in 2021, and $n = 13$ in 2022). Median filtered Cu concentrations are below the criteria in 2021 but above them in 2022.

For both the total and dissolved measurements, over 70% of the water samples contained concentrations of Cu over the MDL. Only dissolved Cu is regulated. All the filtered Cu samples in the 2021 sampling season, except one, were below both the acute and chronic regulatory criteria (Figure 3). However, in the 2022 sampling season (Figure 3), 7 of the 13 samples were above 0.013 mg/L, which is the acute standard. The highest concentration was 0.0329 mg/L. The State regulations for Cu are only applicable to the filtered samples, but for context, unfiltered Cu was above regulatory levels during both sampling seasons. For reference, the UL sediments on average contain 19.5 mg/kg of Cu. We found that unfiltered Cu concentrations are strongly correlated with unfiltered Ni and Fe, with Pearson correlation coefficients (PCC) of 0.901 and 0.752, respectively. This might indicate mineral forms in the suspended solids.

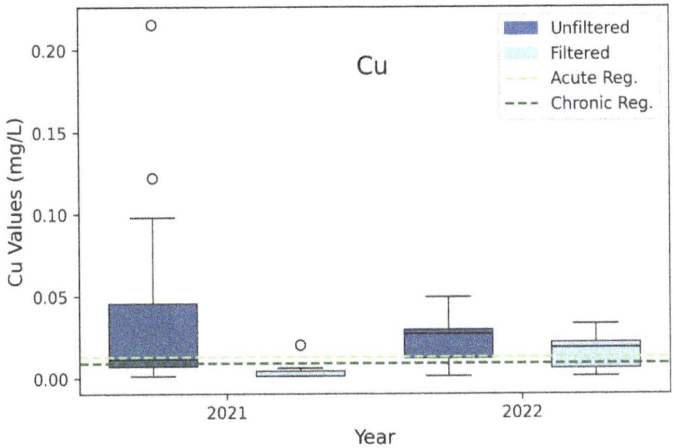

Figure 3. Distributions of Cu grouped by year, 2021 (n = 15) and 2022 (n = 13); and phase, unfiltered (dark blue) and filtered (light blue). The plot includes reference lines for 0.013 mg/L (acute light green) and 0.009 (chronic dark green) regulatory values. The circles are outlier data points.

The elevated levels of Cu may also be related to the use of copper sulfate-based algaecide, which the State began using to treat intense algal blooms in marinas in 2020 [84,85]. Our study location is located less than 1 km (0.6 miles) from Lindon Marina and 2 km (1.25 miles) from American Fork Marina. Lake currents are generally along the north shoreline towards the east, which could carry Cu from the American Fork Marina to our study site. The Utah Department of Environmental Quality (UDEQ) released an interim report on UL marina HAB treatments in which they found that "copper concentration increased considerably" in the days following treatment, but it returned to below toxic levels within a week [84]. The higher concentrations of dissolved Cu in 2022 relative to 2021 may also be due to the lower lake levels during the 2022 sampling season. On average, the lake had about 49 million cubic meters (39,700 ac-ft) less volume in 2022 than in 2021, which may have increased the water column concentrations of several elements [86]. Although the use of algaecide and low lake levels are one potential cause of the elevated levels of Cu observed in both years, they may not be the actual drivers.

Additional research is needed to identify and characterize the source and behavior of these metals. This is important as our results show that levels of dissolved Cu in the lake could be approaching harmful concentrations.

3.4. Zinc (Zn)

Zn (Figure 4) has both toxic effects at both excess and deficient concentrations. It is a micronutrient vital for a healthy ecosystem. Although filter contamination slightly increased concentrations of our filtered Zn samples, both filtered and unfiltered concentrations of Zn were well below the most stringent regulatory standard. More than 80% of Zn samples, both the filtered and unfiltered, measured above the MDL (Figure 2). In 2022, the difference between the total and the dissolved Zn was larger than in 2021, but it remained low. Filtered Zn concentrations, which are lower than those shown because of contamination from the filter, are low enough that it is more likely to be a limiting or colimiting nutrient for algal growth in UL because of deficient concentrations than a toxin due to high concentrations. For reference, Zn concentrations average 145.5 mg/kg in our UL sediment samples.

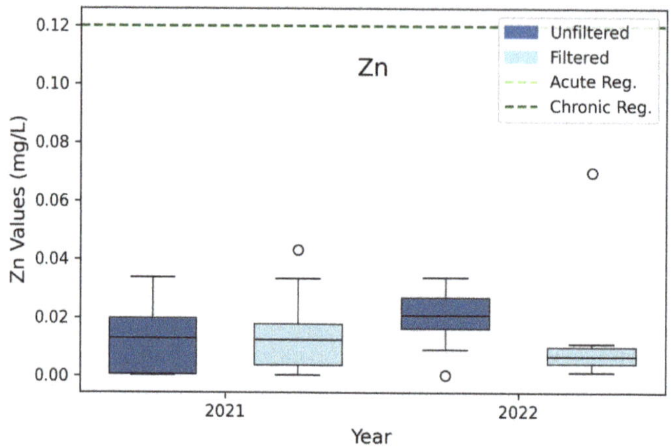

Figure 4. Distributions of Zn grouped by year, 2021 (n = 15) and 2022 (n = 13); and phase, unfiltered (dark blue) and filtered (light blue). The plot includes reference lines for 0.12 mg/L (acute light green) and 0.12 mg/L (chronic dark green) regulatory values. The acute and chronic standards for Zn are the same. Circles are outlier data points; the light green and dark green lines are the same value and only the dark green can be seen.

3.5. Nickel (Ni)

Ni (Figure 5) has adverse effects at both excess and deficient concentrations. It is a micronutrient vital for a healthy ecosystem. Both total and dissolved Ni concentrations in 2021 and 2022 were well below acute regulations, and just below chronic State regulations. Most of the Ni in UL appears to be associated with suspended solids rather than being dissolved in the water column. This matches Figure 2, which shows that ~70% of the unfiltered Ni measurements were over the MDL, compared to only ~20% of the dissolved measurements being over the MDL. These low Ni concentrations suggest that, like Zn, Ni is more likely to be a limiting or colimiting micronutrient because of deficient concentrations than a toxin in UL. Unfiltered Ni is correlated with unfiltered Cu and Fe, with PCCs of 0.901 and 0.735, respectively. This suggests a mineral phase in the suspended solids. For reference, our UL sediments samples average 11.5 mg/kg of Ni.

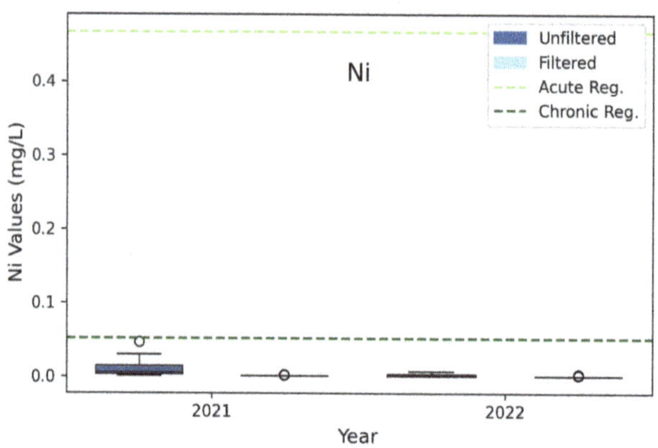

Figure 5. Distributions of Ni grouped by year, 2021 (n = 15) and 2022 (n = 13); and phase, unfiltered (dark blue) and filtered (light blue). The plot includes reference lines for 0.468 mg/L (acute light green) and 0.052 mg/L (chronic dark green) regulatory values. The light blue boxes are the 2nd and 4th from the left, but are so thin they show as black lines. The circles are outliers.

3.6. Aluminum (Al)

Al is a nonessential metal that is toxic at high concentrations [59–61,87]. State regulations state that Al concentrations are based on total recoverable Al, not dissolved. We assume that our unfiltered, digested samples represent total recoverable Al in UL water.

Unfiltered (total) Al median measurements are above both the acute and chronic criteria (Figure 6). Dissolved Al concentrations, however, are well below the criteria. Over 100% and 60% of the unfiltered and filtered Al measurements, respectively, were above the Al MDL (Figure 2).

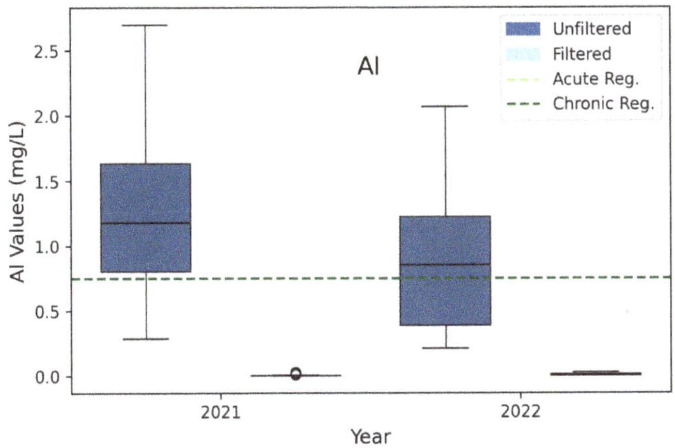

Figure 6. Distributions of Al grouped by year, 2021 and 2022; and phase, unfiltered (dark blue) and filtered (light blue) with reference lines for 0.75 mg/L (acute) and 0.75 mg/L (chronic) regulatory levels. The light blue boxes are so thin they show as black lines. Circles are outlier data points; the light green and dark green lines are the same value and only the dark green can be seen.

The large difference between filtered and unfiltered concentrations indicates that Al is associated with suspended solids in the UL water column. The unfiltered concentrations represent a total digestion of all suspended solids, including clay minerals. Al and Si ratios in the samples are consistent with Al and Mg clay minerals, providing evidence that high Al concentrations are related to suspended clays. We attribute the high total Al concentrations to suspended clay particles in the water column.

As noted earlier, we did not follow regulatory analysis methods. State water quality regulations require measurements of the dissolved (filtered) concentrations of elements, except for Al and P. For Al, the State regulations require "total recoverable" concentrations, which we took to mean unfiltered samples, but we did not use the State-approved method for Al analysis. Figure 6 shows that the unfiltered Al samples, when measured using full digestion and ICP-OES, exceed regulatory criteria in both the 2021 and 2022 seasons for both the acute and chronic regulations. The filtered (dissolved) samples, however, are well below the State regulations for both sampling periods. Our results show that dissolved Al concentrations—the phase likely to be bioavailable—are well below levels of concern. While unfiltered concentrations are above levels of concern, they are likely caused by suspended clays.

For reference, the average concentration of Al in the sediment is 8975.8 mg/kg, or almost 0.9% by mass. This concentration of Al is an order of magnitude higher than the amount of Zn and two orders of magnitude greater than the concentrations of Cu and Ni in the sediments.

Unfiltered Al was strongly correlated with unfiltered Ti, Mn, volatile suspended solids, and total suspended solids, with PCCs of 0.916, 0.750, 0.745, and 0.737, respectively.

3.7. Lead (Pb)

Pb is a nonessential metal that is toxic at high concentrations [59–61,87]. In our samples, dissolved Pb distributions had median values well below the acute standard for both filtered and unfiltered samples. However, the median values were often near or above their chronic State-regulation level (Figure 6). These results are misleading, however, as the acute Pb standard is below the ICP-OES method MDL for Pb. The bottom panel of Figure 7 shows that four dissolved Pb samples were above or at the acute standard, with two samples having values of 0.0072 and 0.0057 and two at the MDL of 0.0045. The remaining data were below the MDL and were set to one-half of the MDL to generate the plots and statistics. All the samples were well below the chronic standard. In 2021 and 2022, ten and three of the unfiltered samples were above the acute standards, respectively. Of the remaining samples, five and ten samples for 2021 and 2022, respectively, were below the MDL. Actual dissolved Pb concentrations could be lower (or higher) than this value. This is a limitation of our study caused by the lack of sensitivity of the ICP-OES compared to the regulatory standard. The EPA standard for Pb analysis uses a more sensitive ICP-MS method, which we did not use.

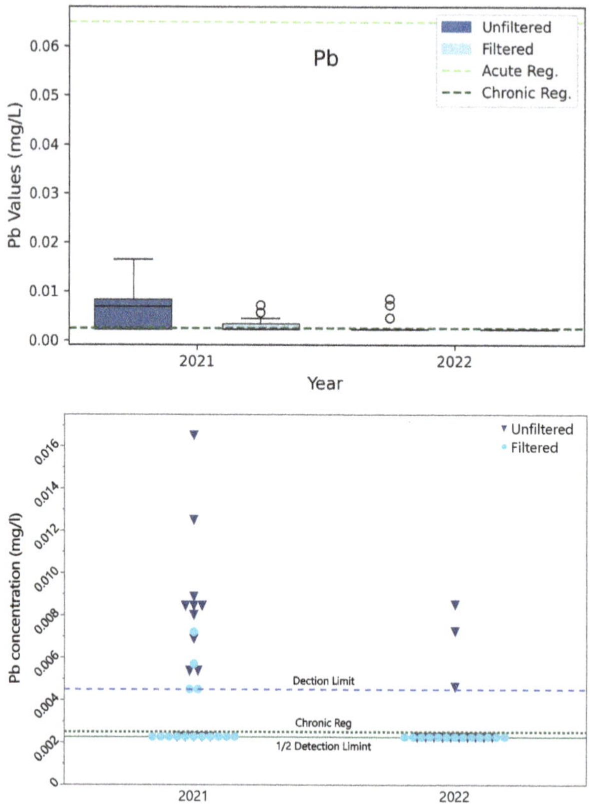

Figure 7. Distributions of Pb grouped by year (**Top** Panel), 2021 (n = 15) and 2022 (n = 13); and phase, unfiltered (dark blue) and filtered (light blue) with reference lines for 0.065 mg/L (acute) and 0.0025 mg/L (chronic) regulatory levels; Pb data, grouped by year (**Bottom** Panel) with lines for the detection limit, the chronic regulatory level, and ½ the detection limit. There are only two data points above detection limits, with two at detection limits. In the top panel, the boxes are so thin they show as black lines. Circles are outlier data points.

The Pb chronic regulation concentration is half of the ICP-OES MDL for Pb. Since we replaced 90% of measurements that were below the MDL with half of the MDL, most of the dissolved Pb measurements are exactly at the chronic regulation. Our data cannot be used to evaluate Pb concentrations relative to the chronic standard.

Figure 2 shows that except for two samples, all the samples with Pb concentrations over the MDL were for unfiltered samples, indicating that the Pb is associated with suspended solids. Both the filtered and unfiltered Pb distributions exhibit large differences between 2021 and 2022, with 2022 exhibiting significantly lower concentrations. We are unaware of any specific cause for this, although it is possible that higher levels of spring runoff in 2021 in comparison to 2022 led to greater inputs of Pb into the lake in 2021.

3.8. Phosphorus (P)

P is a limiting macronutrient in many freshwater lakes [88], but if sufficient P is available, other factors, such as micronutrients or light, can become limiting factors. Both total and dissolved P distributions are above the acute and chronic State regulations at the 25th percentile, with the exception of filtered samples in 2022, where the median is above the limits, but the 25th percentile is below (Figure 8). UL is listed as impaired due to high P concentrations, and the State has begun the process of developing a total maximum daily load (TMDL) for the lake [89]. Our data agree and show that P concentrations in both the filtered and unfiltered samples are above the regulatory levels. The MDL analysis further emphasizes the presence of P, with over 80% of the filtered and all the unfiltered samples over the MDL (Figure 2).

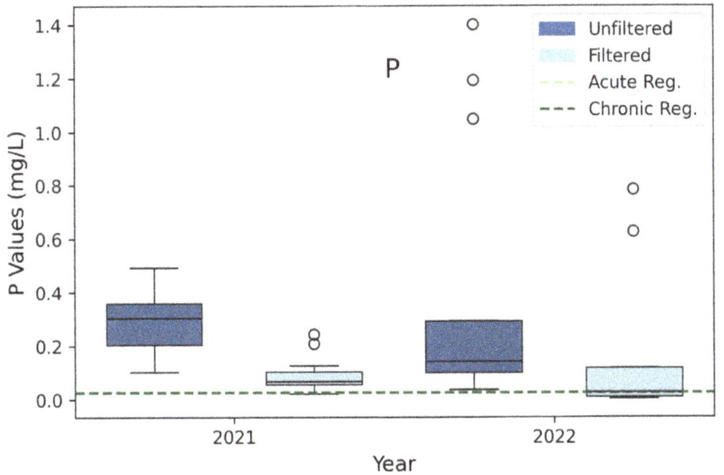

Figure 8. Distributions of P grouped by year, 2021 and 2022; and phase, unfiltered (dark blue) and filtered (light blue) with reference lines for 0.025 mg/L (acute) and 0.025 mg/L (chronic) regulatory levels. Circles are outlier data points; the light green and dark green lines are the same value and only the dark green can be seen.

Filtered and unfiltered P concentrations are strongly correlated, with a PCC of 0.949, but neither filtered nor unfiltered P correlates strongly with any other parameters we measured. The strong correlation between unfiltered and filtered P concentrations indicates that they are dependent on each other. That is, when unfiltered concentrations increase, due to higher turbidity or sediment resuspension, filtered or dissolved concentrations increase as well. These data support the idea that dissolved P concentrations in UL are governed by a sorption process that keeps sorbed (unfiltered) and dissolved (filtered) P in a constant ratio [1]. Because of the large reservoir of P in the UL sediments, water column concentrations are relatively stable, as was shown by Taggart, et al. [1]

Dissolved P concentrations were not correlated with probe measurements of either chlorophyll or phycocyanin, both of which indicate phytoplankton growth. This suggests that phytoplankton growth is not responding directly to bioavailable P concentrations, meaning dissolved P is not a limiting factor for algal growth in the UL water column. This analysis indicated that P concentrations, both filtered and unfiltered, are independent of other processes, which is again consistent with P in UL acting within a sorption-based system, where water column concentrations are in equilibrium with P-rich sediments of geologic origin [1].

3.9. Barium (Ba)

Ba is a potentially toxic metal that is not currently regulated in Utah [80]. Ba occurs naturally and is released by industrial processes, where it becomes mobilized as airborne particles that could enter UL through atmospheric deposition [75]. UL is susceptible to atmospheric deposition, as indicated by several studies [16,17,19]. Based on this, we have chosen to include a discussion on Ba. Figure 9 shows the Ba distributions for both filtered and unfiltered measurements. The Ba distributions for 2021 and 2022 sampling seasons are consistent and emphasize the presence of Ba within the UL ecosystem. There is a difference, but not a large one, between the unfiltered and filtered distributions. This indicates that a significant portion of Ba is dissolved, with some portions associated with suspended solids. About half of the Ba in UL is present in the dissolved phase.

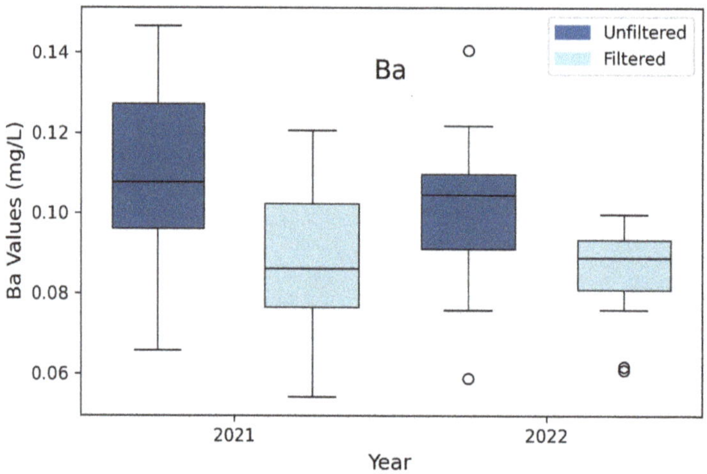

Figure 9. Distributions of Ba grouped by year, 2021 and 2022; and phase, unfiltered (dark blue) and filtered (light blue). Circles are outlier data points.

The U.S. EPA released a factsheet on Ba [90] in which they estimate that between the years of 1987 and 1993, 680 kg (1500 pounds) of Ba was released into the waters of Utah, including UL. The MDL analysis (Section 3.1 and Figure 2) revealed that Ba measurements for both unfiltered and filtered samples were over the MDL 100% of the time. Sediment samples show that there is an average of 179.5 mg/kg in the UL sediments. The Ba boxplot (Figure 9) shows that the median unfiltered and filtered samples for Ba are about the same in both the 2021 and 2022 sampling seasons.

Short-term health effects of Ba include gastrointestinal disturbances and muscular weakness, with long-term effects including high blood pressure for levels above the maximum contaminant limit (MCL) [90]. The U.S. EPA MCL for Ba is 2 mg/L for drinking water—nearly two orders of magnitude higher than any of our measured values in UL [90,91]. Within the State of Utah, Ba is only regulated for the designated use of 1C or "Domestic Source" (which does not apply to Utah Lake) at a maximum dissolved limit of

1 mg/L [80]. The Ba distribution plot (Figure 9) shows that both the unfiltered and filtered measurements are well below the EPA MCL of 2 mg/L for drinking water and the 1 mg/L Utah 1C regulation for domestic source water.

3.10. Arsenic (As)

Both filtered and unfiltered As concentrations were consistently below the most restrictive aquatic use, which in this case was for agriculture such as irrigation, which has values of 0.10 mg/L for both acute and chronic standards (Table 1). Unfiltered concentrations are near the drinking water standards which are 0.01 mg/L, however, Utah Lake water is not current use for domestic supply. As exhibited a seasonal trend in both years of our data (Figure 10). We postulate that As increases in late summer because groundwater inflows with higher concentrations of As become a larger portion of lake inflows at that time of year [92,93]. As is positively correlated with phycocyanin probe measurements and inversely correlated with filtered Al concentrations. We believe these correlations are spurious and are only present because all three variables have seasonal variation. These variations, however, are due to different processes related to seasonal changes, not chemical or ecological processes involving all three parameters. We could not identify any likely physical process that would drive this correlation. While unfiltered As concentrations are well below the most stringent beneficial use requirements for agricultural and aquatic wildlife, they are, on average, slightly above domestic use standards.

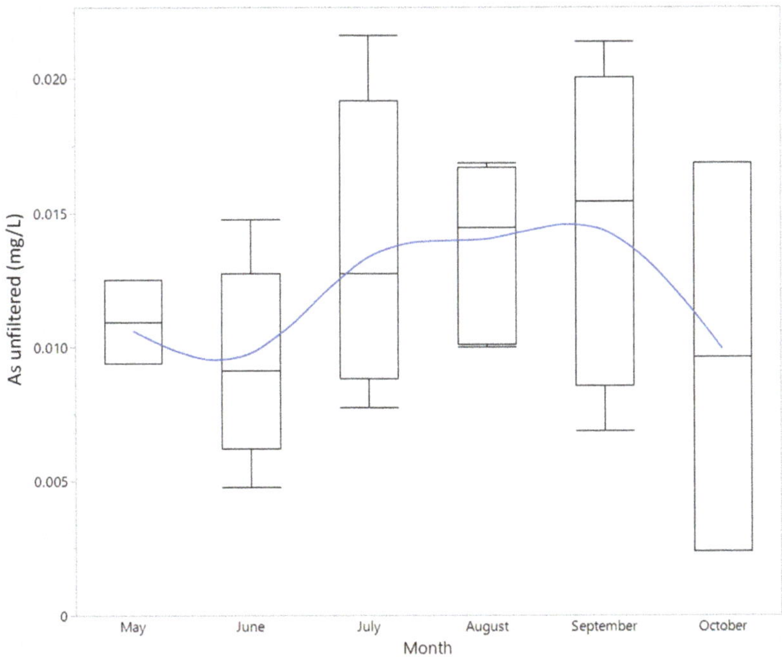

Figure 10. Annual trend in unfiltered (total) arsenic (As). The blue line is a smoothed trend through the mean values.

4. Conclusions

In this study, we used ICP-OES to examine elements in UL water at a ppb level. We found that total recoverable concentrations of Al in UL exceeded regulatory criteria when using our analytical methods, although we attribute these elevated levels to nonreactive, clay-dominated suspended sediment. Concentrations of dissolved Al, which would be bioavailable and have the potential to be toxic, were well below regulatory criteria. We consistently measured concentrations of Cu and P above or very close to regulatory limits. The high concentrations of P in UL are well known, but the elevated levels of Cu are not widely recognized, and as these elements have the potential to impair the beneficial uses of UL, further studies should be conducted. Cu especially should be considered because of an ongoing program using Cu-based algaecide to treat intense algal blooms in UL.

Concentrations of other regulated elements remained well below regulatory criteria, which was interesting because anthropogenic inputs of these elements to UL have been and may still be large—the low concentrations we observed demonstrate the lake's high capacity for absorbing, degrading, and removing pollutants. Additionally, we found no regulations regarding Ba for UL's designated beneficial uses; however, concentrations of Ba are below drinking water limits, despite nearby anthropogenic sources.

We found a strong correlation between unfiltered (total) and filtered (dissolved) P, indicating that the constant resuspension of P-rich lakebed sediments acts as a source of bioavailable P in the water column. This relationship indicates that dissolved P in the water column is in equilibrium with P sorbed onto lakebed sediments and suspended sediments in a sorption-dominated system.

UL, except for Al and P, is well within compliance for all regulated DE elements associated with its designated uses. It may be beneficial for the State to create a separate designated use category for UL, as has been done with the Great Salt Lake. The DE standards drive many regulatory decisions for point sources around UL, and as such, they could be further considered with the uniqueness of UL.

Author Contributions: Conceptualization, R.A.V., K.B.T. and G.P.W.; methodology, R.A.V. and G.P.W.; field data collection, R.A.V., K.B.T., J.B.T., R.L.R., A.C.C., L.M.W. and M.J.O.; software, R.A.V.; validation, J.B.T., R.L.R., A.C.C., A.W.M. and R.B.S.; resources, A.W.M. and G.P.W.; data curation, R.A.V., K.B.T. and A.C.C.; writing—original draft preparation, R.A.V.; writing—review and editing, R.A.V., K.B.T., J.B.T., R.L.R., A.C.C., G.P.W., A.W.M. and R.B.S.; visualization, R.A.V. and G.P.W.; supervision, G.P.W.; project administration, R.A.V., K.B.T. and G.P.W.; funding acquisition, A.W.M. and G.P.W. All authors have read and agreed to the published version of the manuscript.

Funding: This research was funded by Timpanogos Special Service District (TSSD), American Fork, UT.

Data Availability Statement: The data presented in this study are available on request from the corresponding author due to contractual restrictions on data distribution.

Acknowledgments: We acknowledge TSSD management and staff in their support in this work.

Conflicts of Interest: The authors declare no conflict of interest. The funders had no role in the collection, analyses, or interpretation of data or in the decision to publish the results.

References

1. Taggart, J.B.; Ryan, R.L.; Williams, G.P.; Miller, A.W.; Valek, R.A.; Tanner, K.B.; Cardall, A.C. Historical Phosphorus Mass and Concentrations in Utah Lake: A Case Study with Implications for Nutrient Load Management in a Sorption-Dominated Shallow Lake. *Water* **2024**, *16*, 933. [CrossRef]
2. Snow, E. A Preliminary Study of the Algae of Utah Lake. Master's Thesis, Brigham Young University, Provo, UT, USA, 1931; 84p.
3. Harding, W.J. The algae of Utah Lake. Part II. *Great Basin Nat.* **1971**, *31*, 125–134.
4. Rushforth, S.R.; Squires, L.E. New records and comprehensive list of the algal taxa of Utah Lake, Utah, USA. *Great Basin Nat.* **1985**, *45*, 237–254.
5. Squires, L.E.; Rushforth, S.R. Winter phytoplankton communities of Utah Lake, Utah, USA. *Hydrobiologia* **1986**, *131*, 235–248. [CrossRef]
6. Whiting, M.C.; Brotherson, J.D.; Rushforth, S.R. Environmental interaction in summer algal communities of Utah Lake. *Great Basin Nat.* **1978**, *38*, 31–41.
7. Harding, W.J. A preliminary report on the algal species presently found in Utah Lake. *Great Basin Nat.* **1970**, *30*, 99–105.
8. Rushforth, S.R.; St. Clair, L.L.; Grimes, J.A.; Whiting, M.C. Phytoplankton of Utah Lake. *Great Basin Nat. Mem.* **1981**, *5*, 85–100.
9. Squires, L.E.; Whiting, M.C.; Brotherson, J.D.; Rushforth, S.R. Competitive displacement as a factor influencing phytoplankton distribution in Utah Lake, Utah. *Great Basin Nat.* **1979**, *39*, 245–252.
10. Carozzi, A.V. Observations on algal biostromes in the Great Salt Lake, Utah. *J. Geol.* **1962**, *70*, 246–252. [CrossRef]
11. Rushforth, S.R.; Merkley, G.S. Comprehensive list by habitat of the algae of Utah, USA. *Great Basin Nat.* **1988**, *48*, 154–179.

12. Davis, R.; Panja, P.; McLennan, J. Integrated workflow for interpretation of satellite imageries using machine learning to assess and monitor algal blooms in Utah Lake, USA. *Ecol. Inform.* **2023**, *75*, 102033. [CrossRef]
13. Tanner, K.B.; Cardall, A.C.; Williams, G.P. A Spatial Long-Term Trend Analysis of Estimated Chlorophyll-a Concentrations in Utah Lake Using Earth Observation Data. *Remote Sens.* **2022**, *14*, 3664. [CrossRef]
14. Hansen, C.H.; Burian, S.J.; Dennison, P.E.; Williams, G.P. Evaluating historical trends and influences of meteorological and seasonal climate conditions on lake chlorophyll a using remote sensing. *Lake Reserv. Manag.* **2020**, *36*, 45–63. [CrossRef]
15. Hansen, C.H.; Williams, G.P.; Adjei, Z.; Barlow, A.; Nelson, E.J.; Miller, A.W. Reservoir water quality monitoring using remote sensing with seasonal models: Case study of five central-Utah reservoirs. *Lake Reserv. Manag.* **2015**, *31*, 225–240. [CrossRef]
16. Barrus, S.M.; Williams, G.P.; Miller, A.W.; Borup, M.B.; Merritt, L.B.; Richards, D.C.; Miller, T.G. Nutrient Atmospheric Deposition on Utah Lake: A Comparison of Sampling and Analytical Methods. *Hydrology* **2021**, *8*, 123. [CrossRef]
17. Telfer, J.T.; Brown, M.M.; Williams, G.P.; Tanner, K.B.; Miller, A.W.; Sowby, R.B.; Miller, T.G. Source Attribution of Atmospheric Dust Deposition to Utah Lake. *Hydrology* **2023**, *10*, 210. [CrossRef]
18. Abu-Hmeidan, H.Y.; Williams, G.P.; Miller, A.W. Characterizing Total Phosphorus in Current and Geologic Utah Lake Sediments: Implications for Water Quality Management Issues. *Hydrology* **2018**, *5*, 8. [CrossRef]
19. Brown, M.M.; Telfer, J.T.; Williams, G.P.; Miller, A.W.; Sowby, R.B.; Hales, R.C.; Tanner, K.B. Nutrient Loadings to Utah Lake from Precipitation-Related Atmospheric Deposition. *Hydrology* **2023**, *10*, 200. [CrossRef]
20. Olsen, J.M.; Williams, G.P.; Miller, A.W.; Merritt, L. Measuring and calculating current atmospheric phosphorous and nitrogen loadings to utah lake using field samples and geostatistical analysis. *Hydrology* **2018**, *5*, 45. [CrossRef]
21. Bradshaw, J.; Sundrud, R.; White, D.; Barton, J.; Fuhriman, D.; Loveridge, E.; Pratt, D. Chemical response of Utah Lake to nutrient inflow. *J. Water Pollut. Control Fed.* **1973**, *45*, 880–887.
22. Zanazzi, A.; Wang, W.; Peterson, H.; Emerman, S.H. Using Stable Isotopes to Determine the Water Balance of Utah Lake (Utah, USA). *Hydrology* **2020**, *7*, 88. [CrossRef]
23. Strong, A.E. Remote sensing of algal blooms by aircraft and satellite in Lake Erie and Utah Lake. *Remote Sens. Environ.* **1974**, *3*, 99–107. [CrossRef]
24. Strong, A. ERTS-1 observes algal blooms in Lake Erie and Utah Lake. In Proceedings of the NASA. Goddard Space Flight Center Symp. on Significant Results obtained from the ERTS-1, New Carrollton, MD, USA, 5–9 March 1973.
25. Miller, W.; Rango, A. Using heat capacity mapping mission (hcmm) data to assess lake water quality 1. *JAWRA J. Am. Water Resour. Assoc.* **1984**, *20*, 493–501. [CrossRef]
26. Schneider, S.R.; McGinnis, D.F.; Gatlin, J.A. *Use of NOAA/AVHRR Visible and Near-Infrared Data for Land Remote Sensing*; National Earth Satellite Service, Ed.; US Department of Commerce, National Oceanic and Atmospheric Administration: Washington, DC, USA, 1981; Vol. NOAA Technical Report NESS84.
27. Hansen, C.H.; Williams, G.P. Evaluating remote sensing model specification methods for estimating water quality in optically diverse lakes throughout the growing season. *Hydrology* **2018**, *5*, 62. [CrossRef]
28. Hansen, C.H.; Burian, S.J.; Dennison, P.E.; Williams, G.P. Spatiotemporal variability of lake water quality in the context of remote sensing models. *Remote Sens.* **2017**, *9*, 409. [CrossRef]
29. Rivera, S.; Landom, K.; Crowl, T. Monitoring macrophytes cover and taxa in Utah Lake by using 2009-2011 Landsat digital imagery. *Rev. Teledetección* **2013**, *39*, 106–115.
30. Han, Q.; Niu, Z. Construction of the long-term global surface water extent dataset based on water-NDVI spatio-temporal parameter set. *Remote Sens.* **2020**, *12*, 2675. [CrossRef]
31. Seegers, B.N.; Werdell, P.J.; Vandermeulen, R.A.; Salls, W.; Stumpf, R.P.; Schaeffer, B.A.; Owens, T.J.; Bailey, S.W.; Scott, J.P.; Loftin, K.A. Satellites for long-term monitoring of inland US lakes: The MERIS time series and application for chlorophyll-a. *Remote Sens. Environ.* **2021**, *266*, 112685. [CrossRef] [PubMed]
32. Zhu, W.; Zhang, Z.; Yang, Z.; Pang, S.; Chen, J.; Cheng, Q. Spectral probability distribution of closed connected water and remote sensing statistical inference for yellow substance. *Photogramm. Eng. Remote Sens.* **2021**, *87*, 807–819. [CrossRef]
33. Hansen, C.H.; Dennison, P.; Burian, S.; Barber, M.; Williams, G. Hindcasting water quality in an optically complex system. *WIT Trans. Ecol. Environ.* **2016**, *209*, 35–44.
34. Page, B.P.; Kumar, A.; Mishra, D.R. A novel cross-satellite based assessment of the spatio-temporal development of a cyanobacterial harmful algal bloom. *Int. J. Appl. Earth Obs. Geoinf.* **2018**, *66*, 69–81. [CrossRef]
35. Ramsey, R.D.; Falconer, A.; Jensen, J.R. The relationship between NOAA-AVHRR NDVI and ecoregions in Utah. *Remote Sens. Environ.* **1995**, *53*, 188–198. [CrossRef]
36. Hansen, C.; Swain, N.; Munson, K.; Adjei, Z.; Williams, G.P.; Miller, W. Development of sub-seasonal remote sensing chlorophyll-a detection models. *Am. J. Plant Sci.* **2013**, *4*, 40802. [CrossRef]
37. Cardall, A.C.; Hales, R.C.; Tanner, K.B.; Williams, G.P.; Markert, K.N. LASSO (L1) Regularization for Development of Sparse Remote-Sensing Models with Applications in Optically Complex Waters Using GEE Tools. *Remote Sens.* **2023**, *15*, 1670. [CrossRef]
38. Williams, R.; Nelson, S.; Rushforth, S.; Rey, K.; Carling, G.; Bickmore, B.; Heathcote, A.; Miller, T.; Meyers, L. Human-Driven Trophic Changes in a Large, Shallow Urban Lake: Changes in Utah Lake, Utah from Pre-European Settlement to the Present. *Water Air Soil Pollut.* **2023**, *234*, 218. [CrossRef]
39. Nofchissey, S.; Roberts, S.; Hopkinson, J.; McDonald, J.; Emerman, S. Arsenic and other heavy metals in Utah Lake and its tributaries. In Proceedings of the 2014 Spring Runoff Conference, Logan, UT, USA, 1 April 2014.

40. Zhang, X.; Li, B.; Xu, H.; Wells, M.; Tefsen, B.; Qin, B. Effect of micronutrients on algae in different regions of Taihu, a large, spatially diverse, hypereutrophic lake. *Water Res.* **2019**, *151*, 500–514. [CrossRef] [PubMed]
41. Andersen, R.A. (Ed.) *Algal Culturing Techniques*; Elsevier: Boston, MA, USA, 2005.
42. Downs, T.M.; Schallenberg, M.; Burns, C.W. Responses of lake phytoplankton to micronutrient enrichment: A study in two New Zealand lakes and an analysis of published data. *Aquat. Sci.* **2008**, *70*, 347–360. [CrossRef]
43. Norman, L.; Cabanesa, D.J.E.; Blanco-Ameijeiras, S.; Moisset, S.A.M.; Hassler, C.S. Iron Biogeochemistry in Aquatic Systems: From Source to Bioavailability. *CHIMIA* **2014**, *68*, 764. [CrossRef] [PubMed]
44. Krivokapić, M. Study on the Evaluation of (Heavy) Metals in Water and Sediment of Skadar Lake (Montenegro), with BCF Assessment and Translocation Ability (TA) by Trapa natans and a Review of SDGs. *Water* **2021**, *13*, 876. [CrossRef]
45. Rueter, J.G.; Petersen, R.R. Micronutrient effects on cyanobacterial growth and physiology. *N. Z. J. Mar. Freshw. Res.* **1987**, *21*, 435–445. [CrossRef]
46. Bayer, T.K.; Schallenberg, M.; Martin, C.E. Investigation of nutrient limitation status and nutrient pathways in Lake Hayes, Otago, New Zealand: A case study for integrated lake assessment. *N. Z. J. Mar. Freshw. Res.* **2008**, *42*, 285–295. [CrossRef]
47. Dengg, M.; Stirling, C.H.; Reid, M.R.; Verburg, P.; Armstrong, E.; Kelly, L.T.; Wood, S.A. Growth at the limits: Comparing trace metal limitation of a freshwater cyanobacterium (*Dolichospermum lemmermannii*) and a freshwater diatom (*Fragilaria crotonensis*). *Sci. Rep.* **2022**, *12*, 467. [CrossRef]
48. Facey, J.A.; Rogers, T.A.; Apte, S.C.; Mitrovic, S.M. Micronutrients as growth limiting factors in cyanobacterial blooms; a survey of freshwaters in South East Australia. *Aquat. Sci.* **2021**, *83*, 28. [CrossRef]
49. Facey, J.A.; Apte, S.C.; Mitrovic, S.M. A Review of the Effect of Trace Metals on Freshwater Cyanobacterial Growth and Toxin Production. *Toxins* **2019**, *11*, 643. [CrossRef]
50. Shaw-Allen, P.; Sutlerm, G.W. Metals. Available online: https://www.epa.gov/caddis-vol2/metals (accessed on 2 June 2024).
51. Schuler, M.S.; Relyea, R.A. A Review of the Combined Threats of Road Salts and Heavy Metals to Freshwater Systems. *BioScience* **2018**, *68*, 327–335. [CrossRef]
52. Monchanin, C.; Devaud, J.-M.; Barron, A.B.; Lihoreau, M. Current permissible levels of metal pollutants harm terrestrial invertebrates. *Sci. Total Environ.* **2021**, *779*, 146398. [CrossRef] [PubMed]
53. Mogren, C.L.; Trumble, J.T. The impacts of metals and metalloids on insect behavior. *Entomol. Exp. Et Appl.* **2010**, *135*, 1–17. [CrossRef]
54. Rainbow, P.S. Trace metal concentrations in aquatic invertebrates: Why and so what? *Environ. Pollut.* **2002**, *120*, 497–507. [CrossRef] [PubMed]
55. Marschner, H. *Marschner's Mineral Nutrition of Higher Plants*, 3rd ed.; Academic Press: Walhtam, MA, USA, 2011.
56. Gall, J.E.; Boyd, R.S.; Rajakaruna, N. Transfer of heavy metals through terrestrial food webs: A review. *Environ. Monit. Assess.* **2015**, *187*, 201. [CrossRef]
57. Afshan, S.; Ali, S.; Ameen, U.S.; Farid, M.; Bharwana, S.A.; Hannan, F.; Ahmad, R. Effect of different heavy metal pollution on fish. *Res. J. Chem. Environ. Sci* **2014**, *2*, 74–79.
58. Amundsen, P.A.; Staldvik, F.J.; Lukin, A.A.; Kashulin, N.A.; Popova, O.A.; Reshetnikov, Y.S. Heavy metal contamination in freshwater fish from the border region between Norway and Russia. *Sci. Total Environ.* **1997**, *201*, 211–224. [CrossRef] [PubMed]
59. Lucia, M.; André, J.-M.; Gontier, K.; Diot, N.; Veiga, J.; Davail, S. Trace element concentrations (mercury, cadmium, copper, zinc, lead, aluminium, nickel, arsenic, and selenium) in some aquatic birds of the Southwest Atlantic Coast of France. *Arch. Environ. Contam. Toxicol.* **2010**, *58*, 844–853. [CrossRef] [PubMed]
60. Burger, J.; Gochfeld, M. Behavioral impairments of lead-injected young herring gulls in nature. *Toxicol. Sci.* **1994**, *23*, 553–561. [CrossRef]
61. Scheuhammer, A. The chronic toxicity of aluminium, cadmium, mercury, and lead in birds: A review. *Environ. Pollut.* **1987**, *46*, 263–295. [CrossRef] [PubMed]
62. Epstein, E.; Bloom, A.J. *Mineral Nutrition of Plants: Principles and Perspectives*; Sinauer: Sunderland, MA, USA, 1853.
63. Cavet, J.S.; Borrelly, G.P.; Robinson, N.J. Zn, Cu and Co in cyanobacteria: Selective control of metal availability. *FEMS Microbiol. Rev.* **2003**, *27*, 165–181. [CrossRef] [PubMed]
64. Chakraborty, P.; Babu, P.R.; Acharyya, T.; Bandyopadhyay, D. Stress and toxicity of biologically important transition metals (Co, Ni, Cu and Zn) on phytoplankton in a tropical freshwater system: An investigation with pigment analysis by HPLC. *Chemosphere* **2010**, *80*, 548–553. [CrossRef] [PubMed]
65. Moreno-Vivián, C.; Cabello, P.n.; Martínez-Luque, M.; Blasco, R.; Castillo, F. Prokaryotic nitrate reduction: Molecular properties and functional distinction among bacterial nitrate reductases. *J. Bacteriol.* **1999**, *181*, 6573–6584. [CrossRef]
66. Axler, R.; Gersberg, R.; Goldman, C. Stimulation of nitrate uptake and photosynthesis by molybdenum in Castle Lake, California. *Can. J. Fish. Aquat. Sci.* **1980**, *37*, 707–712. [CrossRef]
67. Ali, H.; Khan, E.; Ilahi, I. Environmental Chemistry and Ecotoxicology of Hazardous Heavy Metals: Environmental Persistence, Toxicity, and Bioaccumulation. *J. Chem.* **2019**, *2019*, 6730305. [CrossRef]
68. Whetten, C.L. This Strange Enterprise: Geneva Steel and the American West. Master of Arts Thesis, The University of Utah, Salt Lake City, UT, USA, 2011.
69. Fuhriman, D.K.; Merritt, L.B.; Miller, A.W.; Stock, H.S. Hydrology and Water Quality of Utah Lake. *Great Basin Nat. Mem.* **1981**, *5*, 43–67.

70. United States. Census Bureau. US Census Bureau Publications-Census of Population and Housing. Available online: https://www.census.gov/prod/www/decennial.html (accessed on 1 October 2021).
71. Schoderboeck, L.; Mühlegger, S.; Losert, A.; Gausterer, C.; Hornek, R. Effects assessment: Boron compounds in the aquatic environment. *Chemosphere* **2011**, *82*, 483–487. [CrossRef] [PubMed]
72. Norici, A.; Hell, R.; Giordano, M. Sulfur and primary production in aquatic environments: An ecological perspective. *Photosynth. Res.* **2005**, *86*, 409–417. [CrossRef]
73. Karjalainen, J.; Hu, X.; Mäkinen, M.; Karjalainen, A.; Järvistö, J.; Järvenpää, K.; Sepponen, M.; Leppänen, M.T. Sulfate sensitivity of aquatic organism in soft freshwaters explored by toxicity tests and species sensitivity distribution. *Ecotoxicol. Environ. Saf.* **2023**, *258*, 114984. [CrossRef] [PubMed]
74. Bodzek, M. The removal of boron from the aquatic environment–state of the art. *Desalination Water Treat.* **2016**, *57*, 1107–1131. [CrossRef]
75. Gad, S.C. Barium. In *Encyclopedia of Toxicology*, 3rd ed.; Wexler, P., Ed.; Academic Press: Oxford, UK, 2014; pp. 368–370. [CrossRef]
76. Casbeer, W.; Williams, G.P.; Borup, M.B. Phosphorus Distribution in Delta Sediments: A Unique Data Set from Deer Creek Reservoir. *Hydrology* **2018**, *5*, 58. [CrossRef]
77. Szklarek, S.; Górecka, A.; Wojtal-Frankiewicz, A. The effects of road salt on freshwater ecosystems and solutions for mitigating chloride pollution—A review. *Sci. Total Environ.* **2022**, *805*, 150289. [CrossRef]
78. Maier, K.J.; Knight, A.W. Ecotoxicology of selenium in freshwater systems. In *Reviews of Environmental Contamination and Toxicology: Continuation of Residue Reviews*; Whitacre, D.M., Ed.; Springer: New York, NY, USA, 1994; Volume 197.
79. Diaz, X.; Johnson, W.P.; Naftz, D.L. Selenium mass balance in the Great Salt Lake, Utah. *Sci. Total Environ.* **2008**, *407*, 2333–2341. [CrossRef] [PubMed]
80. Utah Department of Environmental Quality. Standards of Quality for Waters of the State. Available online: https://documents.deq.utah.gov/water-quality/standards-technical-services/DWQ-2021-017555.pdf (accessed on 19 September 2022).
81. Utah Department of Environmental Quality Division of Water Quality. Water Quality Standards-Utah Department of Environmental Quality. Available online: https://deq.utah.gov/water-quality/water-quality-standards (accessed on 19 September 2022).
82. Utah Department of Environmental Quality. Water Qulaity Assessment and Analysis: Utah Lake Water Quality Study. Available online: https://deq.utah.gov/water-quality/water-quality-assessment-and-analysis-utah-lake (accessed on 14 March 2024).
83. Horowitz, A.J.; Elrick, K.A.; Colberg, M.R. The effect of membrane filtration artifacts on dissolved trace element concentrations. *Water Res.* **1992**, *26*, 753–763. [CrossRef]
84. Holcomb, B. *Utah Lake Marina HAB Treatments Evaluation of Treatment Effectiveness 2021 Interim Report Utah Department of Environmental Quality*; Utah Department of Environmental Quality: Salt Lake City, UT, USA, 2021.
85. The Utah Lake Authority. *Utah Lake Authority FY 2023 Annual Monitoring Report*; Utah Division of Environmental Quality: Salt Lake City, UT, USA, 2023.
86. United States Bureau of Reclamation. Historic Data. Available online: https://www.usbr.gov/rsvrWater/HistoricalApp.html (accessed on 15 January 2024).
87. United States Environmental Protection Agency. *2018 Final Aquatic Life Criteria for Aluminum in Freshwater*; US EPA, Environmental Protection Agency, Ed.; US EPA: Washington, DC, USA, 2022.
88. Correll, D.L. The role of phosphorus in the eutrophication of receiving waters: A review. *J. Environ. Qual.* **1998**, *27*, 261–266. [CrossRef]
89. PSOMAS. *Utah Lake TMDL: Pollutant Loading Assessment & Designated Beneficial Use Impairment Assessment*; Utah Department of Environmental Quality, Ed.; Utah Department of Environmental Quality: Salt Lake City, UT, USA, 2007.
90. United States Environmental Protection Agency. Consumer Factsheet on: Barium (Archived). Available online: https://archive.epa.gov/water/archive/web/pdf/archived-consumer-fact-sheet-on-barium.pdf (accessed on 14 March 2024).
91. United States Environmental Protection Agency. National Primary Drinking Water Regulations. Available online: https://www.epa.gov/ground-water-and-drinking-water/national-primary-drinking-water-regulations (accessed on 14 March 2024).
92. Welch, A.H.; Westjohn, D.; Helsel, D.R.; Wanty, R.B. Arsenic in ground water of the United States: Occurrence and geochemistry. *Groundwater* **2000**, *38*, 589–604. [CrossRef]
93. Korte, N.E.; Fernando, Q. A review of arsenic (III) in groundwater. *Crit. Rev. Environ. Control* **1991**, *21*, 1–39. [CrossRef]

water

MDPI

Article

Influence of Anthropogenic and Climatic Factors on the Dynamics of Nitrogen and Phosphorus in an Urbanized River Basin

Qinghuan Zhang [1], Fuzhen Liu [2], Fenfang Tan [3] and Abdul Qadeer [1,*]

1 National Engineering Laboratory for Lake Pollution Control and Ecological Restoration, State Environmental Protection Key Laboratory for Lake Pollution Control, Chinese Research Academy of Environmental Sciences, Beijing 100012, China; zhang.qinghuan@craes.org.cn
2 Changsha Uranium Geology Research Institute, China National Nuclear Corporation, Changsha 410125, China
3 Ecological and Environmental Monitoring Centre of Changsha in Hunan Province, Changsha 410001, China
* Correspondence: dr.aq.geographer@gmail.com

Abstract: The management of and reduction in ammonia nitrogen (NH_3-N) and total phosphorus (TP) in the water environment are crucial for protecting water quality amid rapid urbanization and population growth in highly industrialized regions. Specifically, in the Xiangjiang River Basin, the development of the Chang–Zhu–Tan urban agglomeration resulted in the deterioration of river water quality in the past, where ammonia nitrogen (NH_3-N) and total phosphorus (TP) were the dominant pollutants. This study aims to assess the influence of anthropogenic and climatic factors on the dynamics of nitrogen and phosphorus in an urbanized river basin in the middle Xiangjiang River Basin, China, from 2016 to 2020. This study examines NH_3-N and TP trends and their influencing factors across six tributaries, as well as how their concentrations have changed in response to urbanization and wastewater treatment management. The results reveal that average NH_3-N and TP concentrations decreased from 2016 to 2020 in the urbanized river system. NH_3-N and TP concentrations exhibited positive correlations with the proportion of cropland and negative correlations with population number, percentages of urban lands, and forests. In contrast, the influence of precipitation and streamflow on NH_3-N and TP concentrations was relatively weak. Consequently, agricultural activity was the primary contributing factor to NH_3-N and TP concentrations. Our study also suggests that the government's newly implemented water protection regulations can effectively control pollutant levels in urbanized river basins.

Keywords: urbanized river basin; ammonia nitrogen; total phosphorus; water quality; statistical analyses

1. Introduction

The global hydrological cycle is influenced by various factors, including climate change and anthropogenic activities, and serves as the primary driver of the nutrient cycle within a basin [1–3]. The anthropogenic production of nitrogen (N) and phosphorus (P) for industrial and agricultural development since the 20th century has led to the release of large quantities of these nutrients, which has significantly affected water quality [4]. Surface water pollution caused by N and P poses a serious threat to aquatic habitats [5]. The acceleration of urbanization, characterized by extensive land use changes and rising urban populations, has contributed to environmental issues such as eutrophication in some water bodies [6]. Ammonia nitrogen (NH_3-N) is a common pollutant in aquatic environments and a significant contributor to algal growth [7]. NH_3-N exists in the forms of free ammonia and ammonium ions, while the pH and water temperature are normally in the ranges of 6 to 9 and 5–30 °C, respectively [8]. The primary sources of NH_3-N in the water environment include industrial wastewater, domestic wastewater, and agricultural nonpoint source pollution. Total phosphorus (TP) mainly comes from agricultural runoff, the petrochemical industry, agricultural pesticides, and the food manufacturing industry. The spatial and temporal patterns of nutrients reflect the combined effects of diffuse and point source pollution [9]. Improving water quality in the process of urbanization is crucial for the sustainable development of urban ecosystems [10]. A clear understanding of the water environment and the factors influencing pollutants in industrialized areas is essential for developing effective pollution prevention and control strategies.

Extensive research has been conducted globally on the dynamics of nitrogen (N) and phosphorus (P) in water environments. However, most studies lack a comprehensive analysis of the combined influence of anthropogenic and climatic factors on water quality. For example, in the middle and lower reaches of the Yellow River Basin, industrial discharge, agricultural runoff, and domestic sewage were the main sources of water pollution, particularly concerning N and P [11]. Santy [12] examined the impacts of agricultural activity, industrial sewage, and climate change on water quality in an industrialized region of the Ganga basin, finding that agricultural land use exacerbates water quality degradation. Agricultural nonpoint source pollution, including NH_3-N and TP, places a heavy burden on the water environment [13]. In an agricultural watershed in Central China, N and P concentrations in a pond water were significantly correlated to land cover metrics, such as the proportion of orchard area and the landscape shape index [14]. These examples confirmed the observations of Wang [15] that the spatial distribution of industrialization and population density influences the discharge volumes of industrial and domestic wastewater, leading to varying impacts on surface water quality. The Chinese government has enacted the Action Plan for Prevention and Control of Water Pollution nationwide, which has improved water quality in watersheds [16]. Studies on the combined influence of anthropogenic and climatic factors on the water quality in urbanized watersheds are still lacking.

The Xiangjiang River in Changsha City is situated in a highly industrialized area of the middle Yangtze River, encompassing the Chang–Zhu–Tan urban agglomeration. This region features a diverse and complex terrain, including basins, hills, and mountains, with considerable variations in elevation. As a key component of the middle reaches of the Yangtze River urban agglomeration, it serves as a strategic core area for national economic development and the principal zone for new urbanization [17]. Like other industrialized urban basins, rivers in this area receive various sources of water pollutants, including industrial effluents, domestic sewage, and agricultural runoff [18]. NH_3-N and TP concentrations in some rivers exceed the Grade III standard from 2016 to 2020 (1.0 mg/L for NH_3-N and 0.2 mg/L for TP) [19]. Previous studies in this area have utilized multivariate statistical analyses, such as principal component analysis [20,21] and trend analysis methods [22] to identify water quality patterns. However, there is a lack of research exploring regional differences between tributaries. Furthermore, the Changsha City has enacted water pollution control action in 2016, and the implications of these policies are worthy of investigation. The Action Plan for Prevention and Control of Water Pollution, enacted nationwide, includes strengthening industrial pollution prevention, enhancing urban domestic pollution management, and advancing agricultural and rural pollution control for the treatment of water pollution in key river basins [23]. This paper examines NH_3-N and TP variations in the Xiangjiang River and its six tributaries (i.e., Jinjiang, Liuyang, Laodao, Longwanggang, Shahe, and Weishui rivers) by applying multiple statistical techniques, including Mann–Kendall trend analysis and correlation assessments. We compare the temporal and spatial trends of NH_3-N and TP concentrations in the urbanized river basin from 2016 to 2020, and explore the influence of anthropogenic and climatic factors on these nutrient levels in each tributary catchment. Anthropogenic factors include primary land use types and population number, while climatic factors include precipitation and temperature, with streamflow considered for one tributary.

2. Study Area and Data Collection

2.1. Study Area

The Xiangjiang River flows from south to north through Changsha, the capital city of Hunan Province, China. The total basin area exceeds 10,000 km^2, accounting for 43% of the area of Hunan Province. The Chang–Zhu–Tan urban agglomeration is situated in the middle reaches of the Yangtze River Basin and comprises three cities: Changsha, Zhuzhou, and Xiangtan [17]. The region was initially proposed in 1984 and has experienced rapid industrialization since 2004. Changsha city lies in the lower reach of the Xiangjiang River, where its six tributaries, including Jinjiang, Longwanggang, Weishui, Liuyang, Laodao, and Shahe rivers, enter the main river (Figure 1a). The catchment areas of Longwanggang, Shahe, Jinjiang, Laodao, Weishui, and Liuyang rivers are 179.67 km^2, 226.35 km^2, 777.21 km^2, 2537.38 km^2, 2740.19 km^2, and 4267.52 km^2, respectively. The total area of Changsha city is 11,800 km^2. Land use types in the region include forest, cropland, grassland, water bodies, and urbanized land (impervious area) (Figure 1b). The region is in a humid subtropical monsoon climate zone. Spring and autumn are short, while winter and summer are long. The wet season spans from March to September, while the dry season lasts from October to February of the following year. Historical climate data from 1975 to 2015 show that the average annual temperature was 17.4 °C, and the average annual precipitation was 1403.7 mm [24].

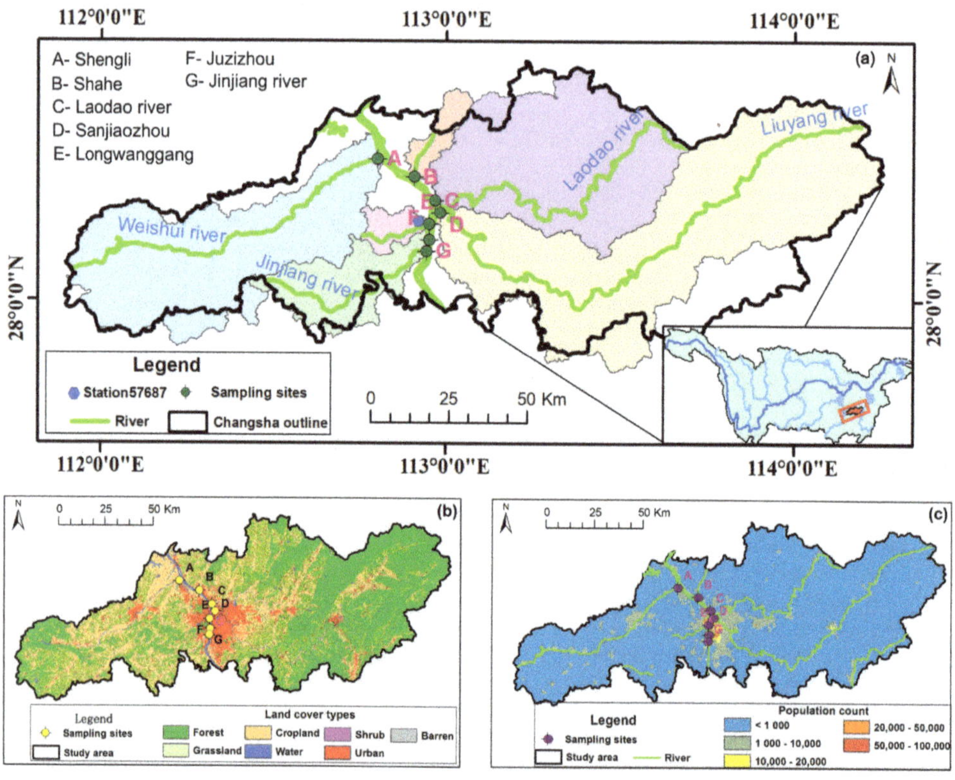

Figure 1. Location map of the (**a**) study area and the sampling sites in Changsha city, with the Xiangjiang River and six tributary rivers; (**b**) land use types in 2020 [25]; (**c**) population number [15].

River systems in this area play a crucial role in sustaining the economy, society, and ecosystem of the metropolitan area. Population number in Figure 1c represents the total number of people per grid cell at a resolution of 30 arc-seconds. Population density per grid cell was utilized as a metric to assess the degree of urbanization intensity [15]. Industrial wastewater and domestic sewage are generally processed in nearby wastewater treatment plants separately and then discharged to rivers. A former study found that water quality was relatively poor because of pollutants associated with industrial activities and household sewage during the 2004–2015 period [22].

2.2. Data Collection

Surface water samples from 2016 to 2020 were collected every month by the Ministry of Ecology and Environment of the People's Republic of China and the Environmental Monitoring Centre of Changsha. Among them, five sampling sites had water quality data from January 2016 to December 2020, while two sampling sites (Longwanggang and Shahe) had water quality data from August 2017 to December 2020. The surface water sampling points are located upstream of the confluence of each tributary river and the Xiangjiang River main stem. The measured water quality parameters included pH, dissolved oxygen (DO), permanganate index (COD_{Mn}), chemical oxygen demand measured by oxidization of dichromate (COD_{Cr}), biological oxygen demand (BOD), ammonia nitrogen (NH_3-N), and total phosphorus (TP). Water pH and DO were measured in situ with a portable multiparameter water quality analyzer (YSI ProQuatro, Columbus, OH, USA), where the detection limit of DO was 0.01 mg/L. TP was determined by the ammonium molybdate spectrophotometric method, with a detection limit of 0.01 mg/L. The minimum values of COD_{Mn} and COD_{Cr} were 0.5 mg/L and 4 mg/L, respectively. BOD was determined after 5 days for dilution and seeding method, where the detection limit was 0.5 mg/L. NH_3-N was measured using the Nessler's reagent spectrophotometry method, which included ammonia nitrogen in the form of free ammonia and ammonium ions. The detection limit was 0.025 mg/L. TP was measured using unfiltered water samples. The Environmental Monitoring Centre of Changsha measured three parameters (NH_3-N, TP, and COD_{Mn}) in Longwanggang and Shahe rivers, and seven parameters (NH_3-N, TP, COD_{Mn}, DO, pH, COD_{Cr}, and BOD) in Jinjiang, Liuyang, Laodao, Weishui, and the Xiangjiang rivers.

The daily precipitation and air temperature data were collected from China Meteorological Data Service Centre at Changsha station (no. 57687 Wangcheng district, Changsha; http://data.cma.cn/en; accessed on 1 January 2023). The daily climatic parameters were aggregated to monthly average values. Digital elevation model (DEM) data were collected from Shuttle Radar Topography Mission V4.1 at a spatial resolution of 90 m (https://www.gscloud.cn/home; accessed on 1 January 2023). The land use datasets at a 30 m spatial resolution were downloaded from Yang and Huang [25]. Urban lands were considered as impervious areas. The datasets of gridded population numbers from 2016 to 2020 were downloaded from the WorldPop Hub (https://hub.worldpop.org/project/categories?id=3; accessed on 1 January 2023). Land use parameters and population number were aggregated to yearly data from 2016 to 2020. We also obtained streamflow data at Liuyang river, which was collected and measured by the Hydrological Bureau of Changsha City. The average slope and elevation of each river catchment were calculated, and there was not a big difference in elevation of each river catchment. In the following analysis, precipitation and temperature represent climatic factors, while main land use types and population number represent anthropogenic factors.

3. Methods

3.1. The Mann–Kendall Test and Correlation Analysis

We applied the Mann–Kendall trend test [26,27] to detect increasing or decreasing trends over time in the data, with a significance level of 0.05. A Z statistic value greater than 1.96 indicates a significant increasing trend at the 0.05 significance level, whereas a Z statistic value less than −1.96 indicates a significant decreasing trend at the 0.05 significance level [28].

Pearson's correlation coefficient (r) was calculated between pollutant concentrations and potential influencing factors [29]. For climatic factors, we calculated the correlation coefficients between monthly precipitation or temperature and monthly NH_3-N or TP concentrations in each river separately. To assess the correlations between different seasons, we separated precipitation and temperature into wet and dry seasons. The wet season spans from March to September, while the dry season is from October to February of the following year. For anthropogenic factors, we calculated the correlation coefficients between annual land use percentages and annual average NH_3-N or TP concentrations in each river. A significant correlation was defined as having a *p*-value of less than 0.05.

We corrected the *p*-values for the correlations between precipitation and NH_3-N or TP in all rivers, and for the correlations between temperature and NH_3-N or TP in all rivers using the Benjamini–Hochberg correction method [30]. For land use types, we corrected the *p*-values for correlations between percentages of each land use type and NH_3-N or TP in all rivers. After corrections, the significant correlations were those with original *p*-values less than 0.01. All statistical analyses in this study were performed using "stats" and "trend" packages in R software version 4.2.3 [31,32].

3.2. The Water Quality Index

The Water Quality Index (WQI) is a mathematical tool used to transform large quantities of water quality data into a single number, thus providing a simple and understandable tool for managers and decision-makers on the quality and possible uses of a given water body [33,34]. WQIs are calculated in two steps. First, the measured concentrations of the selected water quality variables with different units are transformed into unitless subindex values. Second, the values are then used to calculate the WQIs in a simple equation [35]. For the Xiangjiang River and its six tributaries, we calculated the minimum WQI using Equations (1)–(3):

$$C_i = \frac{X_i}{S_i} \tag{1}$$

$$P_i = \frac{C_i}{\sum C_i} \times 100\% \tag{2}$$

$$WQI = \sum_{i=1}^{n} C_i P_i \tag{3}$$

where C_i represents the standardized pollutant concentration for parameter i, X_i is the pollutant concentration for parameter i, S_i represents the national standard water quality levels for parameter i, P_i is the relative weight in percentage assigned to each parameter, and n is the total number of parameters. The water class is defined as excellent when WQI is below 50, good when WQI ranges between 50 and 100, poor when WQI is between 100 and 200, and inferior/unsuitable when WQI is above 200 [36]. COD_{Mn}, NH_3-N, and TP were common variables for the six tributary rivers, and former studies showed that a minimum number of parameters can represent water quality [37]. Thus, we calculated WQI using the three variables, including COD_{Mn}, NH_3-N, and TP.

4. Results

4.1. Time Series of NH_3-N and TP Concentrations in the Rivers and Other Anthropogenic Variables

Concentrations of NH_3-N and TP on a monthly timestep from 2016 to 2020 in the Xiangjiang River and its six tributaries were used to calculate the average values and trends (Table 1). NH_3-N and TP concentrations showed a decreasing trend from 2016 to 2020 in all rivers, except for the Longwanggang River.

Figure 2 shows that NH_3-N and TP concentrations exceeded the Grade III level according to the China Surface Water Environmental Quality Standard in Liuyang, Laodao, and Weishui rivers between 2016 and 2018, while NH_3-N and TP concentrations exceeded the Grade V level in the Longwanggang River between 2018 and 2019. In Jinjiang, Xiangjiang, and Shahe rivers, the two pollutants remained below the Grade III level, indicating that these rivers had better water quality than the other rivers. The national standard levels of NH_3-N and TP concentrations in river water are provided in Table S1. Overall, NH_3-N and TP concentrations in the rivers decreased from 2016 to 2020. The average and standard deviations of each parameter in the rivers are presented in Table S2.

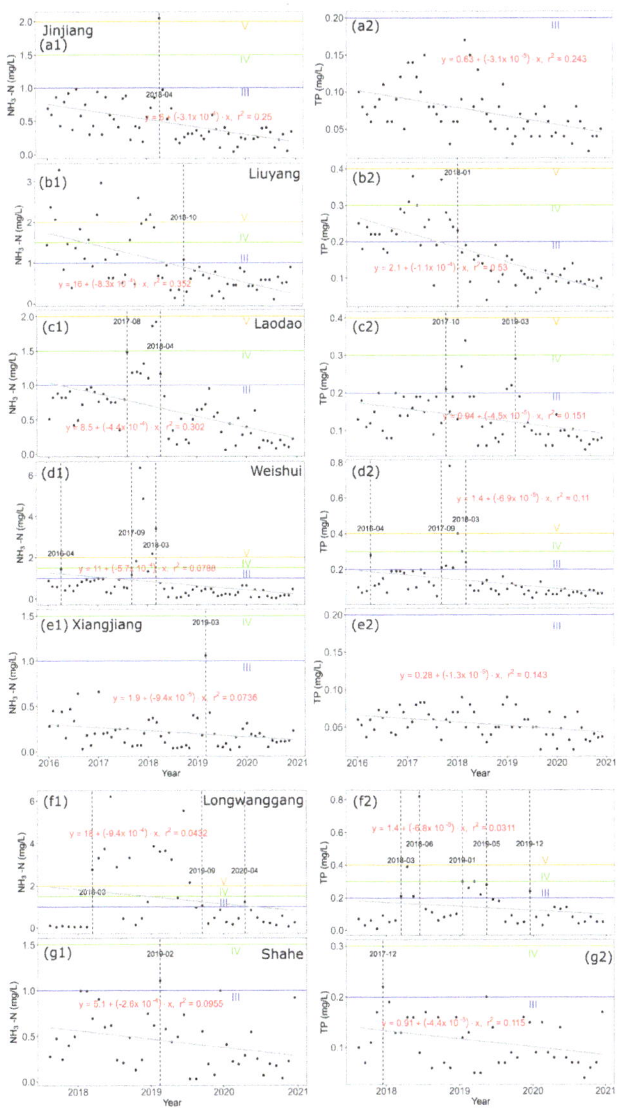

Figure 2. Time series of NH_3-N and TP concentrations in (**a1**,**a2**) Jinjiang, (**b1**,**b2**) Liuyang, (**c1**,**c2**) Laodao, (**d1**,**d2**) Weishui, (**e1**,**e2**) Xiangjiang, (**f1**,**f2**) Longwanggang, and (**g1**,**g2**) Shahe rivers. Colored lines III, IV, and V refer to Grade III, IV, and V national standards for NH_3-N and TP concentrations, respectively. Longwanggang and Shahe rivers have available data from August 2017 to December 2020, so their figures were drawn on a different x-axis compared to other rivers. The vertical dashed lines represent the points in time when NH_3-N or TP levels changed from one grade to another.

Table 1. Average values (mg/L) and Z statistics of pollutant concentrations in the Xiangjiang River and its six tributary rivers.

Parameters		Jinjiang	Longwanggang	Liuyang	Laodao	Weishui	Shahe	Xiangjiang
NH_3-N	Average	0.48	1.38	0.99	0.63	0.75	0.441	0.22
	Std	0.33	1.65	0.74	0.43	1.08	0.30	0.18
	Z statistic	−4.72	−0.47	−4.67	−4.87	−4.52	−2.27	−2.33
	p-value	0.00	0.64	0.00	0.00	0.00	0.02	0.02
TP	Average	0.07	0.14	0.17	0.13	0.13	0.11	0.05
	Std	0.03	0.14	0.08	0.06	0.11	0.05	0.02
	Z statistic	−4.36	−0.89	−6.08	−3.55	−4.10	−2.16	−2.95
	p-value	0.00	0.37	0.00	0.00	0.00	0.03	0.00

Note: std is standard deviation.

Annual values of the three land use types and population number from 2016 to 2020 are presented in Figure 3. The three land use types account for more than 90% of the total area in each catchment. The percentage of urban lands and population number increased (Figure 3a,d), while croplands decreased (Figure 3c) from 2016 to 2020. The percentage of forest areas did not change apparently, as shown in Figure 3b.

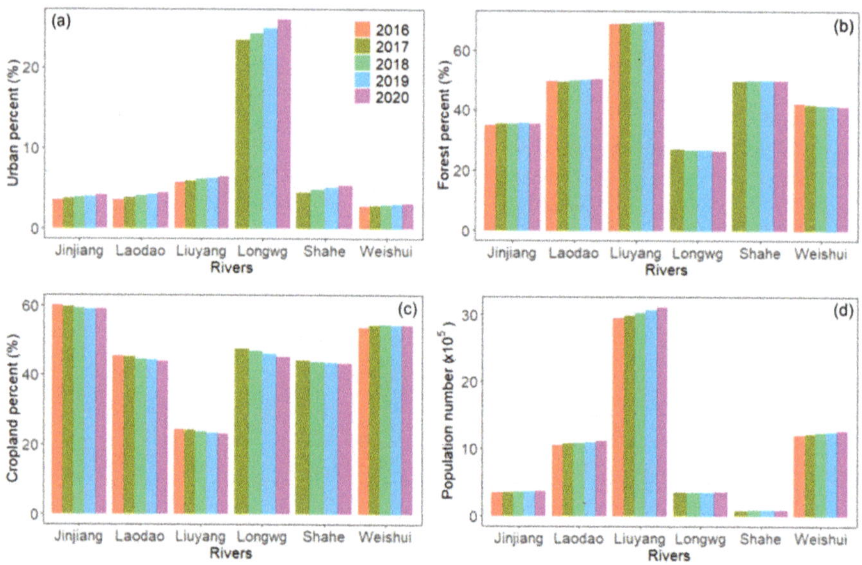

Figure 3. Percent of urban lands (**a**), percent of forest (**b**), percent of croplands (**c**), and population number (**d**) in the six tributary river catchments from 2016 to 2020.

4.2. Correlation Analyses and Water Quality Status

The correlation heatmaps of the seven water quality parameters in the Xiangjiang River, Jinjiang, Liuyang, Laodao, and Weishui rivers are presented in Figure 4. There were only three measured variables in Longwanggang and Shahe rivers, so correlation analysis was not conducted for these two rivers. In the Jinjiang River, COD_{Mn} was significantly negatively correlated to pH and DO; BOD was positively correlated to COD_{Mn}; NH_3-N was positively correlated to COD_{Mn} and COD_{Cr}; and TP was positively correlated to COD_{Cr}, BOD, and NH_3-N. In the Liuyang River, the correlation patterns were similar, except that COD_{Mn} was significantly positively correlated to pH, and TP was significantly negatively correlated to DO. In the Laodao River, TP was significantly positively correlated to DO. In the Weishui River, COD_{Cr} was significantly positively correlated to pH. In the Xiangjiang

River, correlations between the seven parameters were comparatively weak. The data showed that COD_{Mn} was significantly negatively correlated to DO, NH_3-N was positively correlated to DO, and TP was positively correlated to NH_3-N. In general, areas with high NH_3-N and TP also had high COD_{Mn}, COD_{Cr}, and BOD in these rivers.

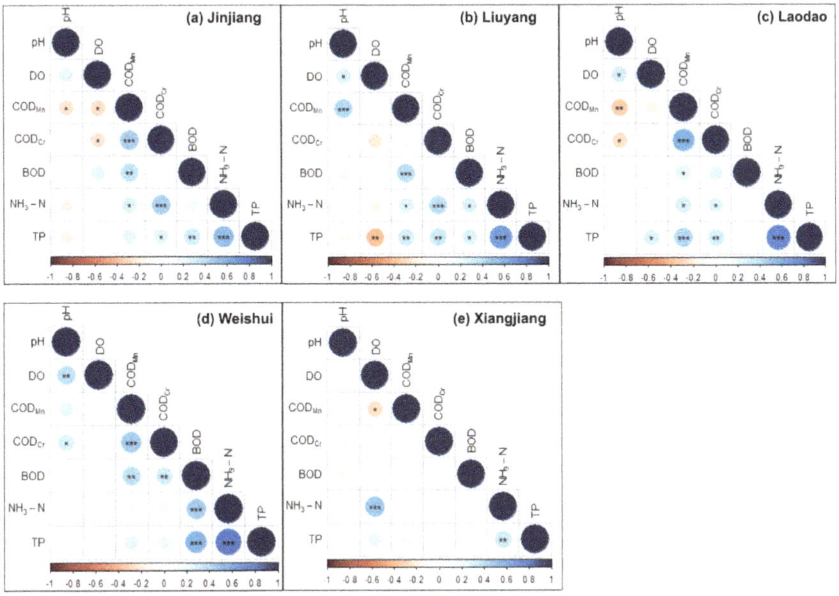

Figure 4. Correlation heatmaps of water quality parameters in the Xiangjiang River and its tributary rivers. In the circles, * represents $p < 0.05$, ** represents $p < 0.01$, and *** represents $p < 0.001$; correlation coefficients close to 0 are shown as blank.

The summary statistics of WQI in the Xiangjiang River and its six tributary rivers are presented in Table 2. In general, the Xiangjiang River exhibited the lowest average WQI (33.86) and, thus, had the best water quality among all the rivers. The Longwanggang River had the highest average WQI and the worst water quality (WQI was 123.29). The water quality of these rivers ranked from best to worst as follows: Xiangjiang, Jinjiang, Shahe, Laodao, Weishui, Liuyang, and Longwanggang rivers. The monthly WQI values from 2016 to 2020 can represent the evolution of water quality (Figure 5). The periods of high WQI values (above 100) overlapped with the periods of high NH_3-N and TP concentrations.

Table 2. Summary statistics of the WQI index in the Xiangjiang River and its six tributary rivers.

Statistics	Jinjiang	Longwanggang	Liuyang	Laodao	Weishui	Shahe	Xiangjiang
Minimum	27.61	26.29	33.90	27.95	40.59	29.03	21.44
Maximum	150.68	514.97	251.21	163.57	508.35	89.79	77.66
Average	56.59	123.29	91.76	70.85	84.72	58.40	33.86
Std	19.09	119.69	49.18	26.21	76.88	19.31	8.19

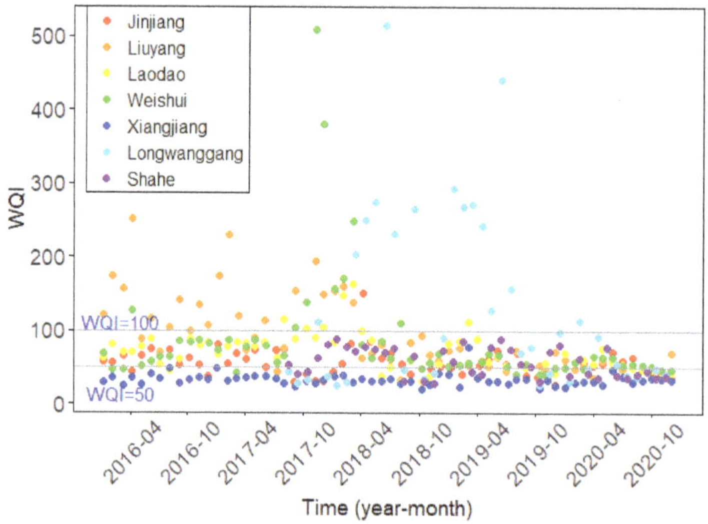

Figure 5. Time series of WQI in the Xiangjiang River and its six tributary rivers.

4.3. Correlations Between NH_3-N and TP Concentrations and Other Factors

We explored the influences of both natural and anthropogenic factors on NH_3-N and TP concentrations by examining their correlation coefficients (r values) with natural and anthropogenic explanatory factors. As the Xiangjiang River is the main stem encompassing the entire basin area, we performed the correlation analyses for the six tributary rivers. In our analyses, natural factors included precipitation, temperature, elevation, and slope. Anthropogenic factors were characterized by population number, the proportion of urban lands, the proportion of forested areas, and the proportion of croplands in each river catchment. In the six tributary river catchments, the primary land use types were croplands and forests, while in Longwanggang, urban lands also constituted a significant proportion.

Correlation coefficients were computed between pollutant concentrations (including NH_3-N and TP) and slope and elevation across the entire study area. The findings indicate that elevation and slope showed no apparent correlations with NH_3-N (with r values of 0.05 and −0.04, respectively, and p-values above 0.1). However, elevation and slope exhibited a weak correlation with TP (with r values of 0.23 and 0.15, respectively, and p-values below 0.05). TP concentrations increased slightly with elevation. The impacts of land use types, population, and climatic factors on NH_3-N and TP concentrations are shown in Figures 6 and 7. In Jinjiang, Laodao, and Liuyang rivers, NH_3-N concentrations were positively correlated with the percentage of croplands (r values were 0.77, 0.90, and 0.98, respectively). The correlation was significant in the Liuyang River, with the corrected p-value below 0.05. NH_3-N concentrations were negatively correlated with the percentage of forested areas in Jinjiang, Laodao, and Liuyang rivers (r values were −0.66, −0.95, and −0.94, respectively). NH_3-N concentrations were negatively correlated with population number and the percentage of urban lands in Jinjiang, Laodao, Liuyang, and Weishui rivers (r values were below −0.74), with significant correlations found in the Liuyang River (the corrected p-value was less than 0.05)

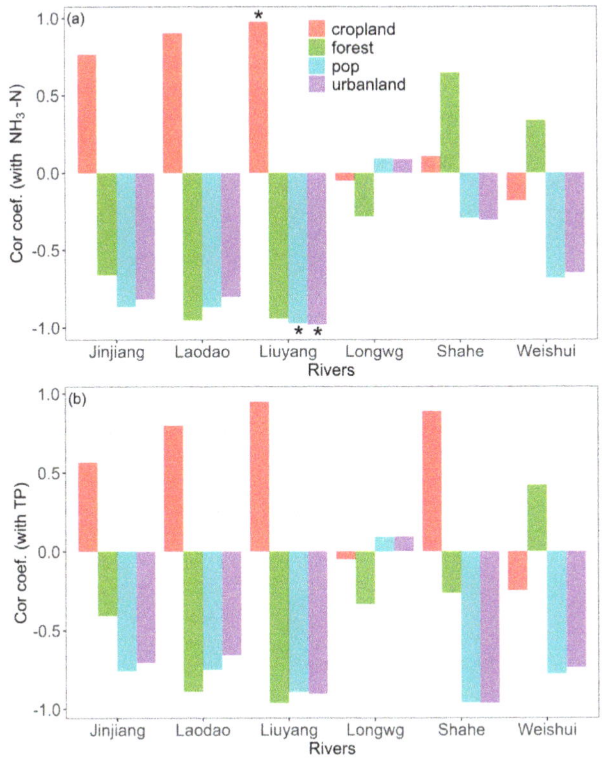

Figure 6. Correlation coefficients between main land use types and population number, and (**a**) NH_3 and (**b**) TP concentrations. * p-value < 0.05, which is corrected using the Benjamini–Hochberg correction method; pop is population number; Longwg represents Longwanggang.

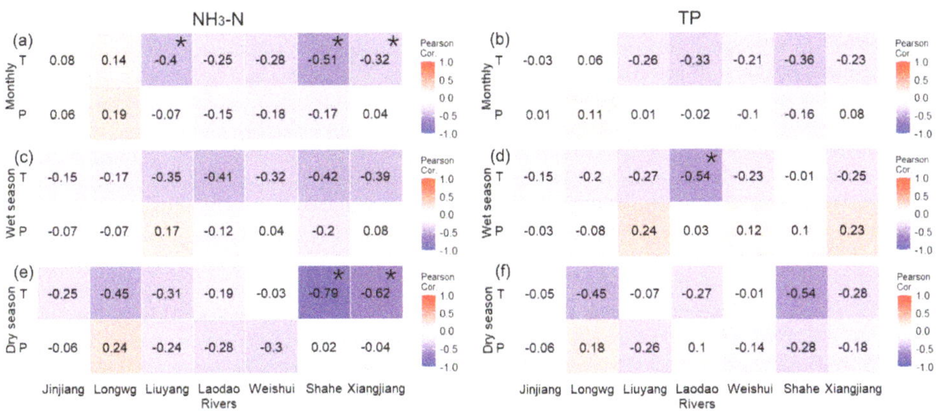

Figure 7. Pearson's correlation coefficients between NH_3-N (left column) and TP (right column) and climatic variables during (**a**,**b**) the whole time period, (**c**,**d**) the wet season, and (**e**,**f**) the dry season in the six rivers. One star represents corrected $p < 0.05$.

TP concentrations were positively correlated with the percentage of croplands in Jinjiang, Laodao, Liuyang, and Shahe rivers (r values were 0.57, 0.80, 0.95, and 0.89, respectively), and negatively correlated with the percentage of forested areas in Jinjiang, Laodao, and Liuyang rivers (r values were −0.41, −0.89, and −0.96, respectively). Moreover, in Jinjiang, Laodao, Liuyang, Shahe and Weishui rivers, TP concentrations were negatively correlated with population number and the percentage of urban lands (r values ranged from −0.96 to −0.66).

The correlation coefficients between monthly average precipitation or temperature and NH_3-N or TP concentrations across the entire time period, for wet and dry seasons, are presented in Figure 7. The correlation coefficients were relatively weaker compared to the relationships between anthropogenic factors and nutrient concentrations. On average, monthly temperature was negatively correlated with NH_3-N or TP concentrations in Liuyang, Laodao, Weishui, and Shahe rivers, and the Xiangjiang River (Figure 7a,b). The correlations with NH_3-N were weak but significant in Liuyang and Shahe rivers, and the Xiangjiang River (corrected p-values were less than 0.05). In the wet season, monthly temperature exhibited negative correlations with NH_3-N concentrations in all rivers, while it showed negative correlations with TP concentrations in all rivers except for the Shahe River (Figure 7c,d). The correlation was significant in the Laodao River, with a corrected p-value less than 0.05. In the dry season, monthly temperature showed negative correlations with NH_3-N concentrations in all rivers except for the Weishui River, and these correlations were significant in Shahe and the Xiangjiang River (corrected p-values were less than 0.05) (Figure 7e,f). Monthly temperature also showed negative correlations with TP concentrations in Longwanggang, Laodao, and Shahe rivers, and the Xiangjiang River. Monthly precipitation showed weak and insignificant correlations with NH_3-N or TP concentrations across all time periods, and wet and dry seasons.

5. Discussions

5.1. Differences in the Trends of NH_3-N and TP Concentrations

In general, NH_3-N and TP concentrations in the rivers decreased from 2016 to 2020. This is consistent with the target discharge permits and reduction action plans for the Xiangjiang River Basin at Grade III level that has been formulated by the Water Pollution Prevention and Control Action Plan of Hunan Province [38]. The action plans include two major parts: (1) The construction or upgrading of several sewage treatment plants in Liuyang and Laodao rivers; (2) Strict environmental management plans, which ensure the minimum ecological water requirement of rivers in the dry season (http://sthjt.hunan.gov.cn/, accessed on 1 January 2023). In addition to the Longwanggang River, the other rivers exhibited significant but slight decreasing trends in NH_3-N and TP concentrations, with p-values less than 0.05. These results are consistent with findings from other studies, which suggest that with proper management and control, water quality can be improved. For example, the total pollutant load control system has been effective in maintaining freshwater quality within Grade III level [39]. A previous study modeling water quality changes in a complex water system in China found that ammonia nitrogen and total phosphorus concentrations could be controlled within the water quality standards after implementing the optimal allocation of pollution loads [40].

5.2. Water Quality Status of the Rivers

In the Yangtze River Basin, COD_{Mn}, NH_3-N, and TP are considered to be the primary parameters affecting water quality [41]. Astaraie-Imani et al. [42] used NH_3-N and DO as water quality indicators and concluded that climate change could exacerbate the deterioration of river water quality. Similarly, a comparative analysis utilizing WQIs in the Kathmandu Valley of Nepal revealed that urban water quality was inferior to rural water quality [43]. Thus, WQIs serve as a useful index for comparing and analyzing the status of river water quality. The average WQIs indicated that the Xiangjiang River main stem had the best water quality, followed by Jinjiang, Shahe, Weishui, and Laodao rivers, which exhibited inferior water quality compared to the former rivers. Liuyang and Longwanggang rivers ranked the lowest in terms of water quality. Ongoing construction activities, along with incomplete infrastructure, such as pipelines and wastewater treatment plants, contributed to the substandard water quality in the Longwanggang river during the study period. In addition to the Longwanggang River, water quality in the other rivers has improved since 2018. The periods of high WQI values also corresponded to the periods of high NH_3-N and TP concentrations.

In the Liuyang River, TP showed a negative correlation with DO, which was different from the correlations observed in the other rivers. This may be related to the fact that the Liuyang River catchment had the highest percentage of forest cover, which could help improving the DO levels [3]. In general, the correlation heatmaps suggested that rivers with high concentrations of NH_3-N and TP also had elevated levels of COD_{Mn}, COD_{Cr}, and BOD in this study area.

5.3. Influencing Factors of NH_3-N and TP Variations

Our results align with the findings from [44], who reported weak correlations between TP and NH_3-N concentrations with precipitation and air temperature in the Yangtze River and other river basins in China. They emphasized that water quality impairments were predominantly influenced by anthropogenic drivers (82.5% for TP and 79.5% for NH_3-N), while natural factors played a secondary role (20.5% for TP and 17.5% for NH_3-N). The negative correlations between NH_3-N or TP concentrations and temperature in some rivers may be attributed to increased biogeochemical processes that consume nutrients. Additionally, we obtained streamflow data from the Langli hydrologic station, which is located before the confluence of the Liuyang River and the Xiangjiang River. This is the only hydrological station with streamflow data that match the time step and time span of the nutrient concentrations among the tributary rivers. The correlation analysis revealed no significant correlations between NH_3-N or TP concentrations and monthly streamflow. For example, Pearson's correlation coefficient between NH_3-N and monthly streamflow was -0.19 (p-value was 0.15), and between TP and monthly streamflow was -0.16 (p-value was 0.23). The correlations were not significant in either the wet or dry seasons. Figure S1 shows that while NH_3-N and TP concentrations decreased over the years, streamflow exhibited a cyclical pattern. Based on the performed analyses, streamflow does not affect nutrient concentrations. If we had obtained high-frequency data of nutrient concentrations, we could have conducted more advanced analyses on this subject. TP concentrations were higher in high-elevation than in low-elevation areas, suggesting an increased chance of erosion/deposition of phosphorus in high-elevation areas. The natural sources of phosphorus include mineral weathering and erosion processes [45]. Animal husbandry and industrial emissions are other sources of atmospheric phosphorus deposition [39], with their transport influenced by air masses [46].

In most of the rivers, NH_3-N exhibited positive correlations with the percentage of cropland and negative correlations with the percentage of forest area, urban land, and population number. The correlations were significant in the Liuyang River. Correlations between TP and land use types showed similar patterns with NH_3-N, except in the Shahe River. These findings suggest that land use types, especially croplands with agricultural activities, were the primary contributors to N and P concentrations during the study period. As a result, NH_3-N and TP concentrations in river water were sensitive to changes in the percentages of cropland and forest areas. The use of pesticides and fertilizers in agriculture can cause leaching of unutilized N and P loads into the water body [18]. In the Yellow River Basin, the reduction in fertilizer applications in recent years has led to a decline in anthropogenic nitrogen and phosphorus inputs [47]. Despite increases in population number and the percentage of urban lands, NH_3-N and TP concentrations decreased, suggesting that the influence of wastewater discharge on water quality did not change significantly. Targeted action plans, such as reducing the discharge of industrial wastewater pollution at the source and improving the treatment rate of industrial wastewater, are effective ways to control water pollution and protect the water environment [48].

Cheng et al. [49] found that population density had a greater impact than industrial structure on water quality at the catchment scale in the Haihe River Basin. Similarly, a recent study by [10] discovered that as urbanization progresses and the proportion of impervious surface area increased, levels of nitrogen and phosphorus pollution in water bodies also increased in the Yangtze River Delta. Their research suggests that expanding urban forests can reduce nitrogen and phosphorus pollution by 37% and 49%, respectively. However, in our study, we found that despite an increase in the proportions of urban land and population number, alongside a decrease in cropland area, concentrations of NH_3-N and TP decreased. This discrepancy is likely due to effective pollution control measures implemented by the Changsha government, as well as the maintenance of forest areas, which have mitigated the expected increase in NH_3-N and TP discharges associated with urban expansion and population growth. In this study, forest areas are located at higher elevations compared to urban and cropland areas, and they primarily receive nitrogen and phosphorus through atmospheric deposition. Basins with larger forested areas showed lower nutrient concentrations, reflecting a reduced presence of anthropogenic activities in comparison to basins with more cropland and urban areas. Urban forests play a crucial role in protecting urban water environments, as they can intercept rainfall, promote the infiltration of surface runoff, and absorb nitrogen and phosphorus in runoff [10].

Furthermore, nitrogen and phosphorus concentrations in the upstream Yangtze River were lower than those in the middle and downstream reaches, reflecting lower population density and agricultural activity, as well as increased pollutant loads [50]. Local governments enacted water pollution prevention and control regulations in 2016, which have been implemented since 2017. These regulations cover the Xiangjiang River and its tributaries, including the implementation of a total pollutant control system, the installation of automatic monitoring equipment for water pollution discharge, the reduction in fertilizer applications, the regular removal of floating objects and water hyacinths from rivers, and the enhancement of effluent quality from wastewater treatment plants. The implementation of effective pollution control measures can help to mitigate the impacts of urbanization on river water quality. For instance, in European rivers, the reduction in nutrient levels has been linked to nitrogen (N) and phosphorus (P) bans implemented by the European Union since the 1980s, which were specifically designed to manage water eutrophication [51]. These measures were part of the European Union Water Framework Directive 2000/60/EC, a significant piece of environmental legislation aimed at restoring Europe's waters, which can serve as a model for future environmental regulations [52]. Such regulations could be adopted in other highly urbanized river systems to improve river water quality.

6. Conclusions

In this study, we used statistical analyses, including Mann–Kendall trend analysis and correlation analysis, to examine variations in NH_3-N and TP concentrations and their influencing factors within an urbanized river system. In general, NH_3-N and TP concentrations decreased from 2016 to 2020, which were consistent with ongoing local governmental pollutant control measures. Agricultural activities, represented by the percentage of croplands, were the primary contributors to pollutant concentrations, while forest areas exhibited lower nutrient input due to the absence of anthropogenic activities compared to croplands and urban areas. On the other hand, natural factors such as precipitation and streamflow showed no significant correlation with NH_3-N and TP concentrations. The increase in temperature may accelerate biogeochemical processes that consume nutrients in water bodies, which warrants further investigation.

Despite rapid urbanization and population growth, water pollutant concentrations decreased, suggesting that local governmental management strategies have been effective. These regulations could be applied to other highly urbanized river systems to protect water quality. Future research should also consider the presence of additional pollutants from agricultural activities, such as pesticides, metals, and antibiotics. These findings provide valuable insights for mitigating water pollution and protecting river ecosystems in urbanized river systems.

Supplementary Materials: The following supporting information can be downloaded at: https://www.mdpi.com/article/10.3390/w16243635/s1. Table S1. Standard values for NH_3-N and TP concentrations of river water according to the China Surface Water Environmental Quality Standard (National Standard of the People's Republic of China, 2002). Table S2. The average value and standard deviation of each parameter. Figure S1. Time series of (a) NH_3-N concentrations and (b) TP concentrations, and (c) average monthly streamflow from 2016 to 2020 in the Liuyang River.

Author Contributions: Q.Z.: Writing—original draft, writing—review and editing, formal analysis, conceptualization, data curation. F.L.: Writing—review and editing. A.Q.: Validation, writing—review and editing. F.T.: Writing—review and editing. All authors have read and agreed to the published version of the manuscript.

Funding: A Government Procurement Contract from Changsha City funded this study (no. CSCG-GK-201909270052-1).

Data Availability Statement: The original contributions presented in the study are included in the article/Supplementary Materials, further inquiries can be directed to the corresponding author.

Acknowledgments: The authors would like to thank the reviewers and editors for their valuable suggestions.

Conflicts of Interest: Fuzhen Liu is now employed by the Changsha Uranium Geology Research Institute. The remaining authors declare that the research was conducted in the absence of any commercial or financial relationships that could be construed as a potential conflict of interest.

References

1. Oki, T.; Kanae, S. Global Hydrological Cycles and World Water Resources. *Science* **2006**, *313*, 1068. [CrossRef]
2. Whitehead, P.G.; Jin, L.; Bussi, G.; Voepel, H.E.; Darby, S.E.; Vasilopoulos, G.; Manley, R.; Rodda, H.; Hutton, C.; Hackney, C.; et al. Water quality modelling of the Mekong River Basin: Climate change and socioeconomic drive flow and nutrient flux changes to the Mekong Delta. *Sci. Total Environ.* **2019**, *673*, 218–229. [CrossRef]
3. Shi, X.; Mao, D.; Song, K.; Xiang, H.; Li, S.; Wang, Z. Effects of landscape changes on water quality: A global meta-analysis. *Water Res.* **2024**, *260*, 1–14. [CrossRef]
4. Xue, J.; Yuan, C.; Ji, X.; Zhang, M. Predictive modeling of nitrogen and phosphorus concentrations in rivers using a machine learning framework: A case study in an urban-rural transitional area in Wenzhou China. *Sci. Total Environ.* **2024**, *910*, 1–21. [CrossRef] [PubMed]
5. Wang, E.; Li, Q.; Hu, H.; Peng, F.; Zhang, P.; Li, J. Spatial characteristics and influencing factors of river pollution in China. *Water Environ. Res.* **2019**, *91*, 351–363. [CrossRef]
6. Xue, J.; Wang, Q.; Zhang, M. A review of non-point source water pollution modeling for the urban-rural transitional areas of China: Research status and prospect. *Sci. Total Environ.* **2022**, *826*, 154146. [CrossRef]
7. Yan, F.; Qiao, D.; Qian, B.; Ma, L.; Xing, X.; Zhang, Y.; Wang, X. Improvement of ccme wqi using grey relational method. *J. Hydrol.* **2016**, *543*, 316–323. [CrossRef]
8. Feng, Y.; Li, N.; Wang, J. Ecological risk evaluation of ammonia nitrogen pollution in China based on the ecological grey water footprint model. *J. Environ. Manag.* **2023**, *347*, 119087.
9. Li, S.; Liu, W.; Gu, S.; Cheng, X.; Xu, Z.; Zhang, Q. Spatio-temporal dynamics of nutrients in the upper Han River basin, China. *J. Hazard. Mater.* **2009**, *162*, 1340–1346. [CrossRef] [PubMed]
10. Wang, R.; Liu, L.; Tao, Z.; Wan, B.; Wang, Y.; Tang, X.; Li, Y.; Li, X. Effect of urbanization and urban forests on water quality improvement in the Yangtze River Delta: A case study in Hangzhou, China. *J. Environ. Manag.* **2024**, *351*, 119980. [CrossRef] [PubMed]
11. Zhao, M.; Wang, S.; Chen, Y.; Wu, J.; Xue, L.; Fan, T. Pollution status of the Yellow River tributaries in middle and lower reaches. *Sci. Total Environ.* **2020**, *722*, 137861. [CrossRef]
12. Santy, S.; Mujumdar, P.; Bala, G. Potential impacts of climate and land use change on the water quality of Ganga River around the industrialized Kanpur Region. *Sci. Rep.* **2020**, *10*, 9107. [CrossRef] [PubMed]
13. Feng, Z.; Zhang, R.; Liu, X.; Peng, Q.; Wang, L. Agricultural nonpoint source pollutant loads into water bodies in a typical basin in the middle reach of the Yangtze River. *Ecotoxicol. Environ. Saf.* **2023**, *268*, 115728. [CrossRef]
14. Xiao, H.; Su, R.; Luo, Y.; Jiang, Y.; Wang, Y.; Hu, R.; Lin, S. Effects of land cover patterns on pond water nitrogen and phosphorus concentrations in a small agricultural watershed in Central China. *Catena* **2024**, *237*, 107800. [CrossRef]
15. Wang, J.; Da, L.; Song, K.; Li, B. Temporal variations of surface water quality in urban, suburban and rural areas during rapid urbanization in Shanghai, China. *Environ. Pollut.* **2008**, *152*, 387–393. [CrossRef]
16. Liu, Q.; Wang, W.; Luo, B.; Wang, K. Analysis on Improvement of Water Environment in Minjiang River Basin by Ten-Action Plan for Prevention and Control of Water Pollution. *Sichuan Environ.* **2020**, *39*, 80–87.

17. Mi, Y.; Li, S.; Wang, Z. Spatial distribution and topographic gradient effects of habitat quality in the Chang-Zhu-Tan Urban Agglomeration, China. *Sci. Rep.* **2024**, *14*, 22563. [CrossRef] [PubMed]
18. Xie, W. Analysis on the current Status and regulatory measures of water pollution in the Xiang River Basin of China. *Nat. Environ. Pollut. Technol.* **2016**, *15*, 1435–1440.
19. *GB3838-2002*; Environmental Quality Standards for Surface Water. National Standard of the People's Republic of China. Ministry of Ecology and Environment of the People's Republic of China: Beijing, China, 2002.
20. Zhang, Q.; Li, Z.; Zeng, G.; Li, J.; Fang, Y.; Yuan, Q.; Wang, Y.; Ye, F. Assessment of surface water quality using multivariate statistical techniques in red soil hilly region: A case study of Xiangjiang watershed, China. *Environ. Monit. Assess.* **2009**, *152*, 123–131. [CrossRef]
21. Zhang, Z.; Tao, F.; Du, J.; Shi, P.; Yu, D.; Meng, Y.; Sun, Y. Surface water quality and its control in a river with intensive human impacts-a case study of the Xiangjiang River, China. *J. Environ. Manag.* **2010**, *91*, 2483–2490. [CrossRef] [PubMed]
22. Duan, W.; He, B.; Chen, Y.; Zou, S.; Wang, Y.; Nover, D. Identification of long-term trends and seasonality in high-frequency water quality data from the Yangtze River basin, China. *PLoS ONE* **2018**, *13*, e0188889. [CrossRef]
23. He, J.; Ma, L.; Wang, D.; Zhao, Y. A Working Drawing of Action Plan for Prevention and Control of Water Pollution: National Plan for Water Pollution Prevention and Control in Key River-Basins during the 13th Five-Year Plan Period. *Environ. Prot.* **2017**, *45*, 7–10.
24. Peng, P.; Liu, B.; Deng, N.; Deng, Y.; Li, X.; Tian, Y.; Ma, F.; Song, Q. Climate change trends and characteristics of Changsha City from 1975 to 2015. *Hunan For. Sci. Technol.* **2021**, *48*, 56–63. (In Chinese)
25. Yang, J.; Huang, X. The 30 m Annual Land Cover Datasets and Its Dynamics in CHINA from 1990 to 2021. 9 August 2022. Available online: https://zenodo.org/record/5816591 (accessed on 20 February 2023).
26. Mann, H.B. Nonparametric tests against trend. *Econometrica* **1945**, *13*, 245–259. [CrossRef]
27. Kendall, M.G. *Rank Correlation Methods*; Charles Griffin: London, UK, 1975.
28. Gocic, M.; Trajkovic, S. Analysis of changes in meteorological variables using Mann-Kendall and Sen's slope estimator statistical tests in Serbia. *Glob. Planet. Change* **2013**, *100*, 172–182. [CrossRef]
29. Bayo, J.; Lopez-Castellanos, J. Principal factor and hierarchical cluster analyses for the performance assessment of an urban wasterwater treatment plant in the Southeast of Spain. *Chemosphere* **2016**, *155*, 152–162. [CrossRef] [PubMed]
30. Benjamini, Y.; Hochberg, Y. Controlling the false discovery rate: A practical and powerful approach to multiple testing. *J. R. Stat. Soc. Ser. B (Methodol.)* **1995**, *57*, 289–300. [CrossRef]
31. Pohlert, T. Trend: Non-Parametric Trend Tests and Change-Point Detection. R Package version 1.1.6. 2023. Available online: https://CRAN.R-project.org/package=trend (accessed on 1 January 2023).
32. R Core Team. *R: A Language and Environment for Statistical Computing*; R Foundation for Statistical Computing: Vienna, Austria, 2024; Available online: https://www.R-project.org/ (accessed on 1 January 2023).
33. Stambuk-Giljanovic, N. Water quality evaluation by index in Dalmatia. *Water Res.* **1999**, *33*, 3423–3440. [CrossRef]
34. Bordalo, A.A.; Nilsumranchit, W.; Chalermwat, K. Water quality and uses of the Bangpakong River (Eastern Thailand). *Water Res.* **2001**, *35*, 3635–3642. [CrossRef]
35. Cude, C.G. Oregon water quality index: A tool for evaluating water quality management effectiveness. *J. Am. Water Resour. Assoc.* **2001**, *37*, 125–137. [CrossRef]
36. Wali, S.U.; Alias, N.B.; Harun, S.B.; Umar, K.J.; Gada, M.A.; Dankani, I.M.; Kaoje, I.U.; Usman, A.A. Water quality indices and multivariate statistical analysis of urban groundwater in semi-arid Sokoto Basin, Northwestern Nigeria. *Groundw. Sustain. Dev.* **2022**, *18*, 100779. [CrossRef]
37. Varol, M.; Karakaya, G.; Alpaslan, K. Water quality assessment of the Karasu River (Turkey) using various indices, multivariate statistics and APCS-MLR model. *Chemosphere* **2022**, *308*, 136415. [CrossRef] [PubMed]
38. People's Government of Hunan Province. Water Pollution Prevention and Control Action Plan. 2016. Available online: http://hunan.gov.cn/hnszf/szf/hnzb/2016/2016nd3q/szfwj_98977/201711/t20171118_4701811.html (accessed on 1 January 2023).
39. Zhang, J.; Ni, S.; Wu, W.; Huang, X.; Jiang, H.; Li, Q.; Wang, J.; Wu, G.; Zorn, C.; Yu, C. Evaluating the effectiveness of the pollutant discharge permit program in China: A case study of the Nenjiang River Basin. *J. Environ. Manag.* **2019**, *251*, 109501. [CrossRef] [PubMed]
40. Chen, L.; Han, L.; Tan, J.; Zhang, F.; Lin, Y. Optimal allocation model of point and non-point pollution loads in river network based on water quality standard of multiple control sections. *Water Resour. Prot.* **2021**, *37*, 128–134.
41. Bai, H.; Chen, Y.; Wang, Y.; Song, Z.; Tong, H.; Wei, Y.; Yu, Q.; Xu, Z.; Yang, S. Contribution rates analysis for sources apportionment to special river sections in Yangtze River Basin. *J. Hydrol.* **2021**, *600*, 126519. [CrossRef]
42. Astaraie-Imani, M.; Kapelan, Z.; Fu, G.; Butler, D. Assessing the combined effects of urbanization and climate change on the river water quality in an integrated urban wastewater system in the UK. *J. Environ. Manag.* **2012**, *112*, 1–9. [CrossRef] [PubMed]
43. Kannel, P.R.; Lee, S.; Lee, Y.-S.; Kanel, S.R.; Khan, S.P. Application of Water Quality Indices and Dissolved Oxygen as Indicators for River Water Classification and Urban Impact Assessment. *Environ. Monit. Assess.* **2007**, *132*, 93–110. [CrossRef]
44. Huang, J.; Zhang, Y.; Bing, H.; Peng, J.; Dong, F.; Gao, J.; Arhonditsis, G.B. Characterizing the river water quality in China: Recent progress and on-going challenges. *Water Res.* **2021**, *201*, 117309. [CrossRef]

45. Gardner, L.R. The role of rock weathering in the phosphorus budget of terrestrial watersheds. *Biogeochemistry* **1990**, *11*, 970110. [CrossRef]
46. Xing, J.; Song, J.; Yuan, H.; Wang, Q.; Li, X.; Li, N.; Duan, L.; Qu, B. Water-soluble nitrogen and phosphorus in aerosols and dry deposition in Jiaozhou Bay, North China: Deposition velocities, origins and biogeochemical implications. *Atmos. Res.* **2018**, *207*, 90–99. [CrossRef]
47. Cui, M.; Guo, Q.; Wei, R.; Chen, T. Temporal-spatial dynamics of anthropogenic nitrogen inputs and hotspots in a large river basin. *Chemosphere* **2021**, *269*, 129411. [CrossRef] [PubMed]
48. Lu, J.; Zhou, N.; Zhou, Z.; Zeng, H. Effect and mechanism of the "Ten-point Water Plan" on the intensity of industrial water pollution. *China Popul. Resour. Environ.* **2021**, *31*, 90–99.
49. Cheng, X.; Chen, L.; Sun, R.; Kong, P. Land use changes and socio-economic development strongly deteriorate river ecosystem health in one of the largest basins in China. *Sci. Total Environ.* **2018**, *616–617*, 376–385. [CrossRef]
50. Liu, S.M.; Zhang, J.; Chen, H.T.; Wu, Y.; Xiong, H.; Zhang, Z.F. Nutrients in the Changjiang and its tributaries. *Biogeochemistry* **2003**, *62*, 1–18. [CrossRef]
51. Ludwig, W.; Bouwman, A.F.; Dumont, E.; Lespinas, F. Water and nutrient fluxes from major Mediterranean and Black Sea rivers: Past and future trends and their implications for the basin-scale budges. *Glob. Biogeochem. Cycles* **2010**, *24*, GB0A13. [CrossRef]
52. Voulvoulis, N.; Arpon, K.D.; Giakoumis, T. The EU Water Framework Directive: From great expectations to problems with implementation. *Sci. Total Environ.* **2017**, *575*, 358–366. [CrossRef]

water

MDPI

Article

Iron Contamination in Groundwater: Risk Assessment and Remediation Techniques in Egypt's New Valley

Ehdaa A. M. Abed [1], Khalid A. N. Alaboudi [2], Mohamed H. H. Abbas [3], Tamer M. S. Attia [4] and Ahmed A. Abdelhafez [1,5,*]

1 Department of Soils and Water, Faculty of Agriculture, New Valley University, El Kharga 72511, Egypt; ehdaaabed94@gmail.com
2 Advanced Agricultural & Food Technologies Institute, King Abdulaziz City for Science and Technology (KACST), P.O. Box 6086, Riyadh 11442, Saudi Arabia; kalaboudi@kacst.gov.sa
3 Department of Soils and Water, Faculty of Agriculture, Benha University, Benha 13518, Egypt; mohamed.abbas@fagr.bu.edu.eg
4 Soils, Water and Environment Research Institute (SWERI), Agricultural Research Center (ARC), Giza 12511, Egypt; tamer_salem43@yahoo.com
5 National Committee of Soil Science, Academy of Scientific Research and Technology, Cairo 12622, Egypt
* Correspondence: ahmed.aziz@agr.nvu.edu.eg

Abstract: This study evaluates water quality (surface and deep wells as well as tap water) within villages of the El-Kharga Oasis (New Valley), focusing on their chemical composition, Fe contents, and potential hazards quantifying human exposure to Fe through different pathways, i.e., ingestion and dermal contact. Generally, the water quality meets the standards of the WHO guidelines for most sites, except for Fe, whose total and soluble contents in surface and ground waters exceed the permissible levels. Fe was higher in surface wells than in the deeper ones. Moreover, total Fe was higher than the permissible level in tap water, indicating potential health hazards for children living therein (hazardous index > 1). Another lab experiment was conducted to assess the efficacy of using dolomite, calcite, glauconite, and biochar for Fe removal from solutions artificially contaminated with Fe at a rate of 1000 mg Fe L^{-1}. Generally, Fe solubility decreased with increasing the pH of media from two to seven. Dolomite exhibited the highest efficiency for removing Fe from the solution for five successive cycles, with slight reductions in efficiency from 100% to 93.67% between the first and fifth cycle. Overall, Fe removal efficiencies followed the order of calcite > dolomite > biochar > glauconite. This sorption fitted the Langmuir isotherm model, and its kinetics (5–20 min) followed a pseudo-second order model. Furthermore, Fe recovery from dolomite was high, ranging from 100%, while decreased slightly to 92.93% in the fifth cycle. Although the results for calcite were also promising, our results showed its higher erodibility rate compared to dolomite. These findings offer valuable insights towards managing water quality and developing solutions for treating contaminated water sources, with a specific emphasis on the efficacy of dolomite and calcite for removing Fe from Fe-contaminated water.

Keywords: water quality; iron removal; sorbent materials; regeneration efficiency; adsorption kinetics

1. Introduction

Global demand for freshwater is ever-increasing to meet rapid population growth and sustain the goals of development [1–3]. In water-stressed countries and regions, groundwater represents an important resource for drinking and agricultural and industrial activities [1,4–6]. Egypt is an example of a water-stressed country, i.e., the water share per capita does not exceed 560 m^3 annually [7]. The New Valley Governorate is one of the promising desert border provinces in Egypt [8], which is marked by vast stretches of the Sahara Desert [9]. This area depends on the Nubian Sandstone Aquifer as the sole source of freshwater [10].

Groundwater quality may generally be affected by various factors, including their contents of potentially toxic elements (PTEs) [5]. Iron (Fe) is among the main contaminants in the groundwater of New Valley (Egypt). Its elevated levels pose significant environmental and health challenges [10], such as gastrointestinal discomfort, organ damage, and increased risk of chronic diseases [11,12]. The World Health Organization (WHO) has, therefore, set standards limiting iron concentrations up to 0.3 mg L^{-1} in drinking water based on considerations of color and staining potential [11]. Additionally, iron-rich water can cause clogging of irrigation systems and impact plant growth and soil health [13,14].

Concentrations and mobility of iron in groundwater are influenced by various factors, including geological substrate, redox conditions, and pH levels [15]. Various technologies were used successfully in remediating iron-contaminated water, including filtration, oxidation, ion exchange, and biological methods [16–18]. Among these techniques, biochar emerges as a safe, cheap, and sustainable approach for removing iron from water via sorption mechanisms [19,20]. This carbon-rich product is produced via the pyrolysis of organic residues under limited oxygen conditions [21]. Because of its porous structure and high CEC; this additive can immobilize PTEs, hence reducing their solubility in water [22].

Two other ecofriendly and economic approaches can help in deironing wastewaters. For example, clay minerals have a high capability for the removal of contaminants that exist in wastewater [23,24], especially the expandable 2:1 ones, such as glauconite [25]. This mineral (glauconite) has high cation exchange capacity and exhibits high chemical and mechanical stability [26]. Also, chemical precipitation of PTEs may effectually downgrade their concentrations in wastewaters [27]. In particular, dolomite and calcite are excellent candidates for the successful removal of PTEs from wastewaters [27–29].

El-Khargais is the biggest oasis of the New Valley, Egypt. Its groundwater contains high levels of Fe, which may pose potential risks to the surroundings [10]. Accordingly, this study aims at (1) monitoring the total and soluble levels of Fe in water samples collected from seven different sites comprising shallow and deep wells across the villages of the El-Kharga Oasis, to estimate health hazards and potential environmental risks associated with using such waters for drinking and irrigation purposes. Tap water was also collected from the same sampling sites for data comparison. Moreover, this study evaluates (2) the efficiencies of using natural safe economic additives, i.e., organic/inorganic adsorbents (biochar and glauconite) and traditional chemical precipitation techniques (calcite and dolomite) for the removal of Fe from aqueous solutions. Sorption isotherms and kinetic models were applied to depict Fe sorption pathways by the investigated materials. Also, the sorbent regeneration after five successive cycles of Fe sorption and desorption was a matter of concern herein. We believe that this integrant study has not yet been performed before in such detail. These objectives are consistent with the Egyptian Sustainable Development 2030 goals, which contribute to sustainable water resource management in the region and ensuring the health and well-being of its inhabitants and the environmental integrity of its ecosystems.

2. Methodology

2.1. Study Location

Water samples were collected from 7 different sites comprising shallow and deep wells across the villages of El-Kharga Oasis (New Valley): El-Monira (7 sites), El-Sherka (9 sites), El-Kharga (9 sites), Porsaied (5 sites), Ganah (5 sites), Bolak (8 sites), and Germashen (6 sites). Additionally, tap-water samples were collected from the abovementioned sites (Figure 1). These samples completely filled sterilized containers, which were tightly closed to avoid Fe oxidation, properly labeled, preserved at 4 °C to stop changes in water chemistry, and transported to the laboratory within 4 h of collection

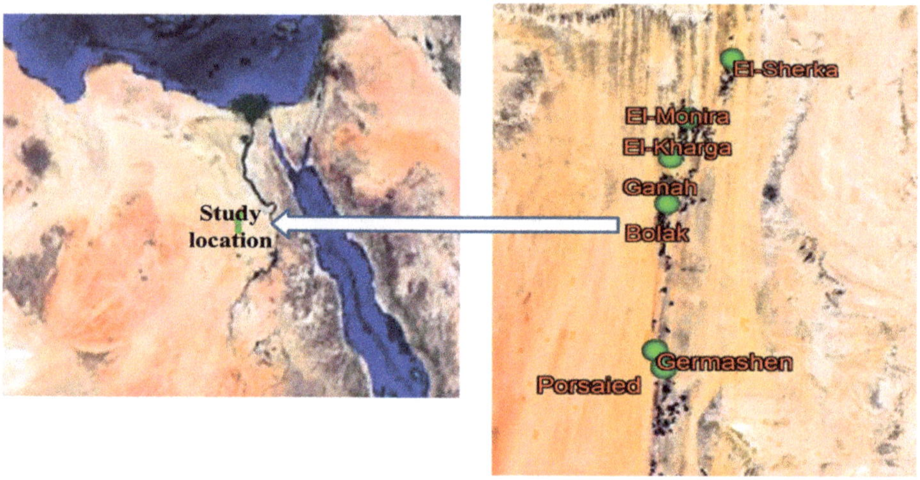

Figure 1. Map of study locations.

2.2. Water Analyses

The water samples underwent a comprehensive chemical and physical analysis following methodologies by Abdelhafez and Li [20]. The key parameters for evaluating water quality in this study were pH, electrical conductivity (Cole-Parmer PC100 pH/Conductivity Meter, Shanghai, China), and turbidity (TB1 Portable Turbidimeter, Velp, Usmate, Italy). For field analysis, the samples were collected after two hours of well operation to ensure representative water quality. Analytical grade reagents (Sigma Aldrich, St. Louis, MO, USA, products of USA) and rigorously acid-washed glassware were used to ensure precise results. All analyses were conducted in triplicate, including blanks and spiked samples as quality controls, with recovery rates between 97.6 and 98.82%, indicating high accuracy and precision. Cation and anion concentrations were assessed per Baird et al. [30]. Soluble and total Fe concentrations were quantified following digestion with concentrated nitric acid by using an Inductively Coupled Plasma instrument (Jobin Yvon Horiba Ultima 2 Inductively Coupled Plasma, Palaiseau, France). The results were then compared to the WHO and FAO guidelines as international water quality benchmarks.

2.3. Risk Assessment

Potential human health hazards were evaluated by quantifying exposure levels to Fe through ingestion and dermal contact. Based on the average daily dose (ADD) and toxicity indices for Fe (Table 1), non-carcinogenic risk assessment was estimated according to Abdelhafez and Li [31] and USEPA [32], and as outlined below. This structured approach incorporating diverse reputable sources ensured a robust evaluation of potential health risks from heavy metal exposure in the studied areas.

$$\text{ADD water ingestion} = \frac{C \times IR \times ED \times EF}{BW \times AT} \tag{1}$$

$$\text{ADD water dermal contact} = \frac{C \times SA \times ABS \times EF \times ED}{BW \times AT} \tag{2}$$

where ADD is the average daily dose of Fe, C is the metal concentration in the water (mg L^{-1}), IR is the ingestion rate per unit time, ED is the exposure duration, EF is the exposure frequency, BW is human body weight, AT is the averaging time, SL is the skin adherence factor, SA is the surface area of contact, and ABS is the absorption factor.

Table 1. Input parameters and toxicity index to characterize the ADD and HI values.

Parameter	Description	Unit	Adult	Children
C	Contaminant concentration in water	$\mu g\ L^{-1}$	--	--
IR	Ingestion rate	$L\ day^{-1}$	2.2	1.8
EF	Exposure frequency	$L\ day^{-1}$	350	
ED	Exposure duration	Years	30.0	6.0
BW	Body weight	Kg	70.0	15
AT	Average time	Days	10,500	2100
SA	Exposure skin area	cm^2	28,000	6600
ET	Exposure time	$h\ day^{-1}$	0.6	1.0
CF	Unit conservation factor	L/cm^3	0.001	
Kp, Dermal permeability coefficient		$cm\ h^{-1}$	0.001	
Fe oral and dermal reference dose		$\mu g\ kg^{-1}\ day^{-1}$	700,000	

The calculated ADD values will then be used to determine the hazard quotient (HQ) and the hazard index (HI), which are as follows:

$$HQ = \frac{ADD}{RfD} \quad (3)$$

$$\mathrm{HI} = \sum HQ \quad (4)$$

where RfD is the reference dose, which is defined as the maximum daily intake of the contaminant without a deleterious health effect [31]. If ADD > RfD, there will be a possible deleterious health impact.

2.4. *Efficiency of Different Sorbent Materials for Fe Removal from Aqueous Solution*

Various materials were evaluated for their efficiencies in removing Fe from aqueous solutions, including biochar, dolomite, calcite, and glauconite. Biochar was produced from date palm waste via pyrolysis at <500 °C [33]. Extensive analyses were also set to characterize the properties of the used sorbents, i.e., pH, EC, organic content, and FTIR bonding structures. Dolomite was processed to 1–3 mm, rinsed, and dried. Its recyclability and hardness avoid the disintegration issues of calcite, enhancing efficiency and sustainability. Calcite and glauconite were carefully collected from a controlled agricultural setting, ensuring authenticity. All materials underwent meticulous preparation for uniformity and accurate comparison.

2.4.1. Preparation of Fe Solutions

Solutions containing iron (Fe) ions (1000 mg Fe L^{-1} stock solution) were prepared for sorption and kinetic experiments using $FeSO_4 \cdot 7H_2O$ (Sigma Aldrich, ACS reagent > 99.9%). This stock was then diluted to prepare a series of solutions with concentrations ranging from 5 to 400 mg L^{-1} Fe using distilled water.

2.4.2. Batch Adsorption Studies

Batch experiments were conducted in the lab to investigate the effects of pH, contact time, and initial Fe concentration on Fe sorption using the materials that were mentioned above, and their efficiencies for removing contaminants from aqueous solutions were evaluated. The experiments utilized a series of 100 mL capped plastic flasks for shaking the samples with a mechanical shaker. The pH of the Fe solutions was carefully adjusted using 0.1 M of either NaOH or HNO_3 to range from 2 to 7, as pH values significantly influence Fe ionization in solutions and its sorption behavior [18].

Time-course experiments were also conducted for periods ranging from 5 to 240 min after adding sorbents at a fixed rate of 1 g L^{-1}. Then, the mixtures were agitated at 120 rpm for each time and centrifuged at 5000 rpm for 10 min to separate the supernatants to determine Fe by using a spectrophotometer using UV2400 dual split-beam UV-VIS. Fe quantification was meticulously carried out in accordance with the ASTM E394 standard test method. This method involved the utilization of the 1,10-Phenanthroline method [34]. The Fe removal efficiency was then calculated using the following equations:

$$\text{Removal efficiency}(\%) = \frac{\text{C0} - \text{Ce}}{\text{Ce}} \times 100 \tag{5}$$

where C0 is the concentration of the applied Fe ions, and Ce is the Fe ions concentration at equilibrium, respectively. Fe ions adsorbed (mg g^{-1}) on the biochar were calculated for each sorbent by the following equation:

$$qe = \frac{\text{C0} - \text{Ce}}{\text{V}} \tag{6}$$

where qe is the amount of adsorbed Fe ions (mg Fe g^{-1}), V is the applied Fe ions solution volume (L), and W is the adsorbent weight (g).

2.4.3. Adsorption Isotherms

Several isotherms were tested to understand the main mechanism beyond Fe removal from aqueous solutions i.e., Freundlich and Langmuir models [20].

$$\text{The Freundlich model is as follows}: \ \text{q}_e = K_F C_e^{\frac{1}{n}} \tag{7}$$

where q_e is adsorbed metal per adsorbent mass, K_F and n are empirical constants, and C_e is the equilibrium concentration. Its linearized form is:

$$\log q_e = \log K + \frac{1}{n} \log C_e \tag{8}$$

The Langmuir model is [35]:

$$q_e = \frac{Q_{\max bC_e}}{1 + bC_e} \tag{9}$$

where Q_{max} is the maximum adsorption capacity and b indicates the binding affinity. Qmax and b were calculated from the linear plot of Ce/qe vs. Ce.

The dimensionless separation factor RL predicts favorable/unfavorable adsorption [36]:

$$RL = \frac{1}{1 + bC_e} \tag{10}$$

RL values imply >1 unfavorable, 1 linear, 0–1 favorable, 0 irreversible.

2.4.4. Adsorption Kinetics Models

Pseudo-first-order and pseudo-second-order models were used to analyze the adsorption kinetics of Fe onto the tested materials [37].

$$\text{The pseudo} - \text{first} - \text{order model is} \log(q_e - q_t) = log q_e - \frac{K_1.ads}{2.303} t \tag{11}$$

where q_e and q_t (mgg^{-1}) are the amount of metal ions adsorbed at equilibrium (mgg^{-1}) and t (mins), respectively, and K_1 is the rate constant of the equation ($mins^{-1}$).

$$Thepseudo - second - order\ is\ \frac{dq}{dt} - \left(q_{eq-}q\right)^2 \tag{12}$$

$$\text{Its integration is } \frac{t}{q} - \frac{1}{K_{2adsqe^2}} + \frac{t}{q_e} \tag{13}$$

where t (min.) is the contact time; q_{eq} (mg g^{-1}) and q (mg g^{-1}) are the amount of Fe ions adsorbed at equilibrium and at any time of reaction t.

2.5. Regeneration

Batch regeneration experiments were conducted on the used materials via 0.1 M of HCl at 1 g L^{-1} to release adsorbed Fe for 5 successive cycles of adsorption–desorption under the optimal pH value to attain successful Fe removal. The desorption rate and removal efficiency were determined to evaluate the sustainable performance of the selected sorbent materials.

2.6. Data Analysis

The statistical analysis involved ANOVA and DMRT for mean separation ($p < 0.05$) to assess differences between treatments. This rigorous methodology enhances scientific precision and reliability.

3. Results and Discussion

3.1. Chemical Analyses of Water

As shown in Table 2, the pH levels were generally within the permissible levels according to the drinking-water guidelines (6.5–8.5) of the WHO [11], where the highest pH values were detected in the El-Sherka drinking water (7.8). The lowest one was in El-Kharga surface 2 (6.42). However, the EC varied considerably, and exceeded the WHO limit (1.50 dS m^{-1}) only at the following sites: El-Sherka surface 1 (1.63 dS m^{-1}) and El-Kharga surface 2 (1.76 dS m^{-1}) and 3 (1.70 dS m^{-1}) [11].

Table 2. Chemical and physical characteristics of tested water samples in the studied locations.

Village	Well	Depth, m	pH	EC dS m^{-1}	Turbidity, NTU
El-Monira	El-Monira 2 old (M1)	706	7.07 ± 0.02	0.60 ± 0.02	16.4 ± 0.12
	El-Monira- country (M2)	550	7.19 ± 0.03	0.69 ± 0.02	2.36 ± 0.31
	El-Monira5 (M3)	304	7.05 ± 0.04	0.67 ± 0.0	7.03 ± 1.80
	El-Monira- Om ELkosour (M4)	290	7.34 ± 0.02	0.46 ± 0.0	5.31 ± 0.10
	El-Monira- 1/9 (M5)	386	7.03 ± 0.0	0.66 ± 0.01	12.55 ± 0.92
	El-Monira-surface 1 (MS1)	<100	6.81 ± 0.01	1.07 ± 0.0	29.55 ± 1.62
	El-Monira- surface 2 (MS2)	<100	7.48 ± 0.11	0.55 ± 0.0	2.19 ± 0.16
	El-Monira- drinking water (MD)	--	7.19 ± 0.02	0.96 ± 0.0	5.90 ± 0.21
El-Sherka	El-Sherka- 55	703	7.15 ± 0.01	0.54 ± 0.0	10.11 ± 0.42
	El-Sherka- 4	696	7.04 ± 0.002	0.51 ± 0.01	3.91 ± 0.11
	El-Sherka- 8	643	7.80 ± 0.001	0.56 ± 0.04	1.28 ± 0.06
	El-Sherka- 21/34	405	7.19 ± 0.00	0.72 ± 0.0	1.40 ± 0.08
	El-Sherka- 5	332	7.08 ± 0.00	0.47 ± 0.0	7.17 ± 1.03
	El-Sherka- 7	302	7.42 ± 0.05	0.723 ± 0.07	0.805 ± 0.06
	El-Sherka- surface 1	<100	7.13 ± 0.02	*1.63 ± 0.0*	21.60 ± 3.40
	El-Sherka- surface 2	<100	7.33 ± 0.00	0.68 ± 0.01	1.80 ± 0.57
	El-Sherka- surface 3	<100	7.22 ± 0.0	0.62 ± 0.0	1.32 ± 0.37
	El-Sherka- drinking water	--	7.78 ± 0.02	0.72 ± 0.39	0.48 ± 0.08
El-Kharga	El-Kharga- 41	762	6.78 ± 0.06	0.36 ± 0.0	16.35 ± 3.61
	El-Kharga- 38	740	6.94 ± 0.06	0.35 ± 0.0	26.90 ± 0.71
	El-Kharga- 12	509	6.95 ± 0.01	0.31 ± 0.02	34.25 ± 1.06
	El-Kharga- 23	322	7.03 ± 0.01	0.44 ± 0.01	28.60 ± 0.71
	El-Kharga- 24	283	6.90 ± 0.08	0.48 ± 0.01	39.05 ± 3.48
	El-Kharga- 26	278	6.82 ± 0.02	0.60 ± 0.01	36.90 ± 0.28
	El-Kharga- surface 1	<100	6.99 ± 0.09	0.51 ± 0.0	24.90 ± 0.14
	El-Kharga- surface 2	<100	6.42 ± 0.03	*1.76 ± 0.02*	69.70 ± 1.98
	El-Kharga- surface 3	<100	6.64 ± 0.03	*1.70 ± 0.01*	80.15 ± 2.47
	El-Kharga- drinking water	--	7.04 ± 0.09	0.82 ± 0.29	0.78 ± 0.07

Table 2. *Cont.*

Village	Well	Depth, m	pH	EC dS m^{-1}	Turbidity, NTU
Porsaied	Porsaied- Nassir new south	573	6.75 ± 0.05	0.80 ± 0.0	34.70 ± 1.27
	Porsaied- Nassir new north	512	6.76 ± 0.04	0.75 ± 0.02	44.852.19
	Porsaied- Nassir tourist	563	7.09 ± 0.04	0.48 ± 0.02	19.10 ± 0.14
	Porsaied- new	323	7.18 ± 0.01	0.32 ± 0.0	6.72 ± 0.0
	Porsaied- surface	<100	6.78 ± 0.05	0.34 ± 0.0	16.45 ± 1.91
	Porsaied- drinking water	--	7.265 ± 0.08	0.34 ± 0.0	0.89 ± 0.29
Ganah	Ganah- 5	800	6.82 ± 0.03	0.54 ± 0.0	42.80 ± 0.57
	Ganah- 2/4	804	6.85 ± 0.02	0.39 ± 0.0	26.35 ± 6.15
	Ganah- 11/16	791	7.43 ± 0.03	0.38 ± 0.0	7.93 ± 0.30
	Ganah- 20 country	453	7.04 ± 0.15	0.61 ± 0.0	25.5 ± 14.42
	Ganah- surface	<100	7.04 ± 0.04	0.35 ± 0.0	17.95 ± 1.20
	Ganah- drinking water	--	7.32 ± 0.04	0.34 ± 0.0	0.77 ± 0.02
Bolak	Bolak- 3 new	813	7.14 ± 0.03	0.47 ± 0.02	10.00 ± 0.57
	Bolak- 25	782	6.75 ± 0.04	0.91 ± 0.04	34.55 ± 2.76
	Bolak- 20	633	6.65 ± 0.06	0.96 ± 0.0	18.53 ± 14.81
	Bolak- 4/23	424	6.71 ± 0.01	0.94 ± 0.0	47.45 ± 2.90
	Bolak- 6	346	6.87 ± 0.01	0.65 ± 0.0	27.6 ± 3.25
	Bolk- palm relief	150	6.76 ± 0.0	0.73 ± 0.03	31.4 ± 2.83
	Bolak- surface 1	<100	6.60 ± 0.30	1.15 ± 0.04	5.82 ± 3.27
	Bolak- surface 2	<100	6.87 ± 0.04	1.00 ± 0.01	12.1 ± 0.71
	Bolak- drinking water	--	6.89 ± 0.01	1.05 ± 0.0	0.82 ± 0.04
Germashen	Germashen- 8	495	6.95 ± 0.02	0.91 ± 0.0	35.1 ± 0.71
	Germashen- 25	554	6.92 ± 0.03	0.84 ± 0.03	30.3 ± 2.69
	Germashen- 1	495	7.09 ± 0.02	0.55 ± 0.02	15.95 ± 0.92
	Germashen- 2G	532	7.10 ± 0.04	0.60 ± 0.0	9.35 ± 3.05
	Germashen- 16	573	6.91 ± 0.06	0.54 ± 0.01	23.3 ± 0.14
	Germashen- surface	<100	6.80 ± 0.0	0.69 ± 0.0	21.15 ± 2.62
	Germashen- drinking water	--	6.89 ± 0.12	0.64 ± 0.01	0.48 ± 0.02
WHO standard (drinking)		--	6.5–8.5	<1.5	--
FAO standard (irrigation)		--	6.5–8.5	<3.0	--

Values higher than the WHO standard are highlighted in italics. Values higher than the FAO standard are underlined. Values higher than both the WHO and FAO standards are highlighted in bold.

Turbidity levels were the highest in El-Kharga surface 2 (69.70 NTU) and the least in Germashen drinking water (0.48 NTU), indicating relative potential contamination occurred at El-Kharga. Overall, the turbidity in many locations exceeded the recommended ones of the WHO (five NTF), and this highlights the need for further treatments to improve water standards for human consumption.

Chemical characterization (Table 3) also showed non-detected carbonate contents in the water samples. Although variable bicarbonate (0.50–2.10 mmol$_c$ L^{-1}) concentrations were detectable among the studied sites, yet, these concentrations were almost less than the permissible levels for drinking purposes.

Sulfate ions exceeded the allowable levels in most sites. In the case of chloride, high contents were detected on the Porsaied surface (9.80 mmol$_c$ L^{-1}). Sodium was also high on the Porsaied surface (7.07 meq L^{-1}), and this result was consistent with prior reports on elevated ion concentrations in surface wells versus deeper ones [38]. Likewise, Ca and Mg contents were higher in surface wells than in deeper ones. Most water sources met Mg^{2+} standards; nevertheless, some exhibited higher Ca^{2+}, Na^+, $SO_4{}^{2-}$, and Cl^- levels, exceeding the limits per the WHO and FAO [11,39]. This indicates the need for targeted interventions, especially for surface wells where Na^+ and Cl^- exceeded recommended levels [38].

Table 3. Chemical characterization of irrigation and drinking-water samples.

Village	Well	HCO_3^-	Cl^-	SO_4^{--}	Na^+	Ca^{++}	Mg^{++}	K^+
					mmol$_c$ L^{-1}			
El-Monira	El-Monira- 2	0.70 ± 0.14	1.40 ± 0.28	**6.48 ± 0.60**	3.07 ± 0.03	2.83 ± 0.60	1.88 ± 0.39	*0.82 ± 0.00*
	El-Monira- 5	0.70 ± 0.14	1.60 ± 0.57	**6.14 ± 0.55**	3.20 ± 0.09	3.43 ± 0.11	1.00 ± 0.35	*0.82 ± 0.00*
	Om eLksor	1.30 ± 0.14	0.80 ± 0.0	*2.74 ± 0.46*	2.20 ± 0.03	1.10 ± 0.42	0.70 ± 0.14	*0.84 ± 0.00*
	Monira Balad	0.80 ± 0.0	1.40 ± 0.28	**4.63 ± 1.29**	3.35 ± 0.00	1.30 ± 0.14	1.40 ± 1.13	*0.78 ± 0.02*
	Monira 1-9	0.50 ± 0.14	1.40 ± 0.28	*3.88 ± 0.19*	3.37 ± 0.03	1.40 ± 0.00	0.20 ± 0.00	*0.81 ± 0.02*
	Monira surface- 1	0.80 ± 0.0	1.20 ± 0.0	**8.25 ± 0.74**	2.98 ± 0.03	*4.33 ± 0.04*	*2.18 ± 0.67*	*0.77 ± 0.00*
	Monira surface- 2	0.90 ± 0.14	1.20 ± 0.57	*3.84 ± 0.93*	3.00 ± 0.06	1.98 ± 0.10	0.43 ± 0.32	*0.54 ± 0.00*
	Monira drinking water	0.80 ± 0.0	2.00 ± 0.57	**4.65 ± 0.74**	4.50 ± 0.03	1.63 ± 0.04	0.28 ± 0.11	*1.05 ± 0.00*
El-Sherka	El-Sherka- 5	0.70 ± 0.14	0.80 ± 0.0	**5.75 ± 0.69**	2.22 ± 0.00	3.03 ± 0.04	1.10 ± 0.71	*0.91 ± 0.13*
	El-Sherka- 55	0.90 ± 0.14	1.20 ± 0.0	**5.65 ± 0.18**	2.76 ± 0.03	3.20 ± 0.07	1.00 ± 0.28	*0.79 ± 0.00*
	El-Sherka- 4	0.60 ± 0.0	1.00 ± 0.28	**5.51 ± 0.04**	2.57 ± 0.00	3.13 ± 0.04	0.60 ± 0.28	*0.82 ± 0.00*
	El-Sherka- 7	1.20 ± 0.0	1.40 ± 0.28	**5.21 ± 0.14**	3.52 ± 0.00	3.15 ± 0.00	0.50 ± 0.42	*0.64 ± 0.00*
	El-Sherka- 8	0.90 ± 0.14	1.20 ± 0.0	**5.18 ± 0.18**	2.44 ± 0.00	3.28 ± 0.04	0.80 ± 0.28	*0.77 ± 0.00*
	El-Sherka- 21-34	1.00 ± 0.57	1.60 ± 0.0	**5.62 ± 0.31**	3.74 ± 0.06	3.30 ± 0.07	0.50 ± 0.14	*0.68 ± 0.02*
	El-Sherka surface- G2	1.00 ± 0.28	3.60 ± 1.13	**9.41 ± 1.38**	5.92 ± 0.25	*5.45 ± 0.21*	1.90 ± 0.99	*0.74 ± 0.00*
	El-Sherka surface- H2	0.80 ± 0.0	1.60 ± 0.0	**5.67 ± 0.67**	3.61 ± 0.00	3.23 ± 0.04	0.70 ± 0.71	*0.54 ± 0.00*
	El-Sherka surface- I 2	2.10 ± 0.14	**8.00 ± 0.0**	**26.65 ± 0.18**	**11.31 ± 0.0**	*14.43 ± 0.53*	**9.40 ± 0.57**	*1.61 ± 0.00*
	El-Sherka drinking water	0.70 ± 0.14	1.60 ± 0.0	**6.65 ± 0.19**	3.55 ± 0.03	3.58 ± 0.04	1.00 ± 0.28	*0.83 ± 0.02*
El-Kharga	El-Kharga- 26	0.90 ± 0.14	1.00 ± 0.28	**5.32 ± 1.36**	1.09 ± 0.00	*3.73 ± 0.04*	1.35 ± 0.4	*1.06 ± 0.02*
	El-Kharga- 12	0.50 ± 0.14	0.80 ± 0.57	*3.65 ± 0.42*	1.13 ± 0.00	2.25 ± 0.00	0.80 ± 0.0	*0.77 ± 0.00*
	El-Kharga- 24	0.80 ± 0.28	1.40 ± 0.28	**4.22 ± 0.89**	1.02 ± 0.03	3.28 ± 0.04	0.98 ± 0.08	*1.15 ± 0.00*
	El-Kharga- 38	0.50 ± 0.14	1.60 ± 0.57	*3.57 ± 0.73*	1.26 ± 0.00	2.38 ± 0.04	1.23 ± 0.04	*0.81 ± 0.02*
	El-Kharga- 41	0.30 ± 0.14	1.00 ± 0.28	*3.57 ± 0.65*	1.24 ± 0.03	2.33 ± 0.04	0.50 ± 0.14	*0.81 ± 0.02*
	El-Kharga- 23	0.70 ± 0.42	1.80 ± 0.28	*3.47 ± 0.34*	0.94 ± 0.03	2.98 ± 0.04	1.00 ± 0.28	*1.06 ± 0.01*
	El-Kharga surface- D2	0.90 ± 0.14	1.20 ± 0.57	**5.03 ± 0.70**	1.11 ± 0.03	3.43 ± 0.04	1.60 ± 0.28	*1.00 ± 0.00*
	El-Kharga surface- E2	0.50 ± 0.14	2.40 ± 0.57	**10.97 ± 0.72**	5.98 ± 0.09	*5.23 ± 0.03*	*2.10 ± 0.14*	*0.56 ± 0.00*
	El-Kharga surface- F2	0.70 ± 0.42	2.60 ± 0.28	**11.01 ± 0.29**	6.18 ± 0.06	*5.10 ± 0.07*	*2.50 ± 0.42*	*0.54 ± 0.00*
	El-Kharga drinking water	0.70 ± 0.14	1.40 ± 0.28	*3.39 ± 0.39*	1.07 ± 0.03	2.38 ± 0.11	1.10 ± 0.14	*0.95 ± 0.04*
Porsaied	Porsaied Nasir new sud	0.50 ± 0.14	1.80 ± 0.28	**6.69 ± 0.59**	2.02 ± 0.03	*4.50 ± 0.21*	1.30 ± 0.65	*1.16 ± 0.01*
	Porsaied Nasir new south	0.40 ± 0.0	1.60 ± 0.0	**6.95 ± 1.26**	1.96 ± 0.00	*4.33 ± 0.04*	1.50 ± 0.09	*1.16 ± 0.02*
	Porsaied tourism	0.90 ± 0.14	1.20 ± 0.0	**5.24 ± 0.55**	1.48 ± 0.00	3.23 ± 0.04	1.70 ± 0.07	*0.93 ± 0.03*
	Porsaied new	0.60 ± 0.0	2.60 ± 0.28	*1.77 ± 0.48*	1.22 ± 0.00	2.48 ± 0.03	0.50 ± 0.14	*0.78 ± 0.02*
	Porsaied surface	0.60 ± 0.0	**9.80 ± 1.98**	**17.35 ± 2.68**	7.07 ± 0.95	*10.88 ± 1.52*	**6.60 ± 1.70**	**3.21 ± 0.49**
	Porsaied drinking water	0.30 ± 0.13	0.80 ± 0.0	*3.76 ± 0.28*	1.24 ± 0.03	2.38 ± 0.04	0.30 ± 0.14	*0.95 ± 0.0*
Ganah	Ganah 5	0.80 ± 0.0	1.23 ± 0.02	**4.99 ± 0.18**	1.61 ± 0.0	3.38 ± 0.04	1.10 ± 0.14	*0.93 ± 0.01*
	Ganah 2-4	0.50 ± 0.14	1.25 ± 0.01	*3.31 ± 0.13*	1.57 ± 0.0	2.50 ± 0.00	0.20 ± 0.00	*0.79 ± 0.00*
	Ganah 11-16	1.30 ± 0.14	1.51 ± 0.08	*2.22 ± 0.25*	1.39 ± 0.06	2.53 ± 0.03	0.30 ± 0.01	*0.81 ± 0.02*
	Ganah 20	0.60 ± 0.0	1.30 ± 0.01	**5.75 ± 1.06**	1.26 ± 0.00	3.68 ± 0.04	1.70 ± 0.09	*1.01 ± 0.02*
	Ganah-surface	0.40 ± 0.0	1.24 ± 0.14	**4.11 ± 1.42**	1.28 ± 0.03	2.55 ± 0.00	1.10 ± 0.03	*0.82 ± 0.04*
	Ganah drinking water	0.60 ± 0.28	1.31 ± 0.02	*3.11 ± 0.05*	1.22 ± 0.00	2.45 ± 0.07	0.40 ± 0.02	*0.95 ± 0.00*
Bolak	Bolak 3 new	0.50 ± 0.14	1.20 ± 0.0	**4.94 ± 0.14**	1.72 ± 0.03	3.13 ± 0.04	1.00 ± 0.28	*0.79 ± 0.00*
	Bolak dates	0.70 ± 0.13	2.20 ± 0.28	**5.57 ± 0.08**	2.07 ± 0.03	*4.48 ± 0.03*	0.80 ± 0.21	*1.13 ± 0.00*
	Bolak 25	0.50 ± 0.12	2.60 ± 0.25	**8.56 ± 0.11**	2.76 ± 0.02	*5.00 ± 0.00*	*2.80 ± 0.00*	*1.10 ± 0.00*
	Bolak 4-32	0.80 ± 0.28	2.40 ± 0.0	**8.23 ± 0.39**	2.89 ± 0.09	*5.03 ± 0.18*	*2.50 ± 0.42*	*1.01 ± 0.02*
	Bolak 6	0.20 ± 0.0	1.80 ± 0.23	**7.01 ± 1.06**	2.11 ± 0.01	*4.03 ± 0.04*	1.90 ± 0.71	*0.97 ± 0.00*
	Bolak 20	0.50 ± 0.11	2.80 ± 0.57	**7.60 ± 2.14**	2.94 ± 0.01	*5.05 ± 0.07*	1.80 ± 0.17	*1.11 ± 0.02*
	Bolak surface 1	0.70 ± 0.10	3.40 ± 0.20	**8.35 ± 0.29**	3.59 ± 0.09	*5.38 ± 0.03*	*2.60 ± 0.00*	*0.88 ± 0.02*
	Bolak surface 2	0.70 ± 0.09	3.20 ± 0.52	**7.47 ± 1.72**	3.20 ± 0.02	*5.03 ± 0.04*	*2.20 ± 0.14*	*0.95 ± 0.18*
	Bolak drinking water	0.90 ± 0.11	3.40 ± 0.21	**7.90 ± 0.68**	2.89 ± 0.03	*5.25 ± 0.14*	*2.70 ± 0.42*	*1.36 ± 0.00*
Germashen	Germashen 8	0.40 ± 0.0	3.00 ± 0.26	**9.90 ± 0.04**	2.94 ± 0.08	*4.98 ± 0.46*	*4.40 ± 0.28*	*0.98 ± 0.05*
	Germashen 25	0.50 ± 0.14	2.80 ± 0.53	**8.68 ± 0.58**	2.65 ± 0.0	*4.63 ± 0.04*	*3.90 ± 0.09*	*0.81 ± 0.02*
	Germashen 1	0.40 ± 0.0	1.60 ± 0.0	**6.07 ± 0.25**	1.98 ± 0.03	3.45 ± 0.0	1.80 ± 0.28	*0.84 ± 0.0*
	Germashen 2 G	0.70 ± 0.13	1.80 ± 0.28	**5.71 ± 0.88**	2.54 ± 0.02	3.50 ± 0.0	1.30 ± 0.42	*0.87 ± 0.0*
	Germashen 16	0.60 ± 0.28	1.60 ± 0.48	**5.71 ± 0.44**	2.28 ± 0.01	3.30 ± 0.0	1.50 ± 0.14	*0.83 ± 0.02*
	Germashen- Surface	0.80 ± 0.25	2.20 ± 0.28	**6.32 ± 0.07**	2.33 ± 0.02	*4.18 ± 0.04*	2.00 ± 0.0	*0.82 ± 0.0*
	Germashen drinking water	0.60 ± 0.20	2.00 ± 0.0	**6.05 ± 0.0**	2.13 ± 0.0	*3.75 ± 0.0*	1.80 ± 0.28	*0.97 ± 0.0*
		6.56	7.05	0.38	8.70	3.7	2.057	0.31
		2.0	<3.0	<4.0	<3.0	20	5.02	2.0

Values higher than the WHO standard are highlighted in italics. Values higher than the FAO standard are underlined. Values higher than both WHO and FAO standards are highlighted in bold.

3.2. Total and Soluble Fe Contents

Significant variations were detectable in the Fe (total and available) concentrations among the seven sites under investigation, where the Fe total and soluble contents were generally higher in surface wells than in deeper ones. Both contained higher Fe contents than the permissible levels. Maybe, anthropogenic activities are beyond such increases. For example, discharges from industrial activities (rich in Fe) found their way into the groundwater and, thus, enriching it with Fe [40]. During the penetration of this water through the soil column, substantial concentrations of Fe were sorbed on soil colloids and, then, subjected to oxidation [41]. For this reason, the total and available concentrations of Fe seemed to be significantly higher in surface-well waters versus the deep ones (Figure 2).

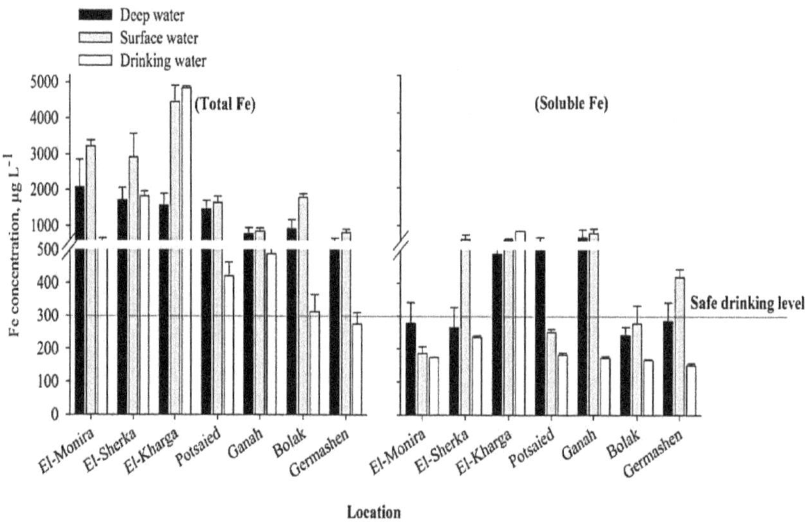

Figure 2. Soluble and total Iron concentrations (μL^{-1}) in the tested water samples.

High total contents of Fe were detected in all water samples of El-Monira, where the highest Fe content was found in its surface wells (3218.75 µg L^{-1}). Nevertheless, soluble Fe seemed to be much lower and did not exceed the permissible level. This might indicate that Fe dominated in these waters mainly in the form of sparingly soluble Fe complexes. Nevertheless, this contaminant is subjected to potential runoff and possesses sediment pollution [42].

Elevated total Fe content was also observed in the El-Sherka surface (2915.00 µg L^{-1}) and deep wells (1821.50 µg L^{-1}), which might be evidence of the presence of anthropogenic activities [11]. In the case of El-Kharga, an alarmingly high total Fe was found in all water samples, exceeding the permissible levels. Likewise, the soluble Fe content was higher than the safe levels in its well samples, while being marginal in drinking water. this could pose long-term health risks [11]. A moderate total Fe level was monitored at Portsaied, particularly within the surface wells (1643.50 µg L^{-1}), and an elevated soluble Fe concentration was detected in its deeper wells (536.8543 µg L^{-1}), which might occur due to geological and/or soil sources [39].

Ganah exhibited high total and soluble Fe contents in both the surface- and deep-well water samples, while the Fe concentration did not exceed the safe level in drinking water, which indicates natural or infrastructure influences.

Relatively moderate total Fe content was found in the Bolak surface water (1778.75 µg L^{-1}). On the other hand, Germashen contained the lowest total Fe, especially in deep wells (548.08 µg L^{-1}), which indicates minimal anthropogenic inputs [43]. Overall, the soluble Fe, but not its total contents, was within the permissible levels in all drinking-water samples. Two possible explanations might explain the increases that occurred in total Fe in drinking water: (1) the water pipers in use were made of Fe and/or (2) there were breaks in the water pipes that allowed the seepage of groundwater. In this concern, high water pressure in water pipes might lessen the seepage of groundwater to the water pipes, but if the water supply in such pipes was not regular, then groundwater might enter within through the pipe breaks and find its way to humans through ingesting water. High levels of Fe in El-Kharga and El-Sherka require urgent intervention. Accordingly, regular monitoring and effective treatment are critical to address Fe issues in water sources.

3.3. Risk Assessment Study

The hazard Index (HI) values of iron (Fe) were assessed in all the water samples according to the calculations of the WHO [44] (Figure 3). The calculated values were generally below the permissible levels for adult consumption; yet, the waters of the surface and deep wells could pose potential hazards for children in nearly all water resources. In this context, El-Monira, El-Sherka, and El-Kharga recorded the highest child HI values, especially within their surface wells. Moreover, El-Kharga contained the highest HI value in the drinking water, suggesting urgent procedures should be taken to lessen these contaminants.

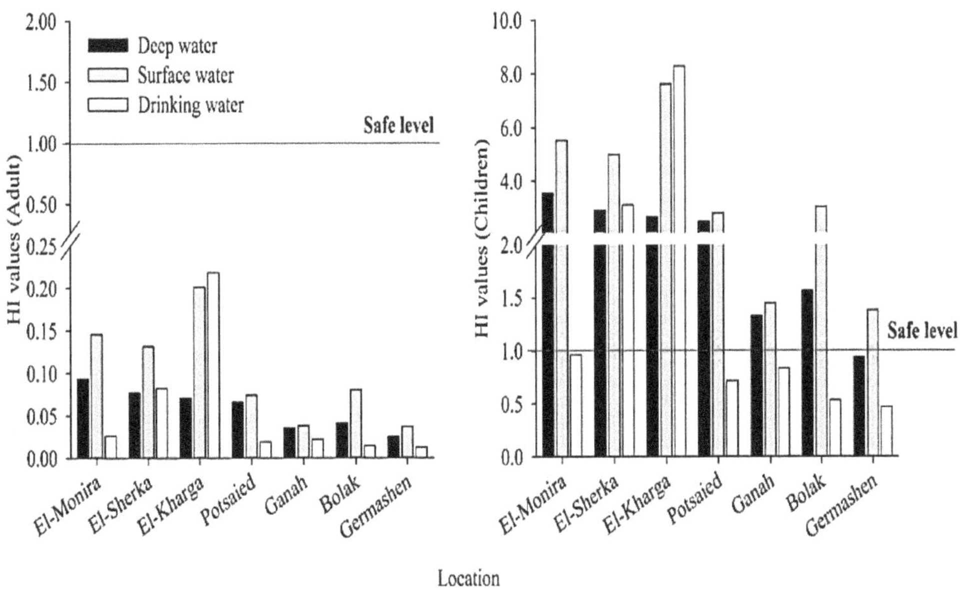

Figure 3. Estimated hazard index values of Fe for adults and children in the tested water samples.

Potsaied, Ganah, and Bolak exhibited moderate HIs in their water resources, while child risk remained unsafe in their wells. Therefore, more precautions should be considered while utilizing these waters. Germashen had the lowest HIs, yet potential child risk still existed in surface wells. This highlights the need for risk management and monitoring to mitigate pediatric Fe-related risks [45]. Children showed consistently higher HI values than adults across the investigated sites and sources. Elevated child HIs in some areas underscore the urgent need for targeted strategies, regular monitoring, and public health interventions to address Fe-related health risks. Maintaining safe Fe limits set by health organizations is crucial for both adult and child well-being.

3.4. Characterization of Calcium Carbonate Soilds and Biochar

3.4.1. Fourier-Transform Infrared Spectroscopy (FTIR) Analysis

Fourier-transform infrared spectroscopy (FTIR) offered insights into the chemical composition of sorbent materials that were used for iron (Fe) removal from the water samples (Figure 4). This technique enables us to select the most efficient and sustainable water remediation approaches [46].

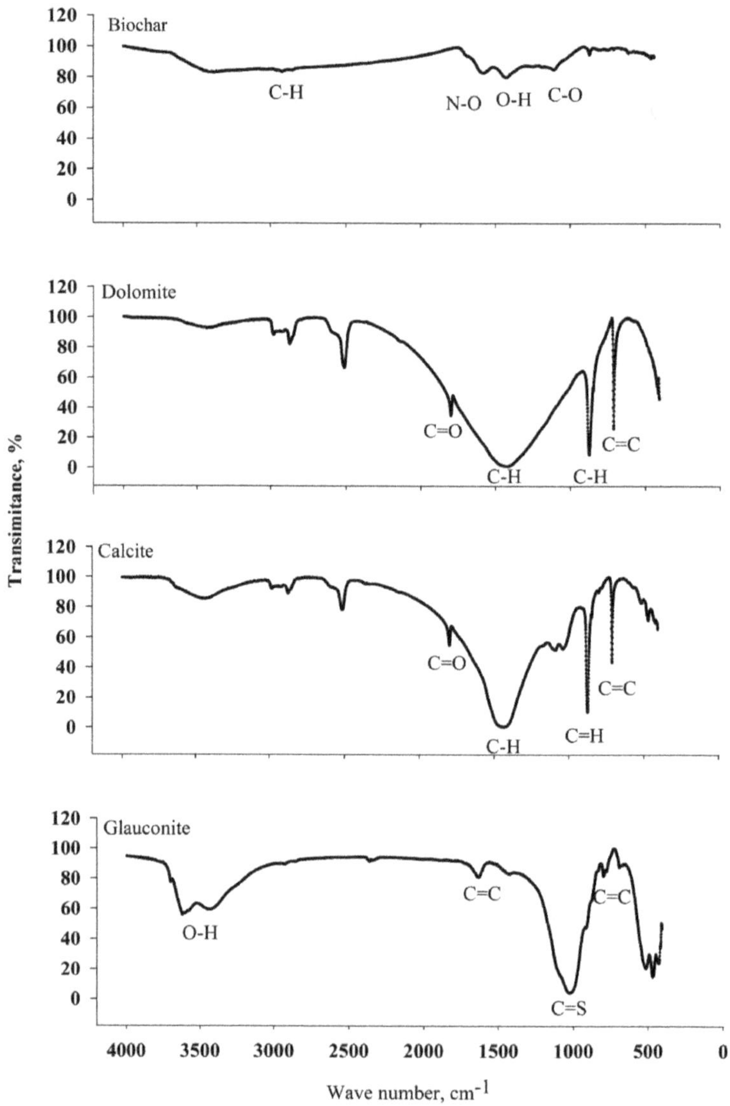

Figure 4. FTIR of the tested sorbent materials.

All calcium carbonate ($CaCO_3$) solids (calcite and dolomite) exhibited a high C=O (carbonyl) group, which exhibits a high affinity for binding metal ions like Fe [47]. Biochar and glauconite showed high levels of functional groups such as C=O (ether) in biochar, N-O (in biochar), and C=C (alkene) in glauconite. In addition, the O-H group (in the two abovementioned additives) demonstrates high hydrophilicity, which facilitates further PTE adsorption on the organic surfaces of biochar [46].

In the case of calcite and dolomite, the C=O group enabled them to form strong adsorption complexes within organic pollutants in the media, particularly potentially toxic elements (PTEs). Accordingly, calcite, dolomite, biochar, and glauconite could be suitable for broader water treatment applications targeting diverse contaminants based on their C=O, C=C, and O-H groups [19].

3.4.2. X-ray Diffraction Analysis (XRD) and Scanning Electron Microscopy (SEM) Analyses of Biochar Material

As shown in Figure 5, the surface of the produced biochar is smooth with a porous structure and high surface area. X-ray diffraction (XRD) analysis revealed that the structure of biochar is basically amorphous in nature but contains strongly conjugated aromatic compounds with some local crystalline structure. The biochar has almost 12 peaks of 2theta, as follows: 28.12, 29.20, 31.14, 35.74, 39.16, 40.14, 42.82, 47.26, 48.28, 50.10, 57.18, and 58.42. Strong cellulose peaks of around 60, 53, 40, and 25 have been reported to gradually lose strength and become wider by increasing charring temperatures of 100° to 300 °C, suggesting a gradual decrease in the cellulose contents of raw biomass due to pyrolysis.

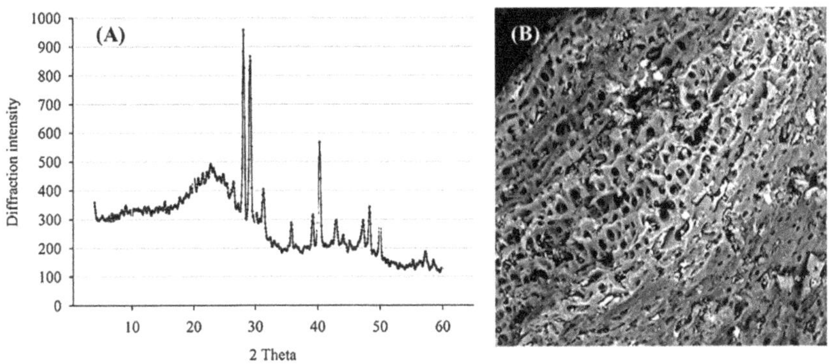

Figure 5. XRD analysis graph (diffraction intensity vs. 2theta) (**A**) and SEM analysis of the produced biochar material (image with a scale of 100 µm) (**B**).

3.5. *Factors Affecting Sorption of Fe by the Investigated Materials*

3.5.1. Sorbent Dosage

For evaluating Fe removal efficiencies by different sorbents, a clear dosage-dependent relationship was observed (Figure 6). Biochar demonstrated a gradual increase in its removal efficiency, starting from 24% at a 0.5 g L^{-1} application dose up to complete removal (100% of Fe) at a dosage of 10 g L^{-1}. These results confirm the findings of Inyang et al. [19] who indicated that the efficiency of biochar for removing PTEs from aqueous solutions was highly influenced by the dose of applied biochar.

Dolomite and calcite were also noted to have remarkable effectiveness in Fe removal from the water media, as these two additives recorded 100% Fe removal using only a dosage of 1 g L^{-1}. Such promising results highlight the strong attributes of dolomite [48] and calcite for binding iron [49]. On the other hand, glauconite exhibited the least efficiency for Fe removal among the tested sorbents, with a maximum removal efficiency of only 66.22%. This efficiency was attained at a 10 g L^{-1} dosage. Probably, it had low ion exchange capacity or trivial surface area constraints in adsorbing iron [29]. Based on the above results, biochar requires higher dosages for achieving high Fe removal efficiency from wastewaters, while dolomite and calcite were effective even at using lower dosages, and this offers practical advantages for water treatment applications.

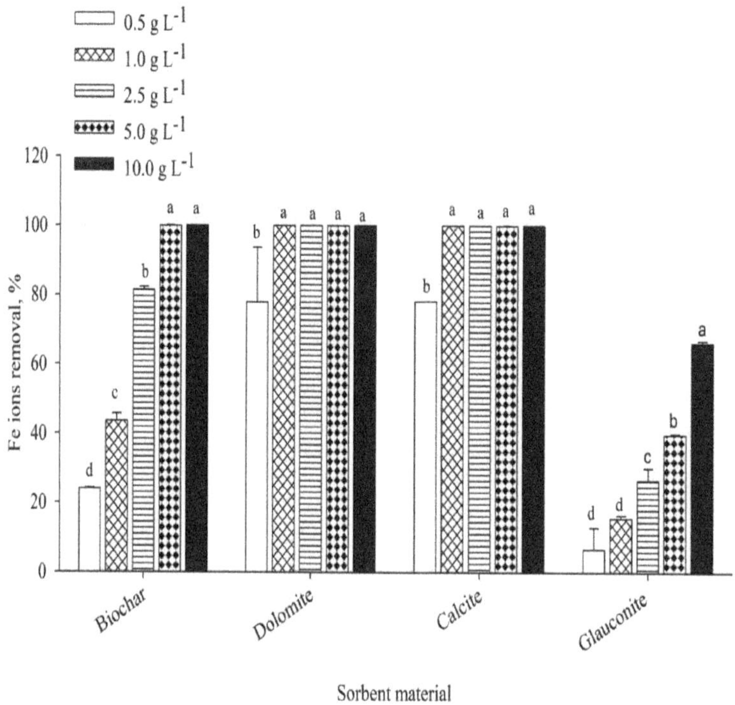

Figure 6. Effect of sorbent dosage on the removal percentage of Fe ions from aqueous solutions. Concentration was 100 mg L^{-1}, and the agitation was carried out at room temperature and 120 rbm. Different letters show significant difference at $p < 0.05$ (values signify means ± SD).

3.5.2. Effect of pH

Figure 7 illustrates that Fe solubility generally decreased because of the increase in the pH of the aqueous solutions within the range of two to seven via adding either diluted hydrochloric acid (HCl) or sodium hydroxide (NaOH). Reduction in Fe solubility with an increasing pH of the media was also noticed by Silva et al. [50]. Additionally, Fe underwent oxidation in the alkaline media to change from Fe^{2+} to Fe^{3+}, which has a lower solubility product and, thus, precipitated as $Fe(OH)_3$ salts [51] to attain a new level of equilibrium in the media [52]. Anyhow, the pH of the media was adjusted to five during the course of this study while testing the removal efficiencies of Fe from contaminated waters by the studied sorbent materials.

3.6. Efficiency of Sorbent Materials for Fe Removal

As shown in Figures 6 and 8, the investigated sorbent materials could be arranged according to their efficiencies for the removal of Fe from aqueous solutions in the following descending order: calcite > dolomite > biochar > glauconite, and the latter exhibited very low removal efficiency for Fe from aqueous solutions. Thus, glauconite was excluded from further investigations, i.e., the isotherm and kinetic studies because of its low capability. The results obtained herein indicate that calcite exhibited higher efficiency than dolomite for Fe decontamination from aqueous solutions; yet, the latter product (dolomite) is more durable upon repeated use compared to calcite, which is highly erodible in acid media, and consequently undergoes significant material breakdown [53]. This makes dolomite a more practical and cost-effective choice for the treatment of irrigation systems.

3.6.1. Adsorption Isotherms

Data on Fe sorption from aqueous solutions were then fitted to both Langmuir and Freundlich isotherm models (Figure 9). Based on the highest r^2 values, Fe sorption data fitted successfully to the Langmuir isotherm model. This suggests the adsorption of a monolayer of iron ions on both biochar and carbonate surfaces [20]. Dolomite and calcite recorded the highest affinity constants (K_L) for iron sorption versus biochar, with mean values of 37.07 and 44.67 mg g^{-1}, respectively, corresponding to only 3.15 mg g^{-1} for biochar. These results are consistent with Walker et al. [54], who found that the adsorption of Fe followed the Langmuir isotherm model, where the higher capacities for Fe sorption were found in carbonate sites versus biochar [54]. Likewise, the maximum adsorption capacity (q_L) was estimated for calcite at 333.33 mg g^{-1}, followed by dolomite at 238.10 mg g^{-1}, and then biochar at 40.00 mg g^{-1} (Figure 8).

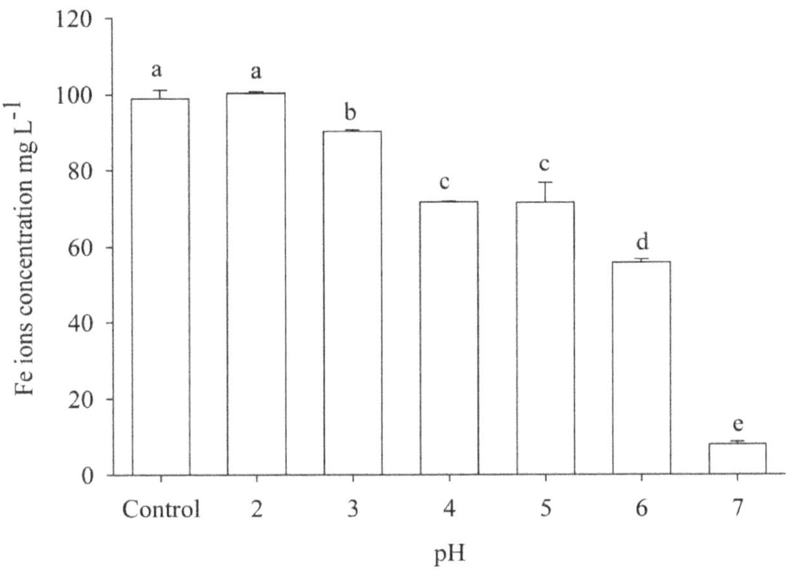

Figure 7. Effect of solution pH on the solubility of Fe (a stock solution of 100 mg L^{-1} Fe was prepared); pH was adjusted by using diluted HCl solution. Different letters show significant difference at $p < 0.05$ (values signify means ± SD).

3.6.2. Kinetics of Fe Sorption

Figure 5 shows the efficiencies of the studied sorbent materials for iron removal from contaminated waters within time intervals ranging from 5 to 240 min. In this aspect, dolomite and calcite demonstrated the highest RE (removal efficiencies). They needed only 120 min to achieve a successful 100% RE. Different mechanisms might explain such high efficiencies, e.g., rapid ion exchange on the carbonate surfaces [48], followed by Fe oxidation and precipitation [29]. Regarding glauconite, its RE capability for removing Fe seemed to be low and could seem nearly constant from the beginning of the investigation (23–25.75%). This result reflects its low capability for ion immobilization [29].

Biochar showed an initial minimal affinity for removing Fe from aqueous solutions for up to 30 min. Then, its RE increased considerably with time, to achieve 89.30% after 240 min of contact with biochar. This indicates that biochar needs a long contact time to attain the efficient removal of Fe from water bodies [19].

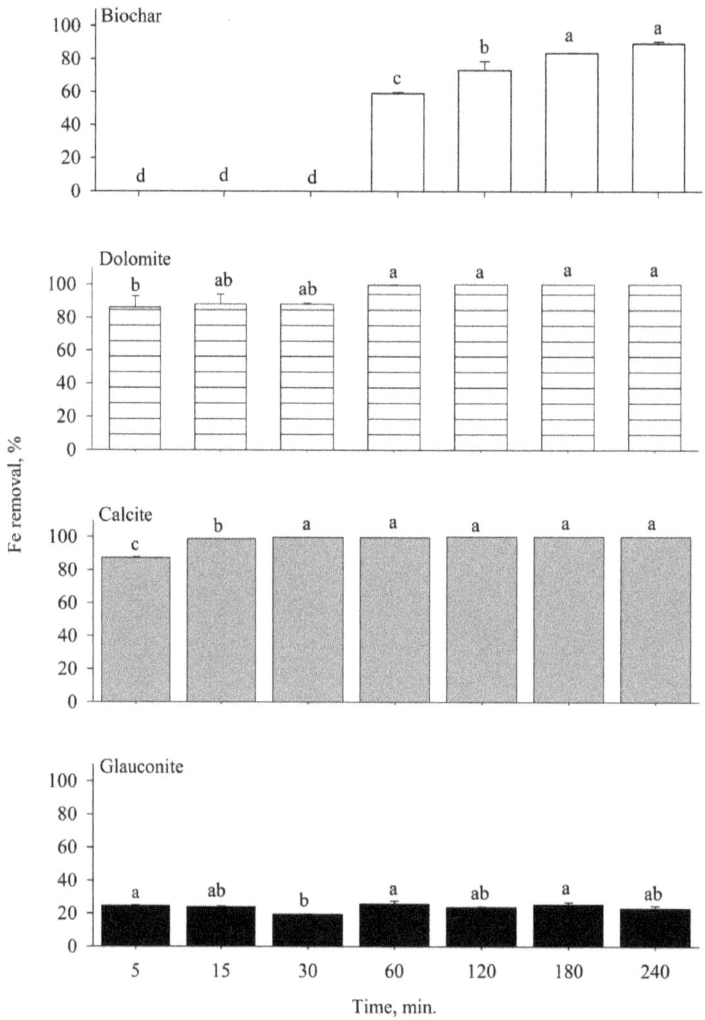

Figure 8. Effect contact time on Fe ions removal by using tested sorbent materials, concentration 100 mg L^{-1}, adsorbent dosage 1.0 g L^{-1}, and the agitation was carried out at room temperature and 120 rbm. Different letters show significant difference at $p < 0.05$ (values signify means ± SD).

Concerning, the kinetics of Fe sorption via calcite, dolomite, and biochar materials, Fe sorption data were well fitted to pseudo-second-order kinetics, as evidenced by the higher r^2 values compared to the values obtained from fitting data to pseudo-first-order kinetics. For example, the pseudo-second-order rate constant (k^2) was 0.000694 g/mg/min (0.694 mg/mg/min) when using biochar for Fe sorption, with an adsorption capacity (q_e) of 106.383 mg/g. Although the adsorption capacity of biochar for Fe could be relatively high, yet, the Fe sorption affinity seemed to be generally very low. Two mechanisms might take place together during the first period of Fe contact (up to 30 min) with biochar, i.e., sorption of Fe on the functional groups of biochar [55], which might lessen considerably the Fe solubility in water. At the same time, electrons were set free to aqueous solutions during the degradation of this organic additive, and these electrons could increase Fe reduction [56] and solubility in water. These two mechanisms, therefore, displayed low-affinity phenomena for Fe removal by biochar from contaminated waters throughout the first time period of contact (up to 30 min). Thereafter (>30 min), easily oxidized organic carbon lessened considerably. Thus, the first mechanism of Fe sorption on biochar surfaces dominated.

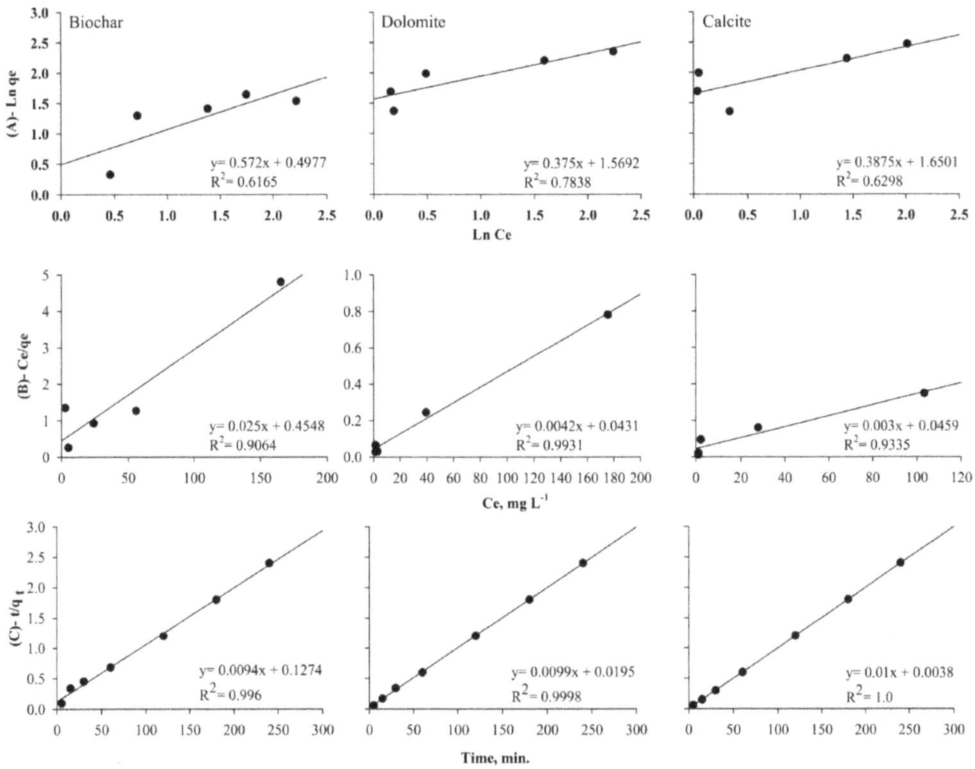

Figure 9. Freundlich (**A**) and Langmuir (**B**) isotherm models and pseudo-second-order kinetics (**C**) for Fe ions adsorption onto biochar, dolomite, and calcite materials.

For dolomite and calcite, the k_2 values were 0.006126 and 0.026316 g/mg/min, respectively, with q_e values around 100 mg g^{-1}. These values characterize the fast and high sorption capacities of Fe by the carbonate minerals. The rapid kinetics and good fits suggest the raw adsorbents are suitable for efficient Fe removal from contaminated waters without modification.

Overall, the fitted sorption and kinetic modeling highlight that the main mechanisms beyond Fe removal by dolomite and calcite from contaminated waters were adsorption on Fe surfaces in the form of a single monolayer, followed by chemisorption, as denoted by the kinetic fittings [57]. Also, our results indicate that co-precipitation of Fe on carbonate surfaces could be more effective for the removal of Fe from contaminated solutions than Fe binding on the functional groups of biochar.

A point to note is that the calculated q_e parameter of the kinetic model aligned closely with the experimental Langmuir maximum adsorption capacity, confirming the applicability of the pseudo-second-order model to interpret adsorption mechanisms [20].

As shown in Table 4. A summary of the adsorption isotherms, showing that the adsorption of iron onto biochar, dolomite, and calcite, is better described by the pseudo-second-order kinetic model, suggesting that chemisorption might be the rate-limiting step in the adsorption process. According to the model, the rate of adsorption relies on the availability of chemisorption sites on the solid adsorbents. It implies that the reaction between the adsorbent and the adsorbate involves a strong chemical bond formation, leading to a more stable adsorption process compared to physisorption, where weak forces dominate. The second-order kinetic model considers the concentration of both the adsorbate and the adsorbent in its equation, indicating a correlated behavior between the two. As a result, the rate of adsorption is not only influenced by the concentration of the adsorbate but also by the surface area and porosity of the adsorbents. This finding suggests that optimizing the chemisorption capacity and characteristics of biochar, dolomite, and calcite can enhance their efficiency as adsorbents for Fe removal in various applications, such as water treatment.

Table 4. Kinetic parameters obtained from pseudo-first order and pseudo-second order for Fe ions adsorption by tested sorbent materials.

Sorbents	Pseudo-1st					Pseudo-2nd				
	Slope	Intercept	R^2	K1/K2	qe (mg g^{-1})	Slope	Intercept	R^2	K1/K2	qe (mg g^{-1})
Biochar	0.0117	1.834	0.914	−0.03	68.23	0.009	0.1274	0.990	0.00	106.38
Dolomite	0.0289	1.5153	0.841	−0.07	32.76	0.01	0.016	0.999	0.006	101.01
Calcite	0.0215	0.729	0.584	−0.05	5.36	0.010	0.0038	1.000	0.026	100

3.6.3. Regeneration

Figure 10 presents the efficacy of the three tested materials—biochar, dolomite, and calcite—in the removal and desorption of iron over five successive cycles of Fe sorption and desorption. Among the sorbents, dolomite exhibited the highest iron removal capacity in the first cycle, with a percentage reaching 100% compared to 86.69% for biochar and 99.93% for calcite. The high removal efficiency of dolomite could be attributed to its unique mineral composition and the porosity of this mineral [58]. The removal efficiency decreased over the successive cycles; nevertheless, the least decline was observed in dolomite, which was only 6.33% by the fifth cycle (RE = 93.67%).

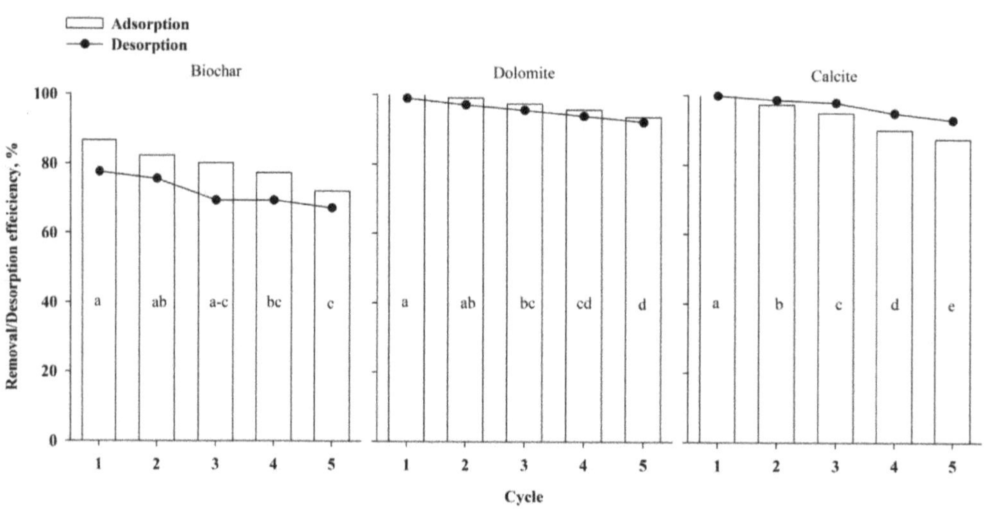

Figure 10. Regeneration of spent biochar, dolomite, and calcite materials and Fe ions desorption. Different letters show significant difference at $p < 0.05$ (values signify means ± SD).

Regarding Fe desorption from the studied materials, dolomite again showed the highest iron recovery, which was 99.97% in the first cycle. This percentage decreased slightly within the proceeding cycles to reach 92.93% by the fifth cycle. Calcite also showed a similar trend as dolomite with a lower adsorption/desorption recovery.

Biochar recorded the lowest removal efficiency, which was 86.69% (cycle 1) and decreased to 72.10% (cycle 5). The desorption recovery of Fe by biochar was also relatively low and ranged from 43.85% in cycle 1 to 38.55% in cycle 5. Biochar is characterized by its relatively high stability, owing to its recalcitrant carbon structure [19]. This stability guarantees that Fe is bound strongly within the pores of biochar, thus lessening Fe desorption. On the other hand, calcite exhibited the greatest weight loss during the regeneration steps reaching 13.67% in cycle 5. Dolomite had a weight loss of 6.23%, while biochar showed the least weight loss at 2.11% in cycle 5. The higher weight loss of calcite indicates high dissolution of the material during its regeneration with diluted HCl, and this could impact its reusability over successive cycles. Overall, dolomite seemed to be the optimum choice for Fe removal from wastewater.

4. Conclusions

This study was conducted to assess water quality in the El-Kharga Oasis (New Valley) by collecting water samples from surface and ground wells from five different villages across this area. Tap water was also collected from the same sites. The results reveal that the water quality of most sites was within the permissible levels of the WHO standards. Nevertheless, some sites exhibited high electrical conductivity (EC) and turbidity, notably in the surface wells of El-Kharga. Elevated levels of total and soluble Fe were detected in both shallow and deep wells of almost all sites, especially in El-Kharga, posing serious health risks, especially for children, as indicated by Hazard Index (HI) values.

Another investigation was carried out to ameliorate artificially contaminated water with Fe at a rate of 1000 mg Fe L^{-1} using four natural safe materials, namely dolomite, calcite, glauconite, and biochar. The efficiencies of the first two additives were much higher than the others, as they both removed successfully 100% of Fe from the water samples within only 120 min, with superiority for dolomite because of its low erodibility rate versus calcite. The regeneration of dolomite via HCl (0.1 M) was also effective enough, exhibiting 99.97% iron recovery. For five successive cycles, dolomite exhibited the highest efficiency for the removal of Fe from aqueous solutions with a slight decrease in removal efficiency from 100% to 93.67% between cycles 1 and 5. Likewise, Fe recovery decreased slightly to 92.93% in the fifth cycle

It seems that the removal of Fe from aqueous solutions by both dolomite and calcite took place through rapid surface Fe adsorption on their surfaces in monolayers, followed by chemisorption, as denoted by the kinetic fittings. Overall, this study highlights the critical need for targeted interventions, continual monitoring, and the adoption of effective sorbents, like dolomite and calcite, to lessen Fe contamination in water and ensure water safety, particularly in high-risk areas.

Author Contributions: All authors contributed to the study's conception and design. Material preparation, data collection, and analysis were performed by E.A.M.A., K.A.N.A., M.H.H.A. and A.A.A. The first draft of the manuscript was written by E.A.M.A., K.A.N.A., M.H.H.A., T.M.S.A. and A.A.A. All authors commented on previous versions of the manuscript. All authors have read and agreed to the published version of the manuscript.

Funding: This study was funded by the Science, Technology, and Innovation Funding Authority (STDF) under grant "Project ID: 44597".

Data Availability Statement: Data will be made available on request.

Conflicts of Interest: The authors declare no competing interests.

References

1. Barbieri, M.; Ricolfi, L.; Vitale, S.; Muteto, P.V.; Nigro, A.; Sappa, G. Assessment of groundwater quality in the buffer zone of Limpopo National Park, Gaza Province, Southern Mozambique. *Environ. Sci. Pollut. Res.* **2019**, *26*, 62–77. [CrossRef] [PubMed]

2. Baggio, G.; Qadir, M.; Smakhtin, V. Freshwater availability status across countries for human and ecosystem needs. *Sci. Total Environ.* **2021**, *792*, 148230. [CrossRef] [PubMed]
3. Lin, J.; Huang, J.; Prell, C.; Bryan, B.A. Changes in supply and demand mediate the effects of land-use change on freshwater ecosystem services flows. *Sci. Total Environ.* **2021**, *763*, 143012. [CrossRef] [PubMed]
4. Krishna, B.; Achari, V.S. Groundwater for drinking and industrial purposes: A study of water stability and human health risk assessment from black sand mineral rich coastal region of Kerala, India. *J. Environ. Manag.* **2024**, *351*, 119783. [CrossRef] [PubMed]
5. Ricolfi, L.; Barbieri, M.; Muteto, P.V.; Nigro, A.; Sappa, G.; Vitale, S. Potential toxic elements in groundwater and their health risk assessment in drinking water of Limpopo National Park, Gaza Province, Southern Mozambique. *Environ. Geochem. Health* **2020**, *42*, 2733–2745. [CrossRef] [PubMed]
6. Abdelbaki, C.; Badr, N.; Mohammedi, H.; Haidara, R.; Belarbi, H. Wastewater Reuse for Agriculture, Qualitative Aspects: A Case Study of Ain Temouchent, Algeria. In *The Water, Climate, and Food Nexus: Linkages, Challenges and Emerging Solutions*; Behnassi, M., Al-Shaikh, A.A., Gurib-Fakim, A., Barjees Baig, M., Bahir, M., Eds.; Springer International Publishing: Cham, Switzerland, 2024; pp. 355–377.
7. Fouad, S.S.; Heggy, E.; Ramah, M.; Abotalib, A.Z.; Palmer, E.M.; Jomaa, S.; Weilacher, U. Egypt's waterways conservation campaigns under growing intrinsic demand and Nile upstream damming. *J. Hydrol. Reg. Stud.* **2023**, *50*, 101537. [CrossRef]
8. El-Shatla, H.; Ahmed, E.A.; Essa, R.; Mohamed, M.F. Economic efficiency and productivity of lifting and distribution methods of groundwater for wheat and tomato in New Valley governorate. *Bull. Natl. Res. Cent.* **2019**, *43*, 157. [CrossRef]
9. Gabr, M.E. Land reclamation projects in the Egyptian Western Desert: Management of 1.5 million acres of groundwater irrigation. *Water Int.* **2023**, *48*, 240–258. [CrossRef]
10. Abdelhafez, A.A.; Abbas, M.H.H.; Kenawy, M.H.M.; Noureldeen, A.; Darwish, H.; Ewis, A.M.G.; Hamed, M.H. Evaluation of underground water quality for drinking and irrigation purposes in New Valley Governorate, Egypt. *Environ. Technol. Innov.* **2021**, *22*, 101486. [CrossRef]
11. WHO. *Guidelines for Drinking-Water Quality*, 4th ed.; Incorporating the 1st Addendum; WHO: Geneva, Switzerland, 2017.
12. Hurrell, R.; Reddy, M.; Cook, J. Inhibition of non-haem iron absorption in man by polyphenolic-containing beverages. *Br. J. Nutr.* **1999**, *81*, 289–295. [CrossRef]
13. Søndergaard, M.; Jensen, J.P.; Jeppesen, E. Role of sediment and internal loading of phosphorus in shallow lakes. *Hydrobiologia* **2003**, *506*, 135–145. [CrossRef]
14. Molot, L.A.; Watson, S.B.; Creed, I.F.; Trick, C.G.; McCabe, S.K.; Verschoor, M.J.; Sorichetti, R.J.; Powe, C.; Venkiteswaran, J.J.; Schiff, S.L. A novel model for cyanobacteria bloom formation: The critical role of anoxia and ferrous iron. *Freshw. Biol.* **2014**, *59*, 1323–1340. [CrossRef]
15. Aladejana, J.A.; Kalin, R.M.; Sentenac, P.; Hassan, I. Assessing the impact of climate change on groundwater quality of the shallow coastal aquifer of eastern dahomey basin, southwestern Nigeria. *Water* **2020**, *12*, 224. [CrossRef]
16. Katsoyiannis, I.A.; Zouboulis, A.I. Application of biological processes for the removal of arsenic from groundwaters. *Water Res.* **2004**, *38*, 17–26. [CrossRef] [PubMed]
17. Aoudj, S.; Khelifa, A.; Drouiche, N.; Hecini, M.; Hamitouche, H. Electrocoagulation process applied to wastewater containing dyes from textile industry. *Chem. Eng. Process. Process Intensif.* **2010**, *49*, 1176–1182. [CrossRef]
18. Fu, F.; Wang, Q. Removal of heavy metal ions from wastewaters: A review. *J. Environ. Manag.* **2011**, *92*, 407–418. [CrossRef] [PubMed]
19. Inyang, M.; Gao, B.; Yao, Y.; Xue, Y.; Zimmerman, A.R.; Pullammanappallil, P.; Cao, X. Removal of heavy metals from aqueous solution by biochars derived from anaerobically digested biomass. *Bioresour. Technol.* **2012**, *110*, 50–56. [CrossRef] [PubMed]
20. Abdelhafez, A.A.; Li, J. Removal of Pb(II) from aqueous solution by using biochars derived from sugar cane bagasse and orange peel. *J. Taiwan Inst. Chem. Eng.* **2016**, *61*, 367–375. [CrossRef]
21. Farid, I.M.; Siam, H.S.; Abbas, M.H.H.; Mohamed, I.; Mahmoud, S.A.; Tolba, M.; Abbas, H.H.; Yang, X.; Antoniadis, V.; Rinklebe, J.; et al. Co-composted biochar derived from rice straw and sugarcane bagasse improved soil properties, carbon balance, and zucchini growth in a sandy soil: A trial for enhancing the health of low fertile arid soils. *Chemosphere* **2022**, *292*, 133389. [CrossRef]
22. Asaad, A.A.; El-Hawary, A.M.; Abbas, M.H.H.; Mohamed, I.; Abdelhafez, A.A.; Bassouny, M.A. Reclamation of wastewater in wetlands using reed plants and biochar. *Sci. Rep.* **2022**, *12*, 19516. [CrossRef]
23. Kutergin, A.S.; Nedobukh, T.A.; Kutergina, I.N. Sorption treatment of surface waters from heavy metals by natural and granulated glauconite. In *AIP Conference Proceedings*; AIP Publishing: Long Island, NY, USA, 2020; Volume 2313. [CrossRef]
24. Martemyanov, D.; Rudmin, M.; Zhuravkov, S.; Korotkova, E.; Godymchuk, A.; Haskelberg, M.; Martemyanova, I.; Chernova, A.; Tyabaev, A.; Artamonov, A.; et al. Application of ural glauconite for groundwater deironing and demanganation. *J. Environ. Sci. Health Part A* **2021**, *56*, 861–866. [CrossRef] [PubMed]
25. Nassar, M.Y.; El-Shahat, M.F.; Osman, A.; Sobeih, M.M.; Zaid, M.A. Adsorptive Removal of Manganese Ions from Polluted Aqueous Media by Glauconite Clay-Functionalized Chitosan Nanocomposites. *J. Inorg. Organomet. Polym. Mater.* **2021**, *31*, 4050–4064. [CrossRef]
26. Abdel-Ghafar, H.M.; Abdel-Aal, E.A.; El Anadouli, B.E. Iron removal from ground water using Egyptian cost-effective clay minerals. *Appl. Chem. Eng.* **2020**, *3*, 23–30. [CrossRef]
27. Hu, H.; Zhang, Q.; Li, X.; Wu, L.; Liu, Y. Efficient heterogeneous precipitation and separation of iron in copper-containing solution using dolomite. *Sep. Purif. Technol.* **2020**, *248*, 117021. [CrossRef]

28. Shah, K.H.; Fahad, M.; Ghazi, Z.A.; Ali, S.; Shahzad, A.; Din, S.U. Optimization, characterization and adsorption properties of natural calcite for toxic As(III) removal from aqueous solutions. *Water SA* **2022**, *48*, 295–303. [CrossRef]
29. Liang-Tong, Z.; Li, Z.; Yuqing, Y.; Na, H.; Bate, B. Investigation of aqueous Fe(III) and Mn(II) removal using dolomite as a permeable reactive barrier material. *Environ. Technol.* **2023**, *44*, 2039–2053. [CrossRef] [PubMed]
30. Baird, R.B.; Eaton, A.D.; Rice, E.W. *Standard Methods for the Examination of Water and Wastewater*, 23rd ed.; American Public Health Association, American Water Works Association, Water Environment Federation: Washington, DC, USA, 2017.
31. Abdelhafez, A.A.; Li, J. Environmental Monitoring of Heavy Metal Status and Human Health Risk Assessment in the Agricultural Soils of the Jinxi River Area, China. *Hum. Ecol. Risk Assess. Int. J.* **2015**, *21*, 952–971. [CrossRef]
32. USEPA. *Exposure Factors Handbook*; Volume I General Factors; USEPA: Washington, DC, USA, 1997; EPA/600/P-96/002Fa.
33. Abdelhafez, A.A.; Zhang, X.; Zhou, L.; Cai, M.; Cui, N.; Chen, G.; Zou, G.; Abbas, M.H.H.; Kenawy, M.H.M.; Ahmad, M.; et al. Eco-friendly production of biochar via conventional pyrolysis: Application of biochar and liquefied smoke for plant productivity and seed germination. *Environ. Technol. Innov.* **2021**, *22*, 101540. [CrossRef]
34. *ASTM E394-22*; Standard Test Method for Iron in TRACE Quantities Using the 1,10-Phenanthroline Method. American Society for Testing and Materials. Available online: https://www.astm.org/e0394-22.html (accessed on 1 December 2023).
35. Joo, J.-H.; Hassan, S.H.A.; Oh, S.-E. Comparative study of biosorption of Zn2+ by Pseudomonas aeruginosa and Bacillus cereus. *Int. Biodeterior. Biodegrad.* **2010**, *64*, 734–741. [CrossRef]
36. Tan, I.A.W.; Ahmad, A.L.; Hameed, B.H. Adsorption isotherms, kinetics, thermodynamics and desorption studies of 2,4,6-trichlorophenol on oil palm empty fruit bunch-based activated carbon. *J. Hazard. Mater.* **2008**, *164*, 473–478. [CrossRef]
37. Ho, Y.S.; McKay, G. The sorption of lead(II) ions on peat. *Water Res.* **1999**, *33*, 578–584. [CrossRef]
38. Sawyer, C.N.; McCarty, P.L.; Parkin, G.F. (Eds.) *Chemistry for Environmental Engineering and Science*, 5th ed.; McGraw-Hill: New York, NY, USA, 2003.
39. Ayers, R.S.; Westcot, D.W. *Water Quality for Agriculture*; Part 1, Rev. 1; FAO Irrigation and Drainage Paper 29; Food and Agriculture Organization of the United Nations: Rome, Italy, 1994. Available online: https://www.fao.org/4/t0234e/t0234e00.htm (accessed on 1 December 2023).
40. Sassane, A.; Touati, M. Metallic groundwater contamination assessment using the heavy metal pollution index: A case study of the Guelma plain alluvial groundwater, Northeast of Algeria. *Sustain. Water Resour. Manag.* **2024**, *10*, 30. [CrossRef]
41. Van Groeningen, N.; ThomasArrigo, L.K.; Byrne, J.M.; Kappler, A.; Christl, I.; Kretzschmar, R. Interactions of ferrous iron with clay mineral surfaces during sorption and subsequent oxidation. *Environ. Sci. Process. Impacts* **2020**, *22*, 1355–1367. [CrossRef]
42. EPA. *National Recommended Water Quality Criteria—Aquatic Life Criteria Table*; EPA: Washington, DC, USA, 2016.
43. Ravikumar, P.; Aneesul Mehmood, M.; Somashekar, R.K. Water quality index to determine the surface water quality of Sankey tank and Mallathahalli lake, Bangalore urban district, Karnataka, India. *Appl. Water Sci.* **2013**, *3*, 247–261. [CrossRef]
44. WHO. *Guidelines for Drinking-Water Quality*, 4th ed.; World Health Organization: Geneva, Switzerland, 2011.
45. USEPA. *EPA's Exposure Factors Handbook (EFH)*; EPA/600/R-09/052F; National Center for Environmental Assessment Office of Research and Development: Washington, DC, USA, 2011.
46. Mohan, D.; Sarswat, A.; Ok, Y.S.; Pittman, C.U. Organic and inorganic contaminants removal from water with biochar, a renewable, low cost and sustainable adsorbent—A critical review. *Bioresour. Technol.* **2014**, *160*, 191–202. [CrossRef]
47. Miretzky, P.; Bisinoti, M.C.; Jardim, W.F. Sorption of mercury (II) in Amazon soils from column studies. *Chemosphere* **2005**, *60*, 1583–1589. [CrossRef]
48. Mayes, S.D.; Calhoun, S.L.; Murray, M.J.; Ahuja, M.; Smith, L.A. Anxiety, depression, and irritability in children with autism relative to other neuropsychiatric disorders and typical development. *Res. Autism Spectr. Disord.* **2011**, *5*, 474–485. [CrossRef]
49. Khoshraftar, Z.; Masoumi, H.; Ghaemi, A. An insight into the potential of dolomite powder as a sorbent in the elimination of heavy metals: A review. *Case Stud. Chem. Environ. Eng.* **2023**, *7*, 100276. [CrossRef]
50. Silva, R.; Cadorin, L.; Rubio, J. Sulphate ions removal from an aqueous solution: I. Co-precipitation with hydrolysed aluminum-bearing salts. *Miner. Eng.* **2010**, *23*, 1220–1226. [CrossRef]
51. Encina, E.R.; Distaso, M.; Klupp Taylor, R.N.; Peukert, W. Synthesis of Goethite α-FeOOH Particles by Air Oxidation of Ferrous Hydroxide Fe(OH)2 Suspensions: Insight on the Formation Mechanism. *Cryst. Growth Des.* **2015**, *15*, 194–203. [CrossRef]
52. Zhu, K.; Hopwood, M.J.; Groenenberg, J.E.; Engel, A.; Achterberg, E.P.; Gledhill, M. Influence of pH and Dissolved Organic Matter on Iron Speciation and Apparent Iron Solubility in the Peruvian Shelf and Slope Region. *Environ. Sci. Technol.* **2021**, *55*, 9372–9383. [CrossRef] [PubMed]
53. Cai, W.K.; Lui, J.H.; Zhou, C.H.; Keeling, J.; Glasmacher, U.A. Structure, genesis and resources efficiency of dolomite: New insights and remaining enigmas. *Chem. Geol.* **2021**, *573*, 120191. [CrossRef]
54. Walker, G.M.; Connor, G.; Allen, S.J. Kinetics of iron (II) removal from aqueous solution using activated dolomite. *Open Chem. Eng. J.* **2007**, *1*, 23–29. [CrossRef]
55. Wang, Y.; Li, H.; Lin, S. Advances in the study of heavy metal adsorption from water and soil by modified biochar. *Water* **2022**, *14*, 3894. [CrossRef]
56. Colombo, C.; Palumbo, G.; He, J.-Z.; Pinton, R.; Cesco, S. Review on iron availability in soil: Interaction of Fe minerals, plants, and microbes. *J. Soils Sediments* **2014**, *14*, 538–548. [CrossRef]

57. Ambaye, T.G.; Vaccari, M.; van Hullebusch, E.D.; Amrane, A.; Rtimi, S. Mechanisms and adsorption capacities of biochar for the removal of organic and inorganic pollutants from industrial wastewater. *Int. J. Environ. Sci. Technol.* **2021**, *18*, 3273–3294. [CrossRef]
58. Kamarzamann, F.F.; Abdullah, M.M.A.B.; Abd Rahim, S.Z.; Abdul Kadir, A.; Jamil, N.H.; Wan Ibrahim, W.M.; Victor Sandu, A. Hydroxyapatite/Dolomite alkaline activated material reaction in the formation of low temperature sintered ceramic as adsorbent materials. *Constr. Build. Mater.* **2022**, *349*, 128603. [CrossRef]

water

MDPI

Article

Potential Climate and Human Water-Use Effects on Water-Quality Trends in a Semiarid, Western U.S. Watershed: Fountain Creek, Colorado, USA

Carleton R. Bern [1,*], Manya H. Ruckhaus [1] and Erin Hennessy [2]

1 U.S. Geological Survey, Colorado Water Science Center, Denver, CO 80225, USA; mruckhaus@usgs.gov
2 U.S. Geological Survey, Colorado Water Science Center, Pueblo, CO 81003, USA; ehennessy@usgs.gov
* Correspondence: cbern@usgs.gov

Abstract: Nutrients, total dissolved solids (TDS), and trace elements affect the suitability of water for human and natural needs. Here, trends in such water-quality constituents are analyzed for 1999–2022 for eight nested monitoring sites in the 24,000 km^2 Fountain Creek watershed in Colorado, USA, by using the weighted regressions on time, discharge, and season (WRTDS) methodology. Fountain Creek shares characteristics with other western U.S. watersheds: (1) an expanding but more water-efficient population, (2) a heavy reliance on imported water, (3) a semiarid climate trending towards warmer and drier conditions, and (4) shifts of water from agricultural to municipal uses. The WRTDS analysis found both upward and downward trends in the concentrations of nutrients that reflected possible shifts in effluent management, instream uptake, and water conservation by a watershed population that grew by about 40%. Selenium, other trace elements, and TDS can pose water-quality challenges downstream and their concentrations were found to have a downwards trend. Those trends could be driven by either a warming and drying of the local climate or decreased agricultural irrigation, as both would reduce recharge and subsequent mobilization from natural geologic sources via groundwater discharge. The patterns illustrate how changes in climate and water use may have affected water quality in Fountain Creek and demonstrate the patterns to look for in other western watersheds.

Keywords: effluent; nitrogen; phosphorus; salinity; selenium; trend analysis; water conservation

1. Introduction

Population growth and a limited water supply are linked challenges that have driven the need for a science-based understanding and the management of water in the western United States [1]. The Fountain Creek watershed, located along Colorado's Front Range, exemplifies such challenges. In Colorado, average annual temperatures have risen by about 1.1 °C in the past 30 years, and although the trends in precipitation are less certain, soil moisture droughts have become more common and most projections of annual streamflow show decreases [2,3]. Local surface water and groundwater resources are relatively limited, and water is imported via several collection and diversion systems. Most drinking water used in Colorado Springs, the main municipality, is imported via pipelines from sources 160 km away in the Rocky Mountain headwaters of the Arkansas and Colorado River Basins [4]. The population in the Fountain Creek watershed has grown by about 40% in the last two decades [5], but human water-use efficiency has increased in that same timeframe. In the city of Colorado Springs, the annual per capita use of water delivered by utilities has decreased from about 0.193 megaliters (mL) per year to 0.106 mL per year between the 2000s and 2021 [6]. Therefore, each resident now uses about 55% as much water as they did about 20 years ago. Finally, the higher economic value associated with urban water use relative to agriculture has driven transfers of water rights in Fountain Creek and the Arkansas River Basin, where it resides, similarly to patterns found across the west [7].

In the Fountain Creek watershed, rapid population growth that started in the 1970s prompted concerns about water quality and quantity [8]. Early studies identified increased concentrations of nutrients and other water-quality components that reflected increasing dominance of effluent in streamflow [8,9]. Subsequent work has addressed a spectrum of issues including controlling excess sediment transport [10], finding that wildlife contributes substantially to upstream fecal contamination [11], documenting declines in some aquatic communities [12,13], and linking salinity, selenium, and uranium loading to groundwater discharge [14]. Storm-generated streamflows during the summer monsoon season have been the focus of targeted sampling efforts that showed how abrupt, rainfall-driven increases in streamflow (stormflow) increase the concentrations of suspended sediments, unfiltered trace elements, and *E. coli* in the creek [15,16]. Stormwater management has been a challenge as the watershed has urbanized, and efforts to improve it are underway [17]. Farther downstream, constituents like total dissolved solids (TDS), a measure of salinity, and selenium are substantial concerns in the Arkansas River, to which Fountain Creek is a tributary [18,19].

The primary goal of this paper is to examine trends in water-quality constituents in Fountain Creek, specifically nutrients, salinity, and selected trace elements. Trends are examined for eight streamgage monitoring locations in the watershed for the period 1999 to 2022 [20]. The timeframe is long enough to yield insights into multi-decadal trends, and the eight locations allow for a spatial analysis of the sources of water-quality constituents. The effects of stormflow sampling and suspended sediments are assessed only for their influence on overall trends across flow conditions. For context on the drivers of water-quality change, changes in water management and trends in local climate and aspects of regional hydrology in unmanaged locations are also assessed. The result is a holistic examination of water-quality change in the watershed in recent decades. An understanding of water-quality trends in Fountain Creek is important to local stakeholders. Through an examination of the trends relative to the drivers, however, Fountain Creek could also provide insights into patterns that are unfolding across the western United States.

2. Materials and Methods

2.1. Watershed Characteristics

Elevations in the Fountain Creek watershed vary from 4302 m in the northwest to about 1413 m at the creek's confluence with the Arkansas River (Figure 1). Terrain at the higher elevations is steep and mountainous with climates typical of the Rocky Mountains in Colorado, including seasonal snowpack accumulations. However, most of the watershed is within the Colorado Piedmont, a region with a semiarid climate which is located below a transition zone at around 2000 m. The major municipality in the watershed, Colorado Springs, is located at around 1800 m elevation and most of the watershed's population lives in the Piedmont region. The higher elevation terrain is underlain primarily by crystalline igneous intrusive and metamorphic rocks, and the Colorado Piedmont terrain is primarily underlain by sedimentary rocks of generally Cretaceous age [21]. Some of the sedimentary rocks, particularly the Niobrara Formation and the lower unit of the Pierre Shale, are known to negatively influence water quality via TDS and trace elements [22].

From 1991 to 2020, mean annual precipitation in Colorado Springs was 40 cm, with the heaviest precipitation being in summer due to thunderstorms (NOAA, [23]). Mean monthly temperatures in the same time period ranged from −0.2 °C in January to 22.4 °C in July (NOAA, [23]). The Köppen Climate Classification for most of the watershed is cold, arid, steppe [24].

Colorado Springs is the major municipality in the Fountain Creek watershed. The population of Colorado Springs was 356,208 in 1999 and grew to 483,953 in 2021, a 36% population increase [5]. Including the other municipalities and unincorporated communities, the population in the Fountain Creek watershed was approximately 421,000 in 1999 and grew to 590,000 by 2021, a 40% population increase [5,25]. Those population estimates excluded Pueblo residents living in the Fountain Creek corridor where it cuts through a narrow section of that municipality near the creek's mouth (Figure 1).

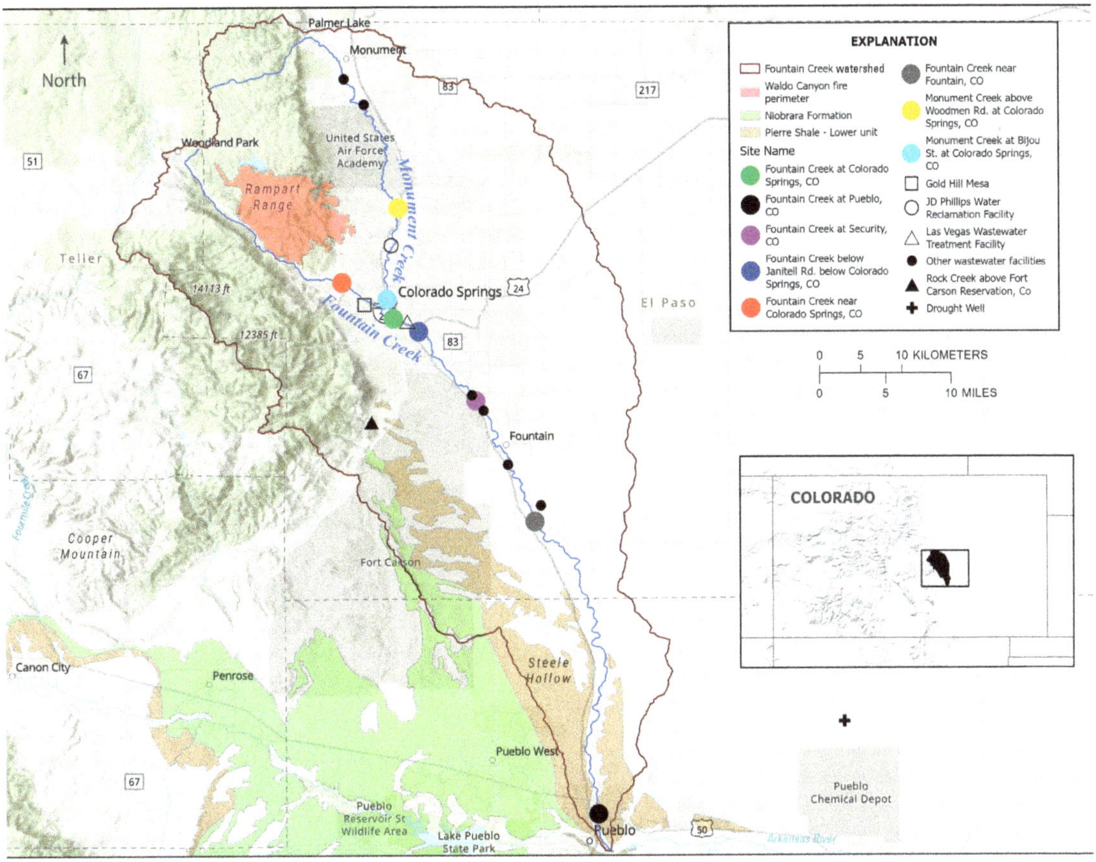

Figure 1. Map of the Fountain Creek watershed and features relevant to the study.

Because the semiarid climate limits natural local water supplies, water is imported for municipal use, and the resulting volumes and locations of treated effluent discharges to Fountain Creek are important to hydrology and water quality. Various changes in treatment processes and effluent compositions have occurred at the numerous wastewater treatment facilities across the watershed over the years (Figure 1). One major change involved the construction of the JD Phillips Water Reclamation Facility which came online in 2008 [26]. The new facility took some of the burden of wastewater treatment for the City of Colorado Springs from the Las Vegas Street Wastewater Treatment Facility, which was and remains the largest such facility in the watershed. In the process, the location of some effluent discharge for the city moved upstream to a location on Monument Creek (Figure 1). A non-potable water system is also operated in the city of Colorado Springs and as much as 5500 mL of reclaimed wastewater was distributed via that system in 2021 [6].

A major wildfire burned in the Fountain Creek watershed in 2012. The Waldo Canyon Fire burned a total of 74 km^2 of the mountainous headwaters primarily of Fountain Creek but also Monument Creek in June and July of that year (Figure 1). Burn severity was classified as 51% high, 20% moderate, and 29% low [27]. Substantial resources were deployed to mitigate runoff effects from the fire, but one flash flooding event occurred in 2012 and another four events occurred in 2013 [28,29].

Eight streamgage monitoring locations were selected to be the focus of the study (Table 1; Figure 1). The criteria for inclusion were continuous monitoring of streamflow and abundance of water-quality data across a variety of constituents. Trend analysis of water-quality constituents focused on the years 1999 to 2022 because of relative data abundance. All water-quality data used in this study are archived in the National Water Information System database and can be accessed using the USGS Station ID [20].

Table 1. U.S. Geological Survey (USGS) streamgage water-quality monitoring sites included in the study along with drainage area and gage elevation information [20].

Short Site Name	USGS Station ID	USGS Station Name	Drainage Area (km^2)	Elevation (m)
Fountain Creek, upstream	07103700	FOUNTAIN CREEK NEAR COLORADO SPRINGS, CO	264	1862
Monument Creek, upstream	07103970	MONUMENT CR ABV WOODMEN RD AT COLORADO SPRINGS, CO	466	1911
Monument Creek at Bijou Street	07104905	MONUMENT CREEK AT BIJOU ST. AT COLO. SPRINGS, CO	609	1823
Fountain Creek at Colorado Springs	07105500	FOUNTAIN CREEK AT COLORADO SPRINGS, CO	1015	1798
Janitell Road	07105530	FOUNTAIN CR BLW JANITELL RD BLW COLO. SPRINGS, CO	1070	1780
Security	07105800	FOUNTAIN CREEK AT SECURITY, CO	1295	1719
Fountain	07106000	FOUNTAIN CREEK NEAR FOUNTAIN, CO.	1740	1632
Pueblo	07106500	FOUNTAIN CREEK AT PUEBLO, CO	2396	1434

2.2. Streamflow and Municipal Water Use Analysis

Streamflow was assessed using data downloaded from the USGS National Water Information System (NWIS) database [20]. Along with mean water year (October to September) flows, mean winter base flows for the period November to February were also calculated, as this period generally represents the base flow season. Unlike native streamflow that originates in the Fountain Creek watershed, transbasin water imports and portions of water rights transferred from irrigation are considered reusable and can be used or consumed to extinction. A transit loss accounting program has been developed to track such reusable water relative to native streamflow [30–32]. Output from this program throughout 2020 was used to assess the amount of reusable water at 6 of the 8 streamgages where output was available [33]. Trends in mean annual streamflows, base flow, and reusable flows were assessed using Kendall's nonparametric test for monotonic trend via the "kendalltrendtest" function from the "EnvStats" package in R v. 4.3.1 [20,34,35]. This same test was used for all trend assessments of non-water-quality parameters in the study.

Daily production volumes of potable water for customers covering 2000 through 2022 and daily effluent discharge volumes data covering 24 July 2010 to 30 September 2022 for the Las Vegas Street Facility and covering 5 May 2008 to 17 November 2021 for the JD Phillips Facility were provided by Annie Berlemann of Colorado Springs Utilities [26]. The proportion of the individual or combined discharges relative to measured streamflow at the appropriate streamgage downstream were calculated along with water year means (Figure 2). Trends were assessed as previously described.

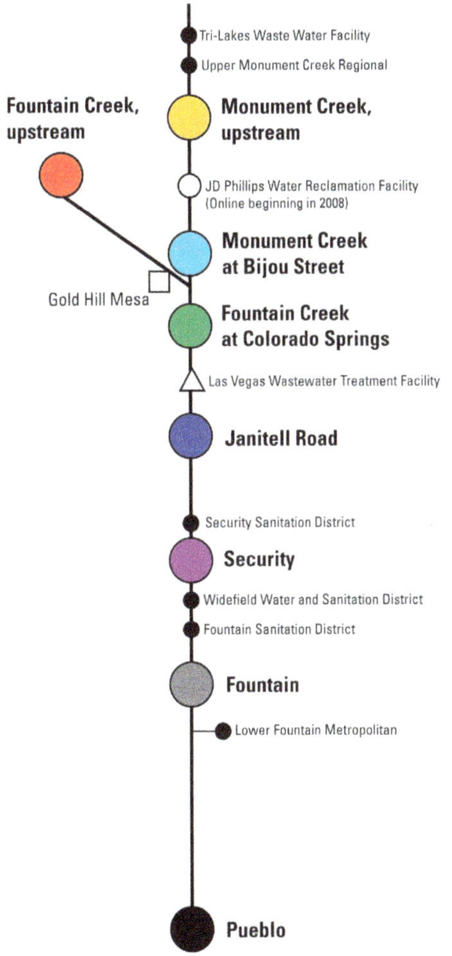

Figure 2. Diagram of the streamgage monitoring locations assessed in this paper relative to the main stems of Fountain and Monument Creeks with the same colors and symbols as Figure 1 and short names from Table 1. Locations where effluent from treatment plants enters the creeks are marked for reference.

2.3. *Climate, Natural Hydrology, and Irrigation Trend Analysis*

As context for water-quality trends, local climate data were analyzed. The timeframe 1980 to 2022 was selected because of the potential lag between patterns in climate and resulting effects on natural hydrology like groundwater recharge, native streamflow, and groundwater contributions to stream base flow. Precipitation totals and maximum daily temperatures were downloaded for stations at the Colorado Springs Municipal Airport and the Pueblo Memorial Airport [36]. These two stations are located in the lower elevation, Colorado Piedmont portion of the watershed where the Cretaceous age sedimentary geology is present. Precipitation was summed across years and maximum temperatures averaged by years.

To evaluate trends in local native surface water hydrology that might relate to climate and be unaffected by imported water, streamflow for the USGS streamgage 07105945 Rock Creek above Fort Carson, Reservation, Co. (Figure 1) was assessed [20]. The gage sits at

1948 m, drains 17.5 km^2, and is underlain primarily by crystalline rock. Just downstream of the streamgage, the creek infiltrates into alluvium that presumably interacts with Fountain Creek through subsurface flow paths. Mean streamflows were calculated by water year for available data which covered 1980 to 2018, and any trends were assessed.

To assess effects of regional climate on recharge of groundwater and subsequent discharge to streams, water level data were analyzed for a continuously monitored "drought well" that monitors shallow groundwater in an upland setting where climate and natural recharge are expected to be the primary controls on water levels ([37] Figure 1). The "drought well" is located outside the Fountain Creek watershed but is subject to similar climate and weather to the Colorado Piedmont in the watershed. The site is part of the USGS Water Mission Area Climate Response Network in South Central Colorado [37]. The well is used in this study because a similar well inside the watershed with a suitably long record and outside the zone affected by the augmented and managed flows in Fountain Creek could not be identified. Water level data were downloaded from the NWIS database for USGS site 382323104200701, SC01906221AAA, Drought Well near Pueblo, Colorado [20]. Water levels were averaged by water year for available data from 2004 to 2022, and the trend was assessed.

To assess how the magnitude of agricultural diversions have changed through time, diverted water volumes were downloaded by irrigation year (November–October) for 29 diversions from Fountain Creek covering the period 1980–2022 [38]. Most irrigation diversions in the watershed occur downstream of the city of Fountain (Fountain site) (Figure 2). Some individual diversions ceased during this time. For the sake of simplicity, trends were assessed using the sum of diversions by irrigation year.

2.4. Water-Quality Trend Analysis

No new water-quality data were collected for this study. Streamflow and water-quality datasets used in the trend analysis were downloaded from the NWIS database [20]. Per the long-term monitoring program, samples had been collected by integrating streamflow using standard techniques such as depth-integrated, multiple vertical profiles across stream channels [39]. Changes to and decisions about which constituents have been monitored and when have been influenced by permits and regulatory requirements. Analyses of water-quality constituents were completed at the USGS National Water Quality Laboratory in Lakewood, Colorado. The quantity of data available for trend analysis varied by site and parameter and year to year with changes in sampling schedules. The average number of annual concentration data points was 8 but ranged from 5 to 13.

The only constituent for which concentrations from NWIS were not directly used in trend analysis was TDS. Specific conductance (SC) is a commonly used proxy for TDS [19,40]. Measurement of SC was more common for site visits than collection of samples for TDS analysis, and four sites had records of continuously monitored SC. Therefore, by translating SC data into TDS concentrations, the quantity of data to support trend analysis could be substantially increased. Existing and new regressions were used to translate SC to TDS (Table S1; [19,40]).

Weighted regressions on time, discharge, and season (WRTDS) [41] were used to describe trends in concentrations and loads of nutrient, salinity, and trace metal constituents from 1999 to 2021 at the 8 selected streamgage monitoring sites. The WRTDS model was run using the default settings, which sets the period of analysis as the water year and the half-window width to 20 years [42]. The analysis was completed using the "EGRET" package in R [34,43]. WRTDS uses discrete water-quality data and daily streamflow data to make daily concentration and load predictions. The method derives coefficients for each combination of streamflow and time, from which fitted values are then calculated using a multiple linear regression model [41]. WRTDS uses time, streamflow, and seasonality as predictor variables for concentration. Because streamflow is often a dominant driver of concentration, concentration–streamflow relationships are defined daily and across the probability distribution of streamflow [44]. A primary output from the model is flow-normalized concentration smoothed by locally weighted scatterplot smoothing (lowess) which minimizes the within-year variability in streamflow and seasonality to yield a concentration estimate that reveals year-to-year trends [41]. The flow-normalization method in WRTDS that assumes stationarity was used because, as described in the results, trends in streamflow were not identified. The flow-normalized output is useful for assessing multi-year trends at individual sites and herein is compared across the 8 streamgage monitoring sites to add a spatial component to the overall analysis. The approach can make WRTDS a powerful tool for deciphering controls on water quality. Because the 2008 opening of the JD Phillips facility could have caused a potential sharp discontinuity in water quality trends, use of the "wall" option in EGRET to divide the 1999–2022 record was explored [45]. However, the "wall" resulted in two time periods that were much shorter than the 20-year minimum for which WRTDS is designed and yielded some unrealistically steep trends [41]. The option was not used in the final analysis.

Trend uncertainty for flow-normalized concentration and flow-normalized load was quantified using a 90 percent confidence interval calculated from the WRTDS Bootstrap Test via the EGRETci package in R [34,46]. The block bootstrap approach estimates Type I error probabilities—the likelihood of detecting a trend where a trend is not present [46]—using a series of Monte Carlo simulations. The bootstrap method produces a set of hypothesis tests and corresponding two-sided p-values to either reject the null hypothesis (trend is more likely than not) or fail to reject the null hypothesis (no trend is more likely than not) using an alpha threshold of 0.1. Bootstrap replicated trend results and confidence bands were obtained via the "wBT" and "ciCalculations" functions, respectively, from the EGRETci package.

The statistical likelihood of water-quality trends is presented as part of the results. The likelihood values and trend directions are computed from the two-sided p-value attained [46]. For a given trend, upward or downward, p-values from 0.95 to 1.0 are considered "highly likely", 0.90 to <0.95 is considered very likely, 0.67 to <0.90 is considered "likely", and 0.50 to <0.67 is considered "uncertain." The EGRET package can also calculate changes in load. Trends in the direction and likelihood of the load closely tracked those for concentration. The Pueblo site is the only location for which load results are presented because of local interest in that site and because mid-record concentration peaks complicate trend results at several other sites.

The initial WRTDS analysis for unfiltered constituents yielded some unusual patterns in flow-normalized concentrations at many sites. Because only the unfiltered constituents were affected, it seems likely that the pattern was driven by high concentrations of suspended sediment in stormflow samples, as suspended sediment in Fountain Creek can commonly host nutrients and trace elements to a greater degree than water [15,16]. Based upon daily hydrographs, most of the samples with the highest concentration of unfiltered constituents came from stormflow samples. In some cases, the peak flows sampled represented one order of magnitude increases in streamflow, but others were more moderate. Daily mean streamflow, the value used as input in WRTDS, and sub-hourly changes in streamflow were separately explored for the purpose of flagging potential high concentration samples that appeared to distort the trends. Those approaches were unsuccessful, possibly because streamflow peaks and peak suspended sediment concentrations are sometimes out of sync [47]. Therefore, a different approach was tested, and a second set of WRTDS analyses were conducted for unfiltered constituents. Z-scores were calculated for all of the concentrations for a given unfiltered constituent by subtracting the mean concentration from the observed concentration and dividing by the standard deviation. A z-score greater than 4 indicated that the sample had a concentration more than 4 standard deviations greater than the mean. Such samples were omitted from the dataset, and then the process was repeated until no more datapoints with z-scores greater than 4 were found. The number of datapoints culled was recorded (Table S2), and a second round of WRTDS analysis was conducted on the remaining data. Differences between the output are discussed in the results.

As context for culling unfiltered constituent data, the relationships between streamflow and turbidity were examined for 17 and 18 June, 2018, for the upstream Fountain Creek site. Turbidity, a common proxy for suspended sediment concentrations, was monitored at this site only for a few months in 2018 [20], which prevented a broader analysis. No discrete water-quality samples overlapped with approved turbidity data. However, the patterns illustrate differences in the timing and magnitude of streamflow versus turbidity responses during stormflow events.

It is worth emphasizing that the intentional sampling of stormflow conditions to different degrees and with differing success across years introduced sampling bias into the dataset compared to regularly scheduled sampling that occurred regardless of streamflow conditions. The bias seems to be driven by the presence of more stormflow samples in the middle of the record compared to the years near the beginning and end. Assessing and attempting to decrease the effects of that bias on the timing and magnitudes of trends was the goal of the data culling procedure described. By doing so, trends during predominant non-stormflow conditions could be assessed.

3. Results

3.1. Streamflow and Municipal Water Use Patterns and Trends

The annual streamflow averages through time illustrate important aspects of Fountain Creek's hydrology. Natural interannual variability yielded coefficients of variation >0.64 at the two most upstream sites in Monument and Fountain Creek, but the coefficients were <0.55 at the Janitell site and downstream because of the moderating effects of imported water and effluent discharge (Figure 3a). The downstream sites gained little additional natural streamflow compared to the combined imported water and natural flows from upstream. Winter base flow variations were even more muted downstream because the lower elevations receive minimal winter precipitation (Figure 3b). The mean annual streamflow and winter base flow show complicated temporal patterns, and no significant trends were found (Figure 3a,b). The exception was Monument Creek at Bijou St. where the significant trends (found $p = 0.04$) may be attributable to the shorter record. The mean annual outflow from the Fountain Creek watershed at Pueblo from 1999 to 2022 was 138,000 mL.

Increased reusable flows were found for all of the sites and across all the years for which data were available (all $p \leq 0.03$) (Figure 3c). These trends illustrate both the increased presence of imported water in Fountain Creek and water rights that allow reuse to extinction. The pattern at the two most upstream sites in Figure 3c reflected the JD Phillips facility coming online in 2008. The annual percentage of reusable water at Fountain Creek at Pueblo varied substantially depending upon the native flows and ranged from 7% to 54%; the average percentage was 30%. The mean reusable flows exceeded the mean native flows from Fountain to Pueblo in 2012 and from the Janitell Road site to Pueblo in 2022. Both 2012 and 2022 were relatively low mean streamflow years (Figure 3a). No significant trends were found for native streamflow from 1999 to 2022.

There was a significant ($p = 0.013$) downward trend in annual potable water production for customers between the water years 2001 and 2022, indicating a 14% decrease (Figure S1). There was no trend in annual effluent discharge from the Las Vegas facility ($p = 0.84$) or both wastewater facilities combined ($p = 0.28$), but the JD Phillips facility ($p < 0.01$) had a 23% increase in effluent discharge between 2009 and 2021 (Figure 4a). The daily percentage of effluent from the JD Phillips facility (Figure 4b) or from both of the facilities combined (Figure 4c) could represent the fact that the streamflow downstream varied. The values >100% at Janitell Road were likely explained by losses of flow to the alluvial aquifer downstream from the JD Phillips discharge point, causing the Janitell Road streamflow to be less than the JD Phillips discharge [48]. Water year averages of the percentage of effluent could in the streamflow for Monument Creek at Bijou Street ranged from 12% to 38% (Figure 4b) and at Janitell Road it ranged from 20% to 59% (Figure 4c), but no significant trends were found (both $p > 0.3$).

3.2. Local Climate, Natural Hydrology, and Irrigation Trends

From 1980 to 2022, a significant downward trend in annual precipitation was found at Colorado Springs ($p = 0.016$, slope = −0.32) but not Pueblo ($p = 0.28$) (Figure 5a,b). Notably though, three years in which there was quite a low total precipitation in Pueblo in that time come in the latter half of the record. Significant upward trends in the annual mean of daily maximum temperature from 1980 to 2022 were found for both Colorado Springs ($p < 0.01$, slope = 0.059) and Pueblo ($p < 0.01$, slope = 0.032) (Figure 5c,d). It can be noted that no recent statewide trends in precipitation have been identified in Colorado, but the amount of precipitation in the 1980s was generally above average and in subsequent decades it was below average; statewide temperatures have trended upward over the same period [49].

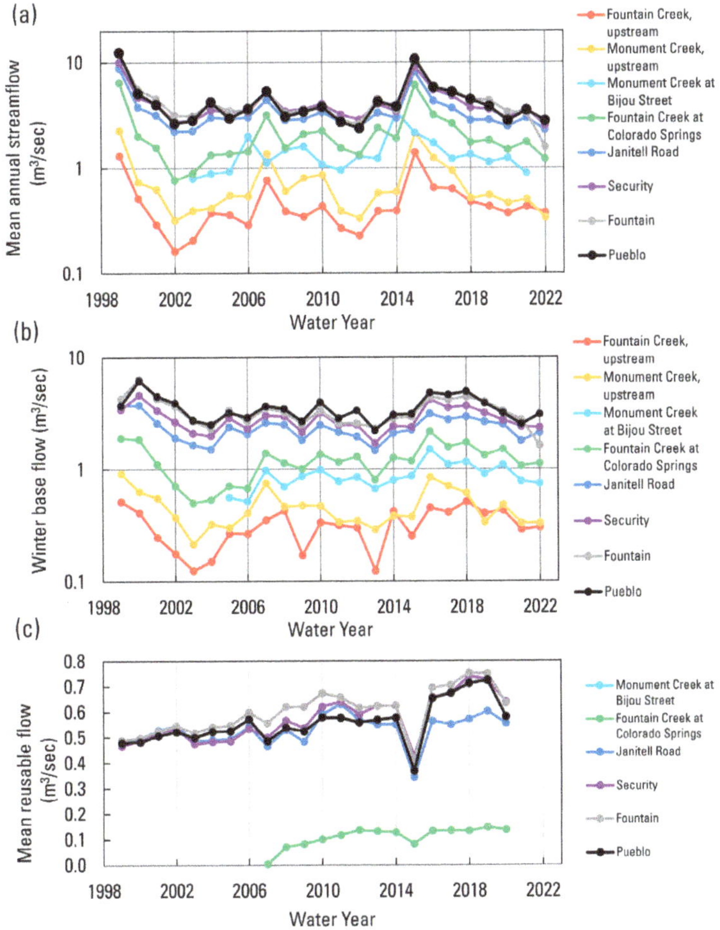

Figure 3. Metrics of streamflow at the 8 streamgages for water years 1999 to 2022: mean annual streamflow (**a**); mean streamflow for the period November to February which represents winter base flow (**b**); and mean flows of reusable water, which is a rough proxy for transbasin imported water (**c**) [20,33]. Note that the Bijou Street site data in (**c**) are overlapped by the next site downstream.

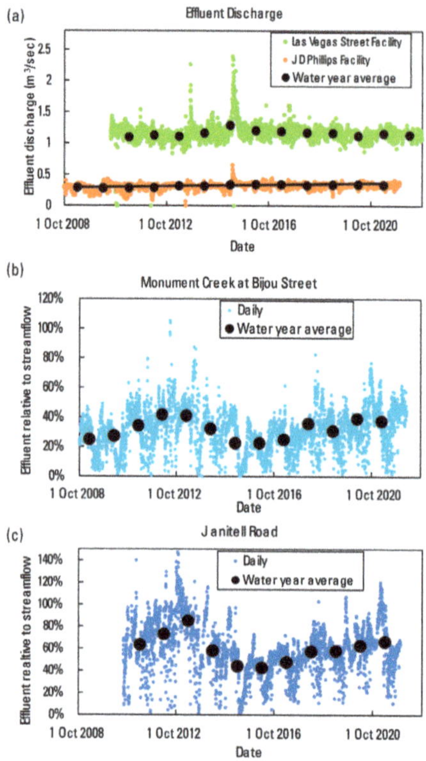

Figure 4. Details of effluent discharge from the two treatment facilities in Colorado Springs: (**a**) effluent discharge rates from the Las Vegas Street and JD Phillips facilities; (**b**) effluent discharge from the JD Phillips facility as potential percentage of streamflow at Monument Creek and Bijou Street; (**c**) and effluent discharge from both facilities as a potential percentage of streamflow at the Janitell Road site [20,26]. The trend line for effluent discharge from the JD Phillips facility is provided for reference.

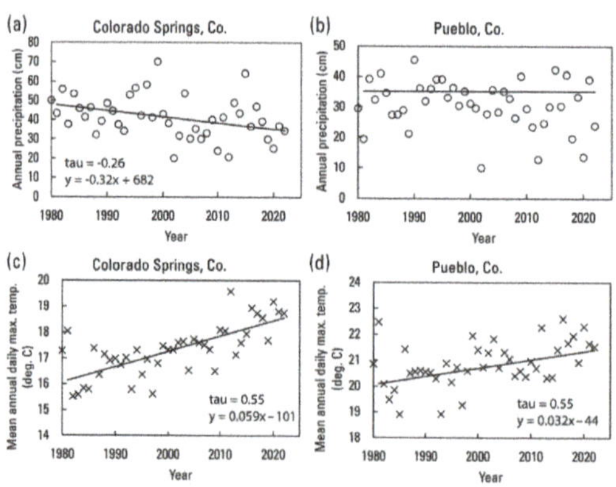

Figure 5. Annual precipitation totals by year for Colorado Springs, Co. (**a**) and Pueblo, Co. (**b**) and mean daily maximum temperatures by year for Colorado Springs, Co. (**c**) and Pueblo, Co. (**d**) [36]. Trend lines are provided for reference along with tau values and equations for significant trends.

The possible effects of warming temperatures and recent below-average precipitation on natural surface water hydrology were assessed in the small Rock Creek watershed, which contains only native flows. There, a downward trend in annual mean streamflow was significant at the 90% level (p = 0.097) (Figure 6a). That trend might have been stronger if not for the presence of high streamflows in 2015, which was a wet year overall (Figures 5a and 6a).

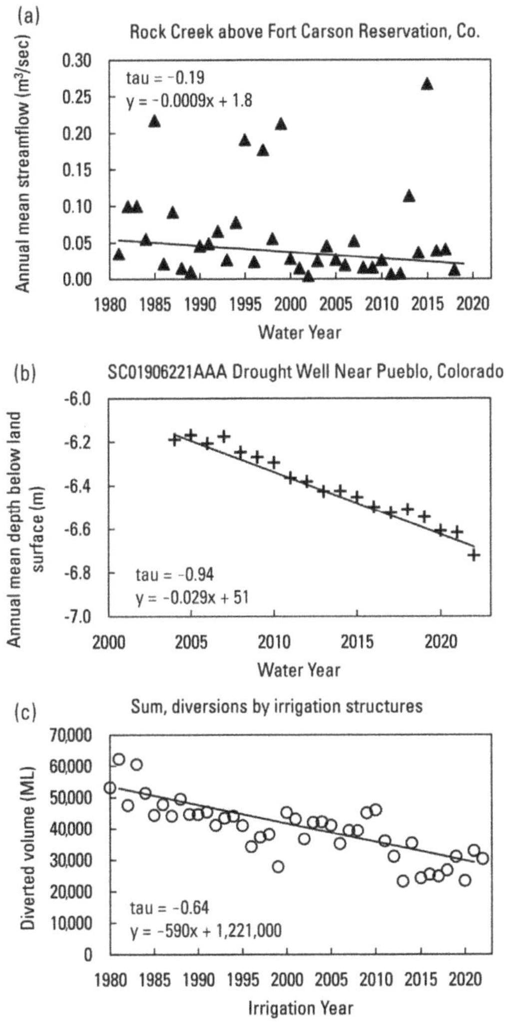

Figure 6. Hydrologic data that yield insight into potential shallow groundwater discharge to Fountain Creek: mean streamflow by year (**a**), mean depth to water below land surface by year for a monitoring well (**b**), and the sum of irrigation diversions from Fountain Creek by irrigation year (**c**) [20,38]. The stream is Rock Creek above Fort Carson Reservation, Colorado. The well is SC01906221AAA Drought Well near Pueblo, Colorado [20]. Trend lines, tau values, and equations provided for reference.

Groundwater responses to potential climate effects in the region were assessed via annual mean depths below land surface for water in SC01906221AAA, Drought Well near Pueblo, Colorado (Figure 6b) [20]. Intra-annual fluctuation was about 0.15–0.2 m per year with seasonal lows in October and highs in May or June, indicating possible natural recharge (data not shown). A significant downward trend in mean annual water levels was found for 2004–2022 ($p < 0.01$) indicating a decline of −0.029 m per year (Figure 6b).

Trends in the combined volumes of water removed from Fountain Creek by 29 irrigation-related diversions were significantly downward from 1980 to 2022 ($p < 0.01$) with a decline of −590 mL/yr/yr (Figure 6c). Diversions ceased for 15 of the ditches between 1980 and 1990. The sum of diverted water decreased by about −24,800 mL in the 42-year period, a decrease of about −46%. From 2016 to 2022, diversions averaged 28,700 mL/yr [38].

3.3. Effects of Stormflow Sampling and Suspended Sediments on Unfiltered Constituent Trends

As noted in the Methods Section, trends in unfiltered constituents exhibited unusual patterns at certain sites when all data were included in the WRTDS analysis. Differing priorities, strategies, and success in targeting stormflows for discrete sampling during the 22-year period analyzed appeared to be responsible, along with a transient effect from the Waldo Canyon Fire particularly at the upstream site in Fountain Creek. Suspended sediment concentrations can strongly influence concentrations of unfiltered constituents, and both suspended sediment concentrations and streamflow can change rapidly in response to stormflows. Continuously monitored turbidity from the Fountain Creek near Colorado Springs site provided an example (Figure 7). During storm events, suspended sediment concentrations, as reflected by turbidity, spiked quickly but unevenly relative to changes in streamflow. Sampling intended to target peak streamflow might have sampled or missed the peak in suspended sediments. This effect complicated the relationship between streamflow and unfiltered constituent concentration in the WRTDS and seemed to strongly influence the trend during periods when high concentrations of suspended sediments were sampled. A further complication was that the WRTDS used daily mean values as the streamflow input [42], and such daily values did not reflect instantaneous streamflow during the short-lived peaks targeted by storm sampling (Figure 7).

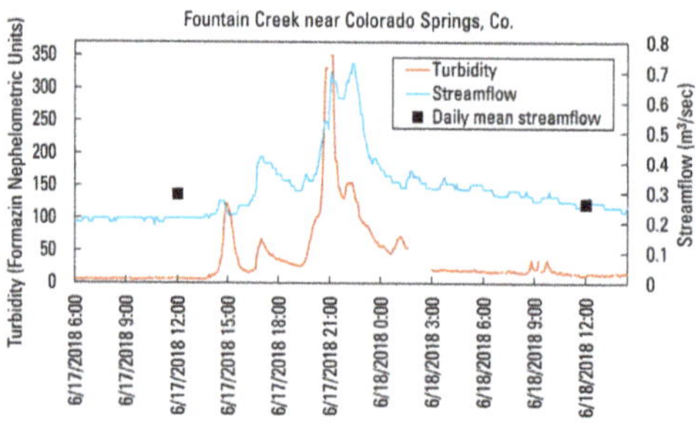

Figure 7. Continuously monitored turbidity (a proxy for suspended sediment concentration) and streamflow for 17 and 18 June, 2018 for Fountain Creek near Colorado Springs, Co. Note the differences in timing and magnitude of response between turbidity and streamflow. Also note how daily mean streamflow (the streamflow parameter used by WRTDS) compares to short duration stormflow [20].

Two examples of how such stormflow samples could have distorted trends were observed for unfiltered phosphorus in the Fountain Creek, upstream, and Pueblo sites. At the first site, the flow-normalized concentration trend was about 0.2 mg/L in 1999, which rose to a peak of about 1.3 mg/L around 2012, and then declined to about 0.2 mg/L by 2021 (Figure 8). The second site showed similar changes. Filtered constituents at these sites did not show dramatic changes in flow-normalized concentration through time.

The concentrations of unfiltered phosphorus in samples from both sites have a strongly skewed positive distribution (Figure 9), even after log transformation, which is one of the steps in WRTDS that compensates for the common situation of skewed concentration data [42]. It seemed that a few very high concentration samples collected in the middle of the period of analysis distorted the flow-normalized concentration trend. For the upstream Fountain Creek site, the few very highest concentrations were in 2013 and 2014, after the Waldo Canyon Fire, and tests removing just those samples diminished but did not remove the pronounced mid-record trend peak seen in Figure 8. The procedure of iteratively culling data points with z-scores >4 (four standard deviations greater than the mean) removed 52 of 192 unfiltered phosphorus concentrations for the Fountain Creek, upstream site (Table S2), which is a large fraction of the available data. In contrast, the z-score culling procedure only removed 12 of the 179 data points for unfiltered phosphorus for the Pueblo site (Table S2). Trends for unfiltered phosphorus at the upstream Fountain Creek site after the z-score culling procedure showed very low concentrations with little change through time (Figure 10a). Comparing the trends in unfiltered phosphorus when using all of the data (Figure 8) to those when using the z-score culled data (Figure 10a) for the Pueblo site, the flow-normalized concentrations at the mid-record peak present in both analyses decreased from almost 1.5 mg/L to less than 0.7 mg/L.

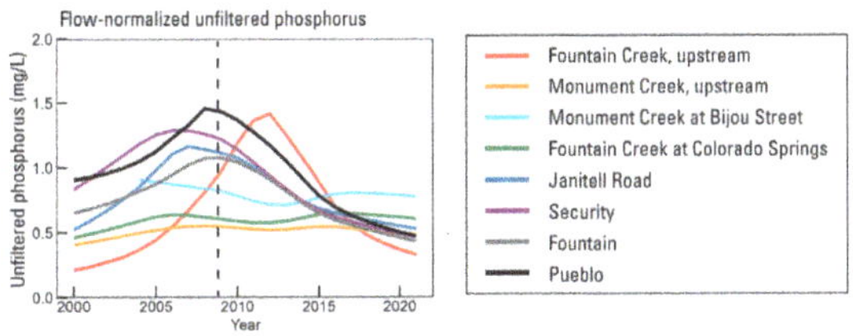

Figure 8. Trends in flow-normalized concentrations of unfiltered phosphorus for the 8 streamgage monitoring sites using all available data [20]. Lines are color coded as in Figures 1 and 2. The trends can be compared to those in Figure 10a where, as described in the text, the z-score-based culling procedure was implemented prior to the WRTDS analysis. For reference, dashed lines indicate when the JD Phillips facility began taking some of the wastewater treatment load from the Las Vegas Street facility.

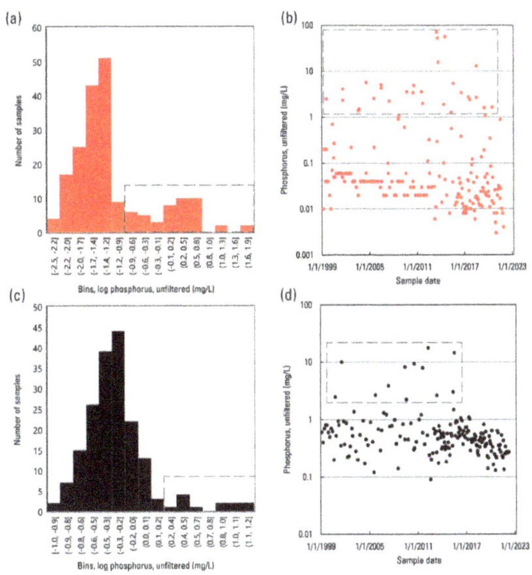

Figure 9. Histogram and all measured concentrations plotted by date of the log of unfiltered phosphorus concentrations [20] for site 07103700, Fountain Creek near Colorado Springs, CO,

(**a**,**b**) and for 07106500 Fountain Creek at Pueblo, CO (**c**,**d**). Note the skewed data distributions even after the log transformation. Also note the timing of highest concentration samples relative to trends in Figure 9. Dashed line boxes indicate the samples removed by the z-score culling procedure in all panels.

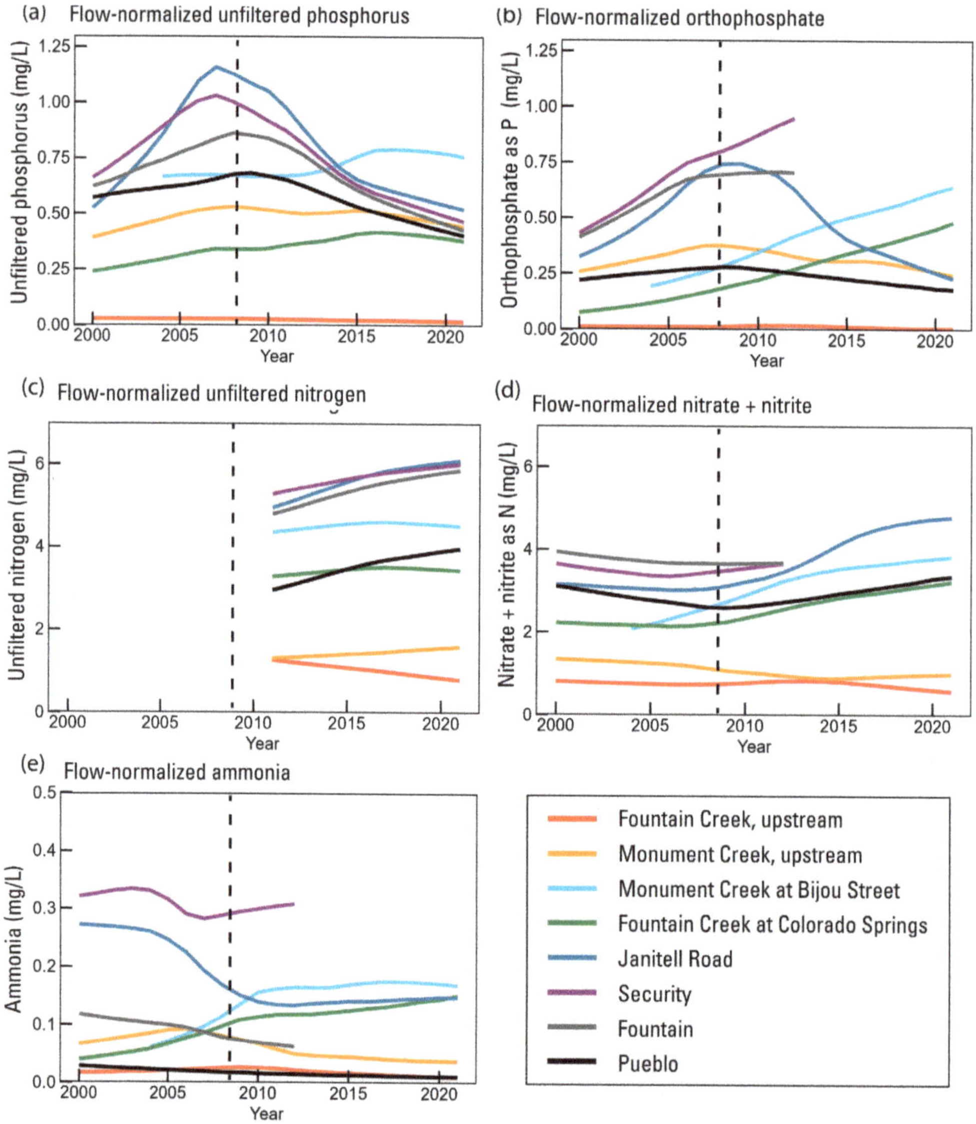

Figure 10. Flow-normalized trends in nutrient concentrations: unfiltered phosphorus as P (**a**), orthophosphate (PO_4^{3-}) as P (**b**), unfiltered nitrogen as N (**c**), nitrate (NO_3^-) and nitrite (NO_2^-) as N (**d**), ammonia (NH_3) as N (**e**) [20]. For reference, dashed lines indicate when the JD Phillips facility began taking some of the wastewater treatment load from the Las Vegas Street facility.

The z-score culling procedure was adopted for trend analysis of all the unfiltered constituents to avoid distorted trends. Data points retained versus culled are listed in Table S2. It is acknowledged that some elements of the effects of stormflows on unfiltered constituents have been omitted from the analysis. The result, however, was that water-quality trends present in the predominant, non-stormflow portion of the hydrograph could be more accurately described.

3.4. Nutrient Trends

Mid-record peaks were a feature of many of the flow-normalized, unfiltered phosphorus, and filtered orthophosphate trends (Figure 10a,b) and for that reason the statistical WRTDS trend likelihood results could be misleading (Table 2 and Table S3). The suggestion of continued increases in orthophosphate at Security and Fountain may be the product of when the monitoring stopped. Flow-normalized trends for both measures of phosphorus peaked around 2008 for the Janitell Road site and those downstream with complete records (Figure 10a,b). The timing of the declines were coincident with the 2008 opening of the JD Phillips facility which began to share the wastewater treatment load carried by the Las Vegas Street facility and shifted some treated water discharge to above the Bijou Street site (Figure 2). Concentrations of unfiltered phosphorus were generally high at Janitell Road and declined in order downstream (Figure 10a). Interestingly, unfiltered phosphorus was high for Monument Creek at Bijou Street before 2008 and no definite response to the new discharge was observed after 2008 (Figure 10a). Notable concentrations of phosphorus were present at the upstream Monument Creek site but not the upstream Fountain Creek site (Figure 10a). Pronounced strong upward trends in orthophosphate were present for the Bijou Street and Fountain Creek sites in Colorado Springs (Figure 10b, Table S3). The suggestions of those trends prior to 2008 might relate to orthophosphate being present at the upstream Monument Creek site. The steady post-2008 increases could be related to discharge from the JD Phillips facility. Distinctions between filtered and unfiltered phosphorus trends may relate to shifts in the proportions the former accounts for within the latter (Figure S2). For the Pueblo site, despite mild mid-record peaks, significant downward trends in both unfiltered phosphorus and orthophosphate were found (Table 2).

Table 2. Trend likelihoods and estimated total changes in concentration and load for Fountain Creek at Pueblo during the period examined. The 90% confidence intervals are given in parentheses. The dataset covers water years 1999 to 2022 and the bootstrap process to estimate likelihoods and changes for water years 2000 to 2021, except as noted. "*" indicates data available only from 2014 to 2022. NA indicates either trend is as likely as not.

Constituent	Concentration Trend Likelihood	Load Trend Likelihood	Total Estimated Concentration Change	Total Estimated Load Change
Phosphorus, unfiltered	Downward trend is highly likely	Downward trend is likely	−0.17 mg/L (−0.27 to −0.03)	−18,810 kg/yr (−55,740 to 2073)
Orthophosphate	Downward trend is highly likely	Downward trend is highly likely	−0.14 mg/L (−0.29 to −0.04)	−15,810 kg/yr (−37,600 to −3310)
Total nitrogen	Upward trend is highly likely	Upward trend is likely	+1.02 mg/L (+0.24 to 1.64)	+96,200 kg/yr (−66,000 to +197,600)
Nitrate plus nitrite	Upward trend is likely	Upward trend is likely	+0.27 mg/L (−0.16 to +0.65)	+45,920 kg/yr (−29,400 to +97,000)
Ammonia and ammonium	Downward trend is highly likely	Downward trend is highly likely	−0.018 mg/L (−0.03 to −0.007)	−2911 kg/yr (−4935 to −622)
Total dissolved solids	Downward trend is highly likely	Downward trend is highly likely	−134 mg/L (−167 to −102)	−16,270 mg/yr (−19,400 to −11,600)
Selenium, unfiltered,	Downward trend is highly likely	Downward trend is highly likely	−16.1 µg/L (−18.5 to −13.9)	−1625 kg/yr (−1930 to −1370)
Selenium, filtered	Downward trend is highly likely	Downward trend is highly likely	−14.9 µg/L (−17.3 to −12.9)	−1362 kg/yr (−1570 to −1140)
Arsenic, unfiltered	Downward trend is highly likely	Either trend is likely as not	−1.8 µg/L (−2.3 to −0.84)	NA
Arsenic, filtered *	Downward trend is highly likely	Downward trend is highly likely	−0.4 µg/L (−0.56 to −0.14)	−67 kg/yr (−92 to −28)
Iron, unfiltered *	Downward trend is likely	Downward trend is likely	−4541 µg/L (−8097 to +2071)	−2227 mg/yr (−4300 to +7803)
Manganese, filtered	Either trend is likely as not	Either trend is likely as not	NA	NA

Data for unfiltered nitrogen only extended back to 2011. Flow-normalized concentrations were lowest at the Fountain Creek and Monument Creek upstream locations (Figure 10c). Downstream, at the two sites within the city of Colorado Springs, concentrations were higher, but trends were not discerned. The concentrations of unfiltered nitrogen were highest in the three sites downstream from the Las Vegas Street facility (Figure 2) and upward trends were likely to very likely (Figure 10c). Downstream at Pueblo, the flow-normalized concentration was lower, but an upward trend was still highly likely (Table 2). Flow-normalized nitrate plus nitrite concentrations had significant upward trends between the Bijou Street and Janitell sites (Figure 10d, Table S3). Recent trends at Fountain and Security were unknown because of a lack of recent data. For the Pueblo site, an overall upward trend in nitrate and nitrite was likely but the trend had both decreased and increased historically (Figure 10d). Pueblo was also the only site where there was an indication that nitrite plus nitrate accounted for more of the unfiltered nitrogen. The concentrations of ammonia and ammonium were comparatively low and showed trends that were localized and divergent among the different sites (Figure 10e). Inflections in many of the trends are found within a couple years of 2008 and seem to indicate effects related to effluent discharge (Figure 10e). Notably, the Security site had the highest flow-normalized concentrations, but the discontinuation of monitoring prevented an assessment of any trends in recent years.

3.5. Salinity-Related Trends

The spatial patterns in the flow-normalized concentrations of TDS showed the that concentrations increased with distance downstream (Figure 11a). One area where a substantial increase in TDS concentrations occurred was in and around the city of Colorado Springs. Additional substantial TDS increases occurred at the sites downstream from the city with a particular increase between the Fountain and Pueblo sites (Figure 11a).

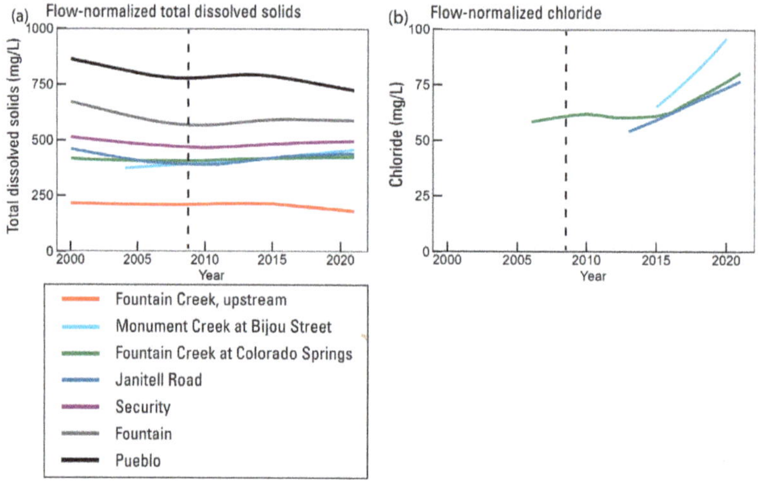

Figure 11. Flow-normalized trends in concentrations of constituents related to salinity: total dissolved solids as determined from measurements of specific conductance and regressions relating those two parameters (**a**), and chloride (**b**) [20]. For reference, dashed lines indicate when the JD Phillips facility began taking some of the wastewater treatment load from the Las Vegas Street facility.

Temporal trends in flow-normalized TDS concentration, both upward and downward, were present in the upper watershed, but the magnitudes were relatively small (Figure 11a, Table S3). Locations where trends were likely and highly likely to be upward were the Bijou Street and Fountain Creek at Colorado Springs sites, respectively. Downward trends were likely for the next two sites downstream. Downward trends were highly likely for Fountain and Pueblo. The bootstrap analysis via WRTDS indicated that flow-normalized TDS concentration at Pueblo had decreased by 134 mg/L and the 90% confidence interval (CI) in the decrease was 102 to 167 mg/L. The estimated change in annual load for TDS was a decrease of 16,000 mg with a 90% CI of 12,000 and 19,000 mg (Table 2). The changes represent approximately 16% decreases for both flow-normalized concentration and load.

Chloride concentrations were not regularly measured at most sites. Upward trends in chloride were significant or highly significant at the three sites where chloride was monitored (Figure 11b, Table S3). There was a subtle upward trend for Fountain Creek at Colorado Springs, CO, before about 2010 and a greater increase after 2015.

3.6. Trace Element Trends

Flow-normalized concentrations of filtered and unfiltered selenium in Fountain Creek show similar patterns spatially and through time (Figure 12a,b). Spatially, there is one marked increase in flow-normalized concentrations of selenium starting around the city of Colorado Springs and another larger increase somewhere between the two most downstream sites (Figure 12a,b). Notably though, trends in the flow-normalized concentrations of unfiltered and filtered selenium were likely or highly likely to be downward at all sites where data were available for 1999 to 2022 (Table S3). At Pueblo, the decreases in unfiltered selenium concentration and load during the period of analysis were 16.1 μg/L (90% CI 13.9 to 18.5 μg/L) and 1625 kg/yr (90% CI 1370 to 1930 kg) (Table S3). The decreases in filtered selenium concentration and load during the period of analysis were 14.9 μg/L (90% CI 12.9 to 17.3 μg/L) and 1362 kg/yr (90% CI 1140 to 1570 kg/yr) (Table S3). The changes represent decreases in the range of 61 to 67%.

The flow-normalized concentrations of filtered and unfiltered arsenic, along with unfiltered iron, showed spatial and temporal patterns that were broadly similar to those for selenium (Figure 12c–e). The concentrations were greater with distance downstream, and there was a marked increase in concentrations between Fountain and Pueblo. Trends were likely to highly likely to be downward at all sites (Table S3). Flow-normalized concentrations of unfiltered lead were distinctly higher at Monument Creek at Bijou St and Security but lower at intervening sites (Figure 12f). Trends for unfiltered lead were likely or highly likely to be downward at all sites except for at upstream Fountain Creek for which either trend was as likely as it was not likely (Table S3). The flow-normalized concentrations of filtered manganese indicated that there were higher concentrations in the upper watershed compared to the lower watershed (Figure 12g). Trends were mixed for filtered manganese, with significant upward trends for both Monument Creek sites and Fountain Creek in Colorado Springs. Many of the trend lines for filtered manganese had an undulating pattern (Figure 12g).

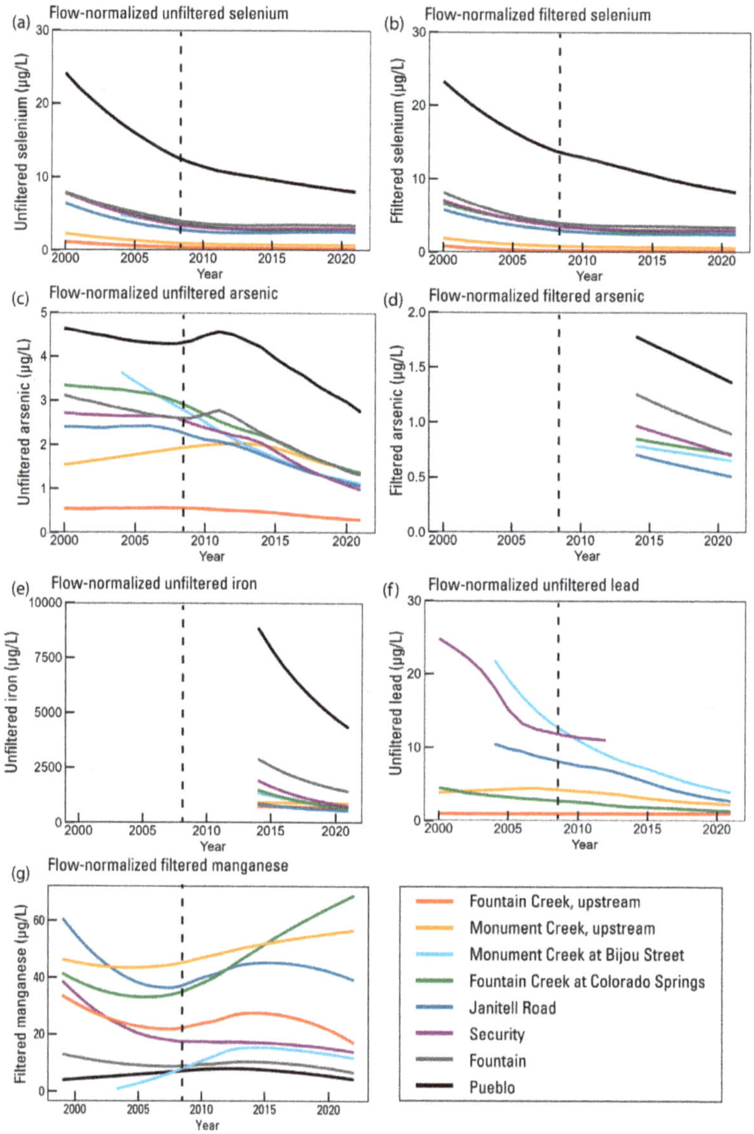

Figure 12. Flow normalized trends in trace element concentrations: unfiltered selenium (**a**), filtered selenium (**b**), unfiltered arsenic (**c**), filtered arsenic (**d**), unfiltered iron (**e**), unfiltered lead (**f**), filtered manganese (**g**) [20]. For reference, dashed lines indicate when the JD Phillips facility began taking some of the wastewater treatment load from the Las Vegas Street facility.

4. Discussion

4.1. Streamflow and Municipal Water-Use Patterns and Trends

The managed nature of streamflow in Fountain Creek was apparent through the decreases in interannual streamflow variability with distance downstream, by the low amount of variability in the winter base flow, which was partially sustained by effluent, and by the substantial volumes of reusable water relative to the total streamflow (Figure 3).

The presence of reusable water, much of it transbasin imports, moderated the inter-annual streamflow variability (Figure 3) compared to a local stream with only native flow (Figure 6). Effluent may have accounted for as much as 59% of streamflow at some locations in some years (Figure 4). Such predominance and flow-moderating effects of effluent in streamflow (Figure 4) are a common occurrence in urban streams in semiarid to arid regions [50,51].

The artificial component of Fountain Creek streamflow is important to consider relative to the human population and water use in the watershed. Historically, across the United States, increases in wastewater generation have closely tracked increases in population [52], but in recent years, the trend has been towards efficiency [53]. The 14% decline in potable water production from 2001 to 2022 (Figure S1), despite a 30% increase in the population of Colorado Springs [5], reflects such a trend. The negligible increase in combined effluent discharge from the Las Vegas Street and JD Phillips facilities (Figure 3c) between 2010 and 2021, despite a roughly 15% increase in the population of Colorado Springs in that time [5], could also reflect increased efficiency. A greater reclamation of water for non-potable uses rather than effluent discharge could account for some of the decrease, but could have only accounted for at most 8% of the potential discharge in 2021 [6]. Per capita effluent discharge in Colorado Springs was about 0.16 mL/yr in 2008 and decreased to 0.12 mL/yr by 2021. For reference, the average urban wastewater generation in North America in 2015 was about 0.23 mL/yr per person [54]. The decline in potable water production in Colorado Springs and the negligible change in effluent discharge might reflect changes in the delivery or sewer systems, but it could also reflect decreases in uses like residential irrigation, with implications for artificial groundwater recharge.

An assessment of per capita outflow from the whole watershed indicates increased water-use efficiency at that scale also. That assessment was made using mean streamflow exiting the watershed at Pueblo of 138,000 mL relative to the watershed population increase from approximately 421,000 in 1999 to approximately 590,000 in 2021. In 1999, outflow was 0.33 mL/yr per capita and it had decreased to 0.23 mL/yr by 2021. That was about a 30% decrease in streamflow per person carrying the sum of human-related water-quality constituents produced in the watershed. These intensifications of human water-use efficiency at the municipal and watershed scales provide context for the water-quality trends.

4.2. *Local Climate, Natural Hydrology, and Irrigation Patterns and Trends*

Warming and drying trends were apparent in the assessment of local climate and natural hydrology. Significant decreases in precipitation were found around Colorado Springs and increases in daily maximum temperatures for both Colorado Springs and Pueblo occurred between 1980 and 2022 (Figure 6). In Rock Creek, a small stream within the watershed that carries only native flows, a significant decrease in streamflow was found over the same period (Figure 6a). The Rock Creek trend was more notable because it and other creeks sink into and recharge the alluvial aquifer along Fountain Creek. At the drought monitoring well slightly outside the study watershed (Figure 1), the shallow groundwater levels had declined significantly which indicated a possible broader pattern of decreased recharge to shallow aquifers in the region (Figure 6b). Drying trends indicate potentially decreased groundwater and resulted in a decreased shallow groundwater discharge to Fountain Creek. Such links between climate, recharge, and groundwater discharge are becoming more commonly identified [55].

The groundwater age near Fountain Creek suggests how quickly groundwater discharge to the creek might respond to the climate trends. The maximum apparent age of groundwater in the alluvium aquifer just south of Colorado Springs was 21.5 years, but most ages were <10 years [48]. Assuming similar ages for general groundwater discharge to Fountain Creek, a response to climate forcing could be possible within a decade [56]. Decreased groundwater discharge has been suggested as driving water-quality trends in other settings [57].

Diversions of water for agricultural irrigation can affect water quality as water is partially consumed by evapotranspiration, mobilizes soluble constituents beneath fields, and re-enters the creek via return flows. Recent irrigation diversions of around 28,700 mL/yr were notable, as the total streamflow out of the watershed at Pueblo is 138,000 mL/yr. Perhaps more important to understanding trends in water quality was the 46% decrease in annual irrigation diversion volume (−24,800 mL) between 1980 and 2022.

4.3. Nutrient Patterns and Trends

The focus of the study was on effluent discharge, but urban nonpoint sources can also contribute substantially to nutrients [58]. Distinguishing between the two was beyond the scope of this study [50]. The most pronounced trends in nutrients in Fountain Creek appeared to be related to the two largest effluent discharges in the watershed (Figure 10). Based upon timing, opening of the JD Phillips facility reduced flow-normalized phosphorus concentrations in much of the watershed (Figure 10a,b), reversing previous upward trends. In more recent years, the Bijou Street site had the highest phosphorus concentrations, but contributions from the upstream Monument Creek site appear to have contributed to that pattern (Figure 10a,b). However, the increasing trends in orthophosphate at the two sites immediately downstream from JD Phillips suggested that this facility is a major driver (Figure 10b). Despite those increases, orthophosphate decreased at Janitell Road later in the record, which may indicate decreased contributions from the Las Vegas Street facility or instream processes consuming the increased orthophosphate from upstream. Instream processes such as uptake by biota like algae, binding to sediments, and other processes serve to remove nutrients from the water column and decrease concentrations with distance downstream [52]. Ratios of filtered nitrogen to filtered phosphorus at Janitell Road have increased in recent years, indicating preferential phosphorus removal (Figure S4). Similar removal may have contributed to the likely to highly likely downward trends at Pueblo that indicated a decreased export of phosphorus from Fountain Creek (Figure S4, Table 2).

Of the nitrogen parameters, trends in ammonia were the most complex and localized (Figure 10e), possibly due to low concentrations, oxidation to nitrate by nitrifying bacteria, and/or rapid uptake by algae and plants during the daytime [59,60]. A notable ammonia source may exist between Janitell Road and Security, a pattern that has been noted previously [16]. In contrast, the other nitrogen constituents showed more straightforward patterns. The highest flow-normalized concentrations of unfiltered nitrogen had a spatial relationship that indicated that the Las Vegas Street facility was a notable source (Figure 10c). Interestingly, concentrations did not decline with distance downstream until somewhere between Fountain and Pueblo, which potentially indicated either minimal instream removal or additional sources between Colorado Springs and Fountain. The highest recent concentrations of nitrate plus nitrite at Bijou Street and Janitell Road, immediately downstream from the two largest effluent discharges, suggested that they were potential sources. However, lower concentrations downstream of those sites at Fountain Creek in Colorado Springs and Pueblo, respectively, pointed to some instream removal (Figure 10d), though not as great as that for phosphorus (Figure S4). The significant or highly significant upward trends for both measures of nitrogen at Pueblo indicated increasing export from the Fountain Creek watershed (Table 2).

The recent downward trends in many phosphorus concentrations and upward trends in many nitrogen concentrations may presumably be attributed to changes in how wastewater was treated prior to effluent discharge. The overall fraction of effluent accounted for in the streamflow changed little (Figure 4b,c). One potential explanation for the upward trends in nutrients is more highly concentrated wastewater entering treatment facilities. Per capita effluent discharge decreased 23% just between 2008 and 2021, from 0.16 mL/yr to 0.12 mL/yr. As domestic water-use efficiency improves, the diluting effect of low-efficiency devices is removed, and wastewater reaching treatment facilities can see dramatic increases in nutrient concentrations [61,62]. A second possible influence could be diminished dilution of effluent in the watershed by urban return flows related to decreased lawn watering and fixing leaking infrastructure. Around Denver, Colorado, the north of Colorado Springs, development greatly influences recharge to groundwater [63] and water from municipal systems accounted for 80% of urban base flow there during the summer [64]. Efficiency-driven reductions in artificial recharge and runoff in the Colorado Springs area may have influenced the water quality patterns observed in Fountain Creek.

4.4. Salinity Related Trends

The decreases in flow-normalized TDS concentration of 134 mg/L and load of about 16,000 mg/yr at Pueblo (Table 2) were important because they indicated decreased TDS contributions from Fountain Creek to the Arkansas River, where TDS presents a major concern [18,19]. The upward trends in TDS concentrations for Monument Creek at Bijou Street and Fountain Creek at Colorado Springs (Table S3) may have been related to effluent discharge from the JD Phillips facility. The increases in TDS in the City of Colorado Springs, compared to sites upstream, were presumably driven by effluent and non-point urban sources. Interestingly, the increased municipal water-use efficiency that may have increased some nutrient concentrations via effluent apparently did not similarly increase TDS, as TDS had a significant downward trend for the Janitell Road site (Table S3; Figure 12a). The pattern could indicate reduced urban return flows.

Spatially, major additional increases in TDS occurred at the three sites farther downstream. As effluent contributions in the lower watershed were small, those increases were presumably driven by either TDS-rich return flows from irrigation or gains from natural geologic sources delivered via groundwater discharge. Both potential sources could enter the creek through similar, diffuse subsurface flow paths.

The 46% decrease in irrigation diversions between 1980 and 2022 could have driven decreased TDS-bearing return flow in the lower watershed. Irrigation-driven flushing of salts from soils in the Arkansas River Valley, downstream from Fountain Creek, mobilizes substantial fractions of TDS [18]. However, unlike the Arkansas River Valley, much of the irrigated agriculture along Fountain Creek occurs on modern alluvium that may contain less salt than upland soils due to recent fluvial re-working [21,65]. An argument against decreased irrigation being a major factor in downward TDS trends is that most irrigation diversions from Fountain Creek are located south of the Fountain site and downward trends in TDS exist at Fountain and the two sites immediately upstream (Table S3). Therefore, decreased agricultural irrigation likely played some role in the downward TDS trends but may not have been the primary driver.

Perhaps a more likely driver of the TDS concentration decreases was decreases in natural discharge of high-TDS shallow groundwater. Such discharges to Fountain Creek may be volumetrically small under the semiarid climate, but concentrations of TDS in shallow groundwater can be substantially greater than those in Fountain Creek [14]. A perspective on the TDS concentrations in shallow groundwater can be gleaned from surface water flows into Fountain Creek south of the Fountain site, which are generally intermittent but may be more perennially connected to Fountain Creek in the subsurface via alluvium. The concentrations of TDS for two streams in the area (07105940 Little Fountain Creek near Fountain, CO; median = 1150 mg/L; range 161–4070 mg/L; n = 10 and 383325104424801 Sand Creek below Fort Carson near Wigwam, CO; median = 740 mg/L; range 301–1980 mg/L; n = 4) indicate that natural water entering Fountain Creek via surface or subsurface flow paths in that region could have high TDS concentrations [20]

Elsewhere in the region, the Niobrara Formation is known to be a disproportionate source of TDS to surface water [22]. Niobrara Formation outcrops are present in the lower watershed, suggesting a role in the substantial TDS gains there (Figure 1). Even where the Niobrara Formation does not outcrop at the surface, it is commonly overlain only by the lower units of the Pierre Shale (Figure 1). Both the Niobrara Formation and the lower units of the Pierre Shale (particularly the Sharon Springs Member) contain abundant pyrite which oxidizes to yield components of TDS [22] and sources therein. Near Pueblo, shallow emplacement of the Niobrara Formation along Fountain Creek indicates that it could be influencing the composition of groundwater that enters the creek (see Figures 4 and 5 in [14]).

The above evidence points to substantial geologic sources of TDS and high TDS concentrations in shallow groundwater in the Fountain Creek watershed. A drier climate, due to reduced recharge, may have played a role in decreasing the discharge of such high-TDS groundwater to the creek. (Figures 6 and 7). Elsewhere in the western United States, in the Upper Colorado River Basin, downward trends for TDS in base flow, in excess of those expected from salinity control projects, have been attributed to climate, landscape changes, or other unidentified processes [57]. Recent drying-driven decreases in groundwater-supported base flow could play a role, but making definitive links between cause and effect can be difficult [57,66,67]. It can be noted that reductions in base flow have been observed in the Upper Rio Grande Basin, roughly 150 km southwest of Fountain Creek and perhaps responding to similar climate forcings as Fountain Creek [68].

The recent subtle increases in TDS concentration at the Bijou Street, Fountain Creek in Colorado Springs, and Janitell sites within Colorado Springs may be influenced by chloride concentrations, which contributes to TDS and saw steep concentration increases (Figure 11a,b). Across the United States, deicing salts have driven steep upward trends in stream chloride concentrations [69]. Effluent can also be a substantial source of chloride and concentrations from the JD Phillips and Las Vegas Street facilities are in the 50 to 60 mg/L range (A. Berlemann, Colorado Springs Utilities, written commun., 2023), quite similar to those in Fountain Creek (Figure 11b). Broader monitoring of chloride along Fountain Creek could help explain these patterns.

4.5. Trace Element Trends

The Colorado chronic wildlife toxicity standard for dissolved selenium is 4.6 μg/L [70]. Flow-normalized concentrations of both filtered and unfiltered selenium were generally above that value (Figure 12a,b). However, the likely downward trends in unfiltered and filtered selenium concentrations and loads were important because of the issues selenium presents downstream [18,19,71].

A volumetrically small groundwater discharge with high selenium concentrations contributes much of the selenium observed in regional surface waters, with geologic units like the Niobrara Formation being the primary source [14,22]. In Fountain Creek, the spatial pattern of larger selenium concentration increases at the southern end of the watershed supports that idea because that is where the Niobrara Formation outcrops (Figures 1 and 12a,b). The lower units of the Pierre Shale (particularly the Sharon Springs Member) also contain abundant selenium and may also be another notable source, as with TDS ([22] Figure 1).

Spatially, filtered and unfiltered selenium, filtered and unfiltered arsenic, and unfiltered iron all increased in concentration in the lower, southern end of Fountain Creek (Figure 12a–e). Like selenium, arsenic and iron are mobilized to surface waters from geologic sources by complex redox processes [72,73]. Similar downstream concentration increases for all three elements appeared to link them and TDS together via shared geologic sources and processes in the lower reaches of Fountain Creek.

Regarding trends through time, selenium, arsenic, and iron all showed downward trends in concentration (Figure 12a–e). Such patterns were similar to those for TDS although TDS can be assumed to have a broader collection of notable sources, including effluent (Figure 12a). In contrast to TDS, the trace elements showed similar downward trends in flow-normalized concentrations across essentially all eight sites. The same effect of drying climate that may have driven downward TDS trends at Pueblo may have driven the trace element trends across the watershed. Climate seems to be a likely driver capable of affecting all trace elements similarly across all 8 sites by decreasing recharge and subsequent discharge of groundwater that heavily influences their concentrations in Fountain Creek.

In a contrasting example, a drier climate was linked to increases in sulfate, manganese, and zinc concentrations in a river in the snow-dominated North Fork of the Snake River elsewhere in Colorado [74]. However, bedrock in that watershed is highly mineralized, and drying was understood to expose more pyrite to oxidation, liberating more trace elements and sulfate through mineral weathering [75]. In Fountain Creek, by contrast, both the trace elements and TDS would seem to exist in forms ready for mobilization, as precipitated salts in the unsaturated zone, ions sorbed on surfaces, or already dissolved in groundwater, and so annual fluxes of natural recharge could be a strong influence on mobilization [18].

Unfiltered lead concentrations in Fountain Creek were generally either low (Fountain Creek, upstream) or significantly trending downward (Figure 12f). Importantly, the higher lead concentrations did not appear to be spatially linked to Gold Hill Mesa (Figures 1 and 2), where a mill that processed ore for gold and silver and generated between 12 and 14 million tons of tailings [76]. Gold Hill Mesa does have a localized influence on the concentrations of arsenic, copper, lead, manganese, selenium, and zinc at a monitoring site on Fountain Creek just before its confluence with Monument Creek [16], a site that was excluded from the present study due to lack of a streamgage. In the present study, the site that had the highest flow-normalized concentrations of filtered manganese and with the steepest upward trend is Fountain Creek in Colorado Springs, which is downstream from Gold Hill Mesa (Figures 1, 2 and 12g). That pattern might indicate increased mobilization of manganese from the tailings, like the drying-induced mineralized-bedrock watershed patterns observed in the Snake River [74]. In other parts of the Fountain Creek watershed, flow-normalized trends of filtered manganese show somewhat undulating patterns (Figure 12g). The cause of those patterns remains unclear. The spatial pattern of lower concentrations downstream may be driven by larger contributions in the upper watershed and removal by oxidation and precipitation with distance downstream [77].

5. Conclusions

Streamflow in Fountain Creek is highly managed, with large fractions of transbasin water and treated wastewater effluent being present. Depending upon location and year, effluent may have accounted for up to 59% of the streamflow. Certain trends in nutrient concentrations between 1999 and 2021 appear to be related to effluent management and major discharges in the city of Colorado Springs. In several cases, phosphorus concentrations have declined substantially from peaks in the middle of the record. Upward trends in orthophosphate are localized and instream processing appears to cause lower concentrations with distance downstream. In contrast, unfiltered nitrogen and nitrate plus nitrite concentrations have significant to highly significant upward trends below major effluent discharge points and towards the mouth of the creek. Upward trends in nutrients may be related to increased human water use efficiency, as deliveries of potable water have been declined and overall effluent discharge has remained essentially flat even as population in Colorado Springs and the larger watershed have increased by about 36% and 40%, respectively. Diminishment of diluting return flows from lawn and park watering and leaking infrastructure could also play a role in nutrient trends.

Salinity, evaluated as TDS, significantly trended slightly upward in parts of the watershed possibly due to chloride from deicing salts or the shift of some effluent discharge to a location farther upstream. However, TDS concentration significantly trends downward at a point downstream from the two major effluent discharges, suggesting that water conservation has not driven increased salinity. Farther downstream, sources of TDS appear related to geology. Significant downward trends in TDS at the mouth of the creek indicate decreases in concentration of 134 mg/L and decreases in load of 16,000 mg/yr, from 1999 to 2021, both representing approximately 16% decreases. The trends might be related to either decreases in agricultural irrigation or decreases in recharge because of recent drying, both of which could decrease the discharge of high-TDS groundwater to the creek. Unfiltered selenium showed highly likely downward trends in concentration and load of 16.1 µg/L and 1625 kg/yr. Selenium, arsenic, and iron concentrations all show significant to highly significant downward trends in streamflow across the watershed. The existence of similar, spatially widespread trends points to a spatially widespread driver, like recent drying-driven decreases in groundwater discharge, but this cannot be confirmed. Nevertheless, downward trends in TDS and selenium are notable because of the challenges these constituents can pose downstream of Fountain Creek in the Arkansas River.

Supplementary Materials: The following supporting information can be downloaded at: https://www.mdpi.com/article/10.3390/w16101343/s1, Table S1 (TDS regressions), Table S2 (data retained after z-score culling), and Table S3 (concentration trend likelihood output from the WRTDS analysis); Figure S1 (potable water production), Figure S2 (ratios of filtered, orthophosphate as P to unfiltered phosphorus), Figure S3 (ratios of filtered nitrate plus nitrite as N to unfiltered nitrogen as P), and S4 (ratios of reactive nitrogen (nitrate plus nitrite) to reactive phosphorus (orthophosphate)).

Author Contributions: Conceptualization, C.R.B.; formal analysis, M.H.R.; data curation, E.H.; writing—original draft preparation, C.R.B.; writing—review and editing, C.R.B., M.H.R. and E.H. All authors have read and agreed to the published version of the manuscript.

Funding: This study was funded by Colorado Springs Utilities and The Colorado Springs Stormwater Enterprise.

Data Availability Statement: Data used in this article regarding potable water deliveries and effluent discharge are available at https://doi.org/10.5066/P97BJ1KK [26]. Data regarding the transit-loss accounting program for streamflow are available at https://doi.org/10.5066/P90J3L5S [33]. All other data used in this article are available from the USGS National Water Information System database at https://doi.org/10.5066/F7P55KJN [20].

Acknowledgments: We thank Sharon Qi for assistance with the map figure and Dustin Ethredge for key knowledge about the Rock Creek streamgage. Ryan Banta, Annie Berlemann, Krystal Brown, Lisa Miller, and Mark Shea contributed to the understanding of Fountain Creek and ideas in this paper through discussion. We thank Natalie Day, Patrick Longley, and several anonymous reviewers' comments that improved the paper. Any use of trade, firm, or product names is for descriptive purposes only and does not imply endorsement by the U.S. Government.

Conflicts of Interest: The authors declare no conflicts of interest. The funders had no role in the design of the study; in the collection, analyses, or interpretation of data; in the writing of the manuscript; or in the decision to publish the results.

References

1. Anderson, M.T.; Woosley, L.H., Jr. *Water Availability for the Western United States—Key Scientific Challenges;* U.S. Geological Survey Circular; U.S. Geological Survey: Reston, VA, USA, 2005; 1261: p. 85 p.
2. Ray, A.J.; Barsugli, J.J.; Averyt, K.B. Climate Change in Colorado: A Synthesis to Support Water Resources Management and Adaptation; Western Water Assessment, Cooperative Institute for Research in Environmental Sciences (CIRES), University of Colorado Boulder 2008; 53p. Available online: https://www.researchgate.net/publication/315476631_Climate_Change_in_Colorado_A_Synthesis_to_Support_Water_Resources_Management_and_Adaptation?channel=doi&linkId=58d159dca6fdcc3fe7852208&showFulltext=true (accessed on 3 May 2014).
3. Gordon, E.; Ojima, D. (Eds.) Colorado Climate Change Vulnerability Study: A Report Submitted to the Colorado Engergy Office; 176p. 2015. Available online: https://climate.colorado.gov/colorado-climate-change-vulnerability-study (accessed on 28 March 2024).
4. Colorado Springs Utilities. *Integrated Water Resources Plan, Final Report;* Colorado Springs Utilities: Colorado Springs, CO, USA, 2017; p. 93 p. Available online: https://www.csu.org/Documents/IWRP.pdf (accessed on 28 March 2024).
5. Colorado Department of Local Affairs. State Demography Office. 2023. Available online: https://demography.dola.colorado.gov/ (accessed on 28 March 2024).
6. Colorado Springs Utilities. 2022 Water Efficiency Plan. 2022. p. 38 p. Available online: https://www.csu.org/Documents/2022WaterEfficiencyPlan.pdf (accessed on 28 March 2024).
7. Howe, C.W.; Goemans, C. Water transfers and their impacts: Lessons from three Colorado water markets. *J. Am. Water Resour. Assoc.* **2003**, *39*, 1055–1065. [CrossRef]
8. Ruddy, B.C. *Water-Quality Variations and Trends in Monument and Fountain Creeks, El Paso and Pueblo Counties, Colorado, Water Years 1976-88;* Water-Resources Investigations Report; U.S. Geological Survey: Reston, VA, USA, 1993; 91-4176: p. 66 p. [CrossRef]
9. Kuhn, G. *Calibration, Verification, and Use of a Steady-State Stream Water-Quality Model for Monument and Fountain Creeks, East-central Colorado;* Water-Resources Investigations Report; U.S. Geological Survey: Reston, VA, USA, 1991; 91-4055: p. 149 p. [CrossRef]
10. Kohn, M.S.; Fulton, J.W.; Williams, C.A.; Stogner, R.W.S. *Remediation Scenarios for Attenuating Peak Flows and Reducing Sediment Transport in Fountain Creek, Colorado, 2013;* U.S. Geological Survey Scientific Investigations Report; U.S. Geological Survey: Reston, VA, USA, 2014; 2014-5019: p. 62 p. [CrossRef]

11. Stoekel, D.M.; Stelzer, E.A.; Stogner, R.W.S.; Mau, D.P. Semi-quantitative evaluation of fecal contamination potential by human and ruminant sources using multiple lines of evidence. *Water Res.* **2011**, *45*, 3225–3244. [CrossRef] [PubMed]
12. Zuellig, R.E.; Bruce, J.F.; Stogner, R.W.S. *Temporal Change in Biological Community Structure in the Fountain Creek Basin, Colorado, 2001–2008*; U.S. Geological Survey, Scientific Investigations Report; U.S. Geological Survey: Reston, VA, USA, 2010; 2010-5103: p. 19 p. [CrossRef]
13. Roberts, J.J.; Bruce, J.F.; Zuellig, R.E. *Changes in Biological Communities of the Fountain Creek Basin, Colorado, 2003–2016, in Relation to Antecedent Streamflow, Water Quality, and Habitat*; Scientific Investigations Report; U.S. Geological Survey: Reston, VA, USA, 2018; 2017-5162: p. 20 p. [CrossRef]
14. Arnold, L.R.; Ortiz, R.F.; Brown, C.R.; Watts, K.R. *Groundwater and Surface-Water Interaction, Water Quality, and Processes Affecting Loads of Dissolved Solids, Selenium, and Uranium in Fountain Creek near Pueblo, Colorado, 2012–2014*; U.S. Geological Survey Scientific Investigations Report; U.S. Geological Survey: Reston, VA, USA, 2016; 2016-5134: p. 78 p.
15. Mau, D.P.; Stogner, R.W.S.; Edelmann, P. *Characterization of Stormflows and Wastewater Treatment-Plant Effluent Discharges on Water Quality. Suspended Sediment, and Stream Morphology for Fountain and Monument Creek Watersheds, Colorado, 1981–2006*; Scientific Investigations Report; U.S. Geological Survey: Reston, VA, USA, 2007; 2007-5104: p. 76 p.
16. Miller, L.D.; Stogner, R.W.S. *Characterization of Water Quality and Suspended Sediment During Cold-Season Flows, Warm-Season Flows, and Stormflows in the Fountain Creek and Monument Creek Watersheds, Colorado, 2007–2015*; Scientific Investigations Report; U.S. Geological Survey: Reston, VA, USA, 2017; 2017-5084: p. 47 p. [CrossRef]
17. Environmental Protection Agency (EPA). *Colorado Springs Agrees to Improve Stormwater Management In Settlement with the United States*; EPA News Release; EPA: Washington, DC, USA, 2020. Available online: https://www.epa.gov/newsreleases/colorado-springs-agrees-improve-stormwater-management-settlement-united-states-0 (accessed on 28 March 2024).
18. Bern, C.R.; Holmberg, M.; Kisfalusi, Z. Salt flushing, salt storage, and controls on selenium and uranium: A 31-year mass-balance analysis of an irrigated, semiarid valley. *J. Am. Water Resour. Assoc.* **2020**, *56*, 647–668. [CrossRef]
19. Miller, L.D.; Watts, K.R.; Ortiz, R.F.; Ivahnenko, T. *Occurrence and Distribution of Dissolved Solids, Selenium, and Uranium In Groundwater and Surface Water in the Arkansas River Basin from the Headwaters to Coolidge, Kansas, 1970–2009*; USGS Scientific Investigations Report; U.S. Geological Survey: Reston, VA, USA, 2010; 2010-5069: p. 59 p. [CrossRef]
20. U.S. Geological Survey (USGS). *USGS Water Data for the Nation*; U.S. Geological Survey National Water Information System database; U.S. Geological Survey: Reston, VA, USA, 2023. [CrossRef]
21. Horton, J.D.; Juan, C.A.S.; Stoeser, D.B. *The State Geologic Map Compilation (SGMC) Geodatabase of the Conterminous United States (ver. 1.1)*; U.S. Geological Survey Data Series; U.S. Geological Survey: Reston, VA, USA, 2017; 1052: p. 46 p. [CrossRef]
22. Bern, C.R.; Stogner, R.W.S. The Niobrara Formation as a challenge to water quality in the Arkansas River, Colorado, USA. *J. Hydrol. Reg. Stud.* **2017**, *12*, 181–195. [CrossRef]
23. NOAA (National Oceanic and Atmospheric Administration). U.S. Climate Normals Quick Access. 23 February 2023. Available online: https://www.ncei.noaa.gov/access/us-climate-normals/ (accessed on 28 March 2024).
24. Peel, M.C.; Finlayson, B.L.; McMahon, T.A. Updated world map of the Koppen-Geiger climate classification. *Hydrol. Earth Syst. Sci.* **2007**, *11*, 1633–1644. Available online: https://www.hydrol-earth-syst-sci.net/11/1633/2007/ (accessed on 28 March 2024). [CrossRef]
25. United States Census Bureau. QuickFacts, United States. 2023. Available online: https://www.census.gov/quickfacts/ (accessed on 28 March 2024).
26. Bern, C.R.; Qi, S.L. *Daily Potable Water Deliveries and Effluent Discharge by Colorado Springs Utilities: 2008–2022*; U.S. Geological Survey Data Release; U.S. Geological Survey: Reston, VA, USA, 2024. [CrossRef]
27. Vanderhoof, M.K.; Burt, C.; Hawbaker, T.J. Time series of high-resolution images enhances efforts to monitor post-fire condition and recovery, Waldo Canyon fire, Colorado, USA. *Int. J. Wildland Fire* **2018**, *27*, 699–713. [CrossRef]
28. Reid, J. Wildfire & Flash Flood Recovery. In *National Association of Counties, Justice & Public Safety Symposium*; National Association of Counties: Washington, DC, USA, 2014. Available online: https://www.naco.org/sites/default/files/documents/Wildfire%20Flash%20Flood%20Recovery%20Jan%202014%20Jim%20Reid.pdf (accessed on 3 May 2024).
29. Kinoshita, A.M.; Chin, A.; Simon, G.L.; Briles, C.; Hogue, T.S.; O'dowd, A.P.; Gerlak, A.K.; Albornoz, A.U. Wildfire, water, and society: Toward integrative research in the Anthropocene. *Anthropocene* **2016**, *16*, 16–27. [CrossRef]
30. Kuhn; Samuels, E.L.; Bemis, W.D.; Steger, R.D. *Descriptions of the Program Changes (1989–97) User Manual for a Transit-Loss Accounting Program Applied to Fountain Creek between Colorado Springs and the Arkansas River, Colorado*; U.S. Geological Survey Open-File Report; U.S. Geological Survey: Reston, VA, USA, 1998; 97-637: p. 39 p.
31. Kuhn; Krammes, G.S.; Beal, V.J. *Description and User Manual for a Web-Based Interface to a Transit-Loss Accounting Program for Monument and Fountain Creeks, El. Paso and Pueblo Counties, Colorado*; U.S. Geological Survey Scientific Investigations Report; U.S. Geological Survey: Reston, VA, USA, 2007; 2007-5028: p. 36 p.
32. Colarullo, S.J.; Miller, L.D. *Upgrades to a Fortran Program for Estimating Stream Transit Losses of Reusable Water, El Paso and Pueblo Counties, Colorado*; U.S. Geological Survey, Scientific Investigations Report; U.S. Geological Survey: Reston, VA, USA, 2018; 2018-5163: p. 21 p. [CrossRef]
33. Hennessy, E.K.; Miller, L.D.; Bern, C.R.; Qi, S. *Daily Subreach Output from the Transit-Loss Accounting Program for Monument and Fountain Creeks, Colorado, 1988–2022*; U.S. Geological Survey Data Release; U.S. Geological Survey: Reston, VA, USA, 2024. [CrossRef]

34. R Core Team. *R: A Language and Environment for Statistical Computing;* R Foundation for Statistical Computing: Vienna, Austria, 2024. Available online: https://www.R-project.org/ (accessed on 28 March 2024).
35. Millard, S.P. EnvStats: An R Package for Environmental Statistics. 2013. Available online: https://link.springer.com/book/10.1007/978-1-4614-8456-1 (accessed on 28 March 2024).
36. NOAA (National Oceanic and Atmospheric Administration). Climate Data Online. 9 March 2023. Available online: https://www.ncei.noaa.gov/cdo-web/ (accessed on 28 March 2024).
37. U.S. Geological Survey. *USGS National Ground-Water Monitoring Network;* U.S. Geological Survey: Reston, VA, USA, 2023. Available online: https://cida.usgs.gov/ngwmn/provider/USGS/ (accessed on 28 March 2024).
38. Colorado Department of Water Resources. Colorado's Decision Support Systems—Structures. 2023. Available online: https://dwr.state.co.us/Tools/Structures (accessed on 28 March 2024).
39. U.S. Geological Survey. *National Field Manual for the Collection of Water-Quality Data;* U.S. Geological Survey Techniques of Water-Resources Investigations, Book 9, Chaps A1–A10. Variously Dated; U.S. Geological Survey: Reston, VA, USA, 2015. [CrossRef]
40. Ortiz, R.F. *Methods to Identify Changes in Background Water-Quality Conditions Using Dissolved-Solids Concentrations and Loads as Indicators, Arkansas River and Fountain Creek in the Vicinity of Pueblo, Colorado;* Scientific Investigations Report; U.S. Geological Survey: Reston, VA, USA, 2004; 2004-5024: p. 20 p. Available online: https://pubs.er.usgs.gov/publication/sir20045024 (accessed on 28 March 2024).
41. Hirsch, R.M.; Moyer, D.L.; Archfield, S.A. Weighted Regressions on Time, Discharge, and Season (WRTDS), with an Application to Chesapeake Bay River Inputs. *J. Am. Water Resour. Assoc.* **2010**, *46*, 857–880. [CrossRef]
42. Hirsch, R.M.; De Cicco, L.A. User guide to Exploration and Graphics for RivEr Trends (EGRET) and dataRetrieval: R packages for hydrologic data. In *U.S. Geological Survey Techniques and Methods 4A-10;* U.S. Geological Survey: Reston, VA, USA, 2015.
43. Hirsch, R.M.; DeCicco, L.; Murphy, J. Exploration and Graphics for RivEr Trends (EGRET). 2023. Available online: https://pubs.usgs.gov/tm/04/a10/ (accessed on 28 March 2024).
44. Choquette, A.F.; Hirsch, R.M.; Murphy, J.C.; Johnson, L.T.; Confesor Jr, R.B. Tracking changes in nutrient delivery to western Lake Erie: Approaches to compensate for variability and trends in streamflow. *J. Great Lakes Res.* **2019**, *45*, 21–39. [CrossRef]
45. Hirsch, R.M.; DeCicco, L. Guide to EGRET 3.0 enhancements. 2018. Available online: https://rconnect.usgs.gov/EGRET/articles/Enhancements.html (accessed on 28 March 2024).
46. Hirsch, R.M.; Archfield, S.A.; de Cicco, L.A. A bootstrap method for estimating uncertainty of water quality trends. *Environ. Model. Softw.* **2015**, *73*, 148–166. [CrossRef]
47. Sansalone, J.J.; Cristina, C.M. First flush concepts for suspended and dissolved solids in small impervious watersheds. *J. Environ. Eng.* **2004**, *130*, 1301–1314. [CrossRef]
48. Newman, C.P.; Paschke, S.P.; Keith, G. Natural and anthropogenic geochemical tracers to investigate residence times and groundwater-surface water interactions in an urban alluvial aquifer. *Water* **2021**, *13*, 871. [CrossRef]
49. Lukas, J.; Gordon, E. Colorado's Climate: Past and Future. In *Colorado Climate Change Vulnerability Study;* Gordon, E., Ojima, D., Eds.; University of Colorado: Boulder, CO, USA; Colorado State University: Fort Collins, CO, USA, 2015; pp. 9–14.
50. Schliemann, S.A.; Grevstad, N.; Brazeau, R.H. Water quality and spatio-temporal hot spotsin an effluent-dominated urban river. *Hydrol. Process.* **2020**, *35*, e14001. [CrossRef]
51. Wolfand, J.M.; Sytsma, A.; Hennon, V.L.; Stein, E.D.; Hogue, T.S. Dilution and pollution: Assessing the impacts of water reuse and flow reduction on water quality in the Los Angeles River Basin. *ACS EST Water* **2022**, *2*, 1309–1319. [CrossRef]
52. Carey, R.O.; Migliaccio, K.W. Contribution of wastewater treatment plant effluents to nutrient dynamics in aquatic systems: A review. *Environ. Manag.* **2009**, *44*, 205–217. [CrossRef]
53. Chinnasamy, C.V.; Arabi, M.; Sharvelle, S.; Warziniack, T.; Furth, C.D.; Dozier, A. Characterization of municipal water uses in the continguous United States. *Water Resour. Res.* **2021**, *57*, e2020WR028627. [CrossRef]
54. Qadir, M.; Drechsel, P.; Jiménez Cisneros, B.; Kim, Y.; Pramanik, A.; Mehta, P.; Olaniyan, O. Global and regional potential of wastewater as a water, nutrient and energy source. *Nat. Resour. Forum* **2019**, *44*, 40–51. [CrossRef]
55. Kløve, B.; Ala-Aho, P.; Bertrand, G.; Gurdak, J.J.; Kupfersberger, H.; Kværner, J.; Muotka, T.; Mykrä, H.; Preda, E.; Rossi, P.; et al. Climate change impacts on groundwater and dependent ecosystems. *J. Hydrol.* **2014**, *518*, 250–266. [CrossRef]
56. Ficklin, D.L.; Robeson, S.M.; Knouft, J.H. Impacts of recent climate change on trends in baseflow and stormflow in United States watersheds. *Geophys. Res. Lett.* **2016**, *43*, 5079–5088. [CrossRef]
57. Rumsey, C.A.; Miller, M.P.; Schwarz, G.E.; Hirsch, R.M.; Susong, D.D. The role of baseflow in dissolved solids delivery to streams in the Upper Colorado River Basin. *Hydrol. Process.* **2017**, *31*, 4705–4718. [CrossRef]
58. Pilone, F.G.; Garcia-Chevesich, P.A.; McCray, J.E. Urban Drool Water Quality in Denver, Colorado: Pollutant Occurrences and Sources in Dry-Weather Flows. *Water* **2021**, *13*, 3436. [CrossRef]
59. Martí, E.; Feijoó, C.; Vilches, C.; Ferreiro, N.; Gantes, P.; Ranieri, C.; Torremorell, A.; Castro, M.C.R.; Gultemiriam, M.L.; Giorgi, A.; et al. Diel variation of nutrient retention is associated with metabolism for ammonium but not phosphorus in a lowland stream. *Freshw. Sci.* **2020**, *39*, 268–280. [CrossRef]
60. Brion, N.; Billen, G. Wastewater as a source of nitrifying bacteria in river systems: The case of the River Seine downstream from Paris. *Water Res.* **2000**, *34*, 3213–3221. [CrossRef]

61. Bailey, O.; Zlatanovic, L.; van der Hoek, J.P.; Kapelan, Z.; Blokker, M.; Arnot, T.; Hofman, J. A Stochastic Model to Predict Flow, Nutrient and Temperature Changes in a Sewer under Water Conservation Scenarios. *Water* **2020**, *12*, 1187. [CrossRef]
62. Lee, M.; Tansel, B.; Balbin, M. Urban Sustainability Incentives for Residential Water Conservation: Adoption of Multiple High Efficiency Appliances. *Water Resour. Manag.* **2013**, *27*, 2531–2540. [CrossRef]
63. Paschke, S. *Groundwater Availability of the Denver Basin Aquifer System, Colorado*; Professional Paper, U.S. Geological Survey: Reston, VA, USA, 2011; 1770: p. 274 p. [CrossRef]
64. Fillo, N.K.; Bhaskar, A.S.; Jefferson, A.J. Lawn Irrigation Contributions to Semi-Arid Urban Baseflow Based on Water-Stable Isotopes. *Water Resour. Res.* **2021**, *57*, e2020WR028777. [CrossRef]
65. Colorado Division of Water Resources. Colorado's Decision Support Systems, Stations—Current and Historical. 2021. Available online: https://dwr.state.co.us (accessed on 16 August 2021).
66. Miller, O.L.; Miller, M.P.; Longley, P.C.; Alder, J.R.; Bearup, L.A.; Pruitt, T.; Jones, D.K.; Putman, A.L.; Rumsey, C.A.; McKinney, T. How Will Baseflow Respond to Climate Change in the Upper Colorado River Basin? *Geophys. Res. Lett.* **2021**, *48*, e2021GL095085. [CrossRef]
67. Rumsey, C.A.; Miller, O.; Hirsch, R.M.; Marston, T.M.; Susong, D.D. Substantial Declines in Salinity Observed Across the Upper Colorado River Basin During the 20th Century, 1929–2019. *Water Resour. Res.* **2021**, *57*, e2020WR028581. [CrossRef]
68. Rumsey, C.A.; Miller, M.P.; Sexstone, G.A. Relating hydroclimatic change to streamflow, baseflow, and hydrologic partitioning in the Upper Rio Grande Basin, 1980 to 2015. *J. Hydrol.* **2020**, *584*, 124715. [CrossRef]
69. Corsi, S.R.; De Cicco, L.A.; Lutz, M.A.; Hirsch, R.M. River chloride trends in snow-affected urban watersheds: Increasing concentrations outpace urban growth rate and are common among all seasons. *Sci. Total. Environ.* **2015**, *508*, 488–497. [CrossRef]
70. CDPHE (Colorado Department of Public Health and the Environment). Regulation No. 32—Classifications and Numeric Standards for Arkansas River Basin. 2016. Available online: https://cdphe.colorado.gov/recent-epa-actions (accessed on 28 March 2024).
71. Bailey, R.T.; Romero, E.C.; Gates, T.K. Assessing best management practices for remediation of selenium loading in groundwater to streams in an irrigated region. *J. Hydrol.* **2015**, *521*, 341–359. [CrossRef]
72. Malakar, A.; Singh, R.; Westrop, J.; Weber, K.A.; Elofson, C.N.; Kumar, M.; Snow, D.D. Occurrence of arsenite in surface and groundwater associated with a perennial stream located in Western Nebraska, USA. *J. Hazard. Mater.* **2021**, *416*, 126170. [CrossRef] [PubMed]
73. Herath, I.; Vithanage, M.; Bundschuh, J.; Maity, J.P.; Bhattacharya, P. Natural Arsenic in Global Groundwaters: Distribution and Geochemical Triggers for Mobilization. *Curr. Pollut. Rep.* **2016**, *2*, 68–89. [CrossRef]
74. Todd, A.S.; Manning, A.H.; Verplanck, P.L.; Crouch, C.; McKnight, D.M.; Dunham, R. Climate-Change-Driven Deterioration of Water Quality in a Mineralized Watershed. *Environ. Sci. Technol.* **2012**, *46*, 9324–9332. [CrossRef] [PubMed]
75. Manning, A.H.; Verplanck, P.L.; Caine, J.S.; Todd, A.S. Links between climate change, water-table depth, and water chemistry in a mineralized mountain watershed. *Appl. Geochem.* **2013**, *37*, 64–78. [CrossRef]
76. Beach, R.A.; U.S. Geological Survey. Mineral Resources on-line SpatialData: Gold Hill Mesa Tailings. 2006. Available online: https://mrdata.usgs.gov/mrds/show-mrds.php?dep_id=10311068 (accessed on 28 March 2024).
77. Harvey, J.W.; Fuller, C.C. Effect of enhanced manganese oxidation in the hyporheic zone on basin-scale geochemical mass balance. *Water Resour. Res.* **1998**, *34*, 623–636. [CrossRef]

water

MDPI

Article

Simultaneous Synthesis of Single- and Multiple-Contaminant Water Networks Using LINGO and Excel Software

Abeer M. Shoaib [1], Amr A. Atawia [1], Mohamed H. Hassanean [1], Abdelrahman G. Gadallah [2,3] and Ahmed A. Bhran [2,*]

[1] Petroleum Refining and Petrochemical Engineering Department, Faculty of Petroleum and Mining Engineering, Suez University, Suez 43512, Egypt; a.shoaib@suezuni.edu.eg (A.M.S.); amr.atawia@gmail.com (A.A.A.); mhassanean@yahoo.com (M.H.H.)
[2] Chemical Engineering Department, College of Engineering, Imam Mohammad Ibn Saud Islamic University (IMSIU), Riyadh 11432, Saudi Arabia; agadallah@imamu.edu.sa
[3] Chemical Engineering Department, National Research Center, Cairo 11241, Egypt
* Correspondence: aabahran@imamu.edu.sa; Tel.: +966-58-077-2698

Abstract: Controlling the distribution of water and wastewater between industrial processes is vital to rationalize water usage and preserve the environment. In this paper, a mathematical technique is proposed to optimize water–wastewater networks, and a nonlinear program is introduced to minimize the consumption of freshwater and, consequently, the flowrate of wastewater discharge. A general mathematical model, able to handle industrial plants containing up to eight sources and eight sinks, is developed using LINGO optimization software to facilitate dealing with complex case studies. The introduced model can handle single-contaminant networks as well as multiple-contaminant ones. The optimal water network is synthesized through two steps; the first step involves the introduction of the case study data into the developed mathematical model. The second step considers using the optimal solution produced after running the developed LINGO model as feed data for a pre-designed Excel sheet able to deal with these results and simultaneously draw the optimal water–wastewater network. The proposed mathematical model is applied to two case studies. The first case study includes actual data from four fertilizer plants located in Egypt; the water resources and requirements are simultaneously integrated to obtain a sensible cutting in both freshwater consumption (lowered by 52.2%) and wastewater discharge (zero wastewater discharge). The second case study regards a Brazilian petrochemical plant; the obtained results show noticeable reductions in freshwater consumption by 12.3%, while the reduction percentage of wastewater discharge is 4.5%.

Keywords: wastewater networks; multiple contaminants; mathematical model; optimization

1. Introduction

The careful use of water in mass-exchange networks between processes or plants will reduce both freshwater consumption and wastewater discharge. Various mathematical programming-based methodologies were presented in the last decades for minimizing both freshwater consumption and wastewater discharge. A relaxed linear model aiming to design an optimal wastewater distribution network was firstly proposed in 1998 by Galan and Grossmann; they used different treatment technologies to decrease the contaminant levels in the network and achieve the maximum reuse of wastewater streams in the plant [1]. Gabriel and El-Halwagi introduced a linear program aiming to decrease the concentrations of contaminants and maximize the reuse and recycle of wastewater through the optimal design of a water interception network [2], while a mixed integer linear program, directed to design water–wastewater networks, was presented by Faria and Bagajewicz for the best handling of water allocation problems [3]. A mixed integer nonlinear programming (MINLP) model was proposed by Galan and Grossmann to design a wastewater network that included five contaminants with thirteen types of technologies to decrease the concentration of the contaminants [4]; their proposed model was applied on a mining plant in Turkey, and an optimal wastewater network with minimum cost was designed. Many other researchers have optimized water networks by MINLP, but for different applications, such as Buabeng and Majozi [5] who aimed to maximize the reuse of wastewater by regeneration–reuse and regeneration–recycling using a hollow-fiber reverse-osmosis membrane to minimize the cost of freshwater consumption. A MINLP model was also proposed by Padron et al. for the design of a wastewater network that included treatment units to decrease the concentration of contaminants and achieve the minimum consumption of freshwater [6]; they presented a case study of Mexico City to show the validity and effectiveness of their presented model. Many other mathematical methodologies have been introduced over the years. Nejad et al. [7] presented a mathematical model and applied it in a case study of a Tehran oil refinery to minimize water and wastewater flowrates; their study included three parameters of water quality, i.e., chemical oxygen demand, suspended solids and hardness. Hansen et al. [8] introduced a mathematical program directed to minimize freshwater consumption between several operations, including cooling systems and washing processes; they applied their proposed model in a petrochemical plant. The priority of source–sink matches was determined by using a ranking matrix technique [9]; three case studies of single and multiple contaminants were introduced to show the applicability of that technique. A closed-loop supply chain was proposed by Mohammad et al. [10] to control the shortage of water in an Azerbaijan city in Iran by using a Social Engineering Optimizer (SEO); their model showed a lowering in freshwater consumption. Arola et al. [11] presented a mathematical approach to maximize the reuse of wastewater by decreasing the concentration of contaminants such as phosphorous, chemical oxygen demand (COD) and total oxygen carbon (TOC) in wastewater; the study included a membrane bioreactor performing a regeneration process to decrease the concentration of contaminants. A mathematical technique was suggested by Chin et al. [12] to optimize the total cost of freshwater consumption and design an optimal water integration network in multiple-contaminant systems. Tuba et al. [13] presented a review of mathematical programming-based methodologies used for water minimization in industrial processes. Ho et al. [14] described a fuzzy multi-objective optimization to minimize freshwater consumption in oil industries in the presence of a single contaminant. A mathematical program was suggested by Liang et al. [15] to allocate wastewater and minimize the consumption of freshwater in a water allocation system in the Zhanghe River basin. Gajendra et al. [16] used the GAMS software (version 23.5.2) to allocate water and wastewater in water networks; the proposed model was formulated as linear Equations.

Graphical-based techniques were also considered in several studies to improve the distribution of water and wastewater in water networks; Irene et al. [17] used a graphical technique to minimize waste flowrates and fresh resources in the design of an inter-plant resource conservation network. Sahu et al. [18] introduced a graphical method to add a treatment unit to minimize the concentrations of contaminants and maximize the reuse of wastewater, while minimizing the costs of treatment. Another graphical technique was proposed by Farrag et al. [19] for the analysis and design of wastewater networks using the composition driving forces; they studied the effect of the mass load of contaminants on the maximum mass recovery. Yamin et al. [20] proposed a graphical technique based on six scenarios in the design of wastewater networks; the technique was applied to agricultural irrigation systems. A study of a water distribution system was presented by Praint et al. [21] to minimize the consumed freshwater in Bota the River area; this study used a graphical technique. The pinch technique was presented also to minimize wastewater discharge from industrial plants. Many other pinch-based techniques were introduced in the last decades to solve the problem of wastewater network optimization and minimize freshwater use and wastewater discharge in industrial plant operations [22–30]. However, pinch and graphical methods are only suitable for simple case studies; when a plant includes a high number of streams, it is difficult to optimize a water–wastewater network via graphical methods. Another group of research works used combined pinch and mathematical programming techniques to overcome the limitations of graphical methods in optimizing water–wastewater networks [31–35].

To date, no simple ready model capable of managing simple as well as complex cases, having a large number of streams and contaminants, and simultaneously drawing the optimum water–wastewater network is available. In this paper, a generalized mathematical model is proposed to minimize both freshwater consumption and the flowrate of wastewater discharge; nonlinear equations based on overall mass balance and component mass balance between several sources (up to eight sources) and several sinks (up to eight sinks) are formulated to be solved by LINGO software (version 14); Excel software (version 2010) is used to draw the water–wastewater network with the automated technique. Two case studies involving a single and multiple contaminants are presented to show the effectiveness of the suggested model in minimizing the flowrates of total freshwater consumption and wastewater discharge in the network. In the future, the proposed model can be applied to agriculture networks to maximize the reuse of irrigation water and minimize freshwater consumption in the mixing units that decreases the concentration of contaminants. Furthermore, the proposed model can be applied to drinkable water networks.

2. Methods

We considered a set of source streams, including up to eight sources, with flowrate F_{SK}, where K is the source number in the network design. It was assumed that each source contained multiple contaminants (up to three contaminants) with concentrations X_{SKA}, X_{SKB}, and X_{SKC}. The contaminants' concentration limits in the discharged wastewater with flowrate $g_{k\text{-waste}}$ were $X_{SAwaste}$, $X_{SBwaste}$, and $X_{SCwaste}$.

We considered a set of sink streams, including up to eight sinks, with flowrates G_Q, where Q is the sink number in the network design; each sink was assumed to contain up to three contaminants with concentrations Z_{QAin}, Z_{QBin}, and Z_{QCin}. Freshwater with flowrate F_{WQ} was distributed to each sink according to the mass load limit, with concentrations X_A, X_B, and X_C. Figure 1 presents the flowchart of the proposed model. The positive or zero values of the outlet flowrates for the sources and the inlet flowrates of the sinks are presented in the proposed model as assumptions; also, we assumed positive contaminant inlet concentrations for the sinks and outlet concentrations for the sources in the proposed model.

The procedure for obtaining the optimum design of a water–wastewater network is shown in Figure 1; the overall mass balance was applied for each source stream with flowrate F_{SK}, which was equal to the sum of the flowrates from that source to the sinks ($g_{K\text{-}Q}$) and of that of the portion of that stream which could be discharged as waste ($g_{K\text{-}Waste}$).

The overall mass balance was applied to each sink; the inlet flowrate for each sink is G_Q, which depends on the flowrate of freshwater fed to each sink (F_{WQ}) and the flowrate from each source to that sink, $g_{K\text{-}Q}$. The component mass balance of three contaminants (A, B, and C) was applied to each sink, and the concentration limits of the contaminants A, B, and C in each sink were Z_{QAin}, Z_{QBin}, and Z_{QCin} respectively. The freshwater flowrate (F_{WQ}) was determined assuming that the freshwater was mixed with the feed water in each sink to optimize the inlet mass load.

The overall mass balance was applied to determine the total waste discharge flowrate (G_{Waste}), which depended on the discharge waste flowrate from each source ($G_{K\text{-}Waste}$); the component mass balance was applied to the discharged wastewater, with the contaminants A, B, and C having the concentrations $X_{SAwaste}$, $X_{SBwaste}$, and $X_{SCwaste}$, respectively. The contaminant concentration limits in discharged wastewater are established by environmental laws.

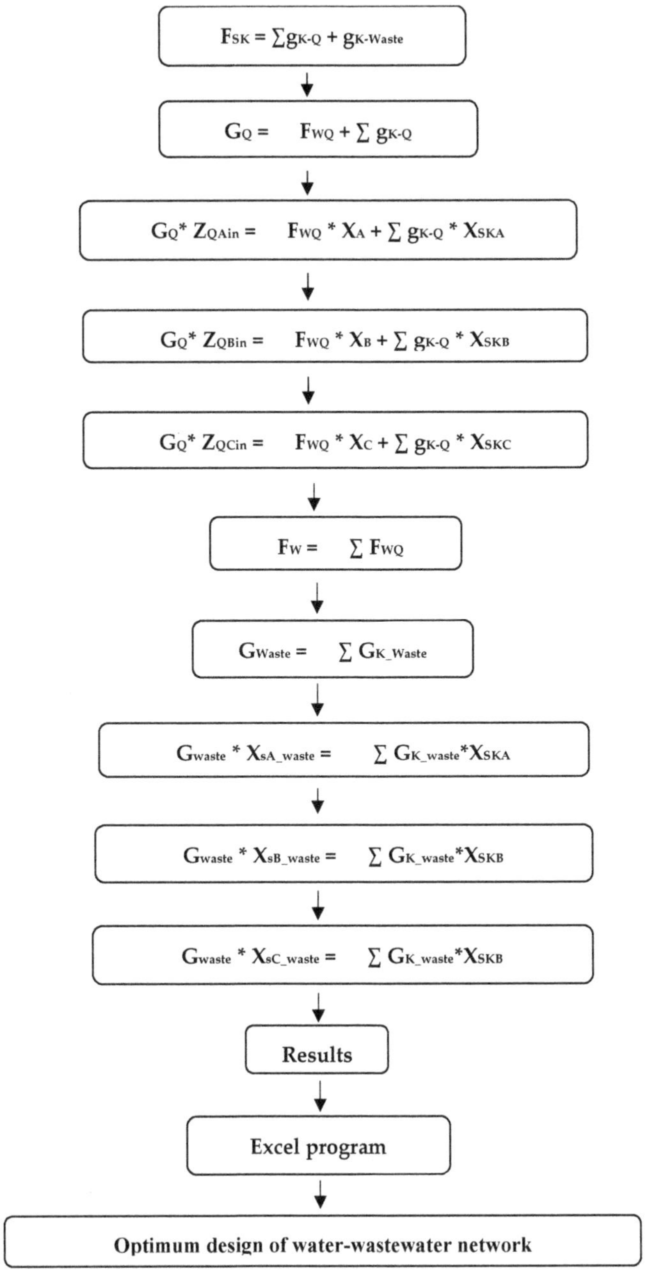

Figure 1. Procedure for achieving the optimum design of a water–wastewater network.

3. Case Studies

The proposed mathematical model was applied to two case studies to show its effectiveness in minimizing freshwater consumption and reducing the total wastewater discharge flowrate. The following subsections describe in detail the considered case studies.

3.1. Case Study 1

The first case study in this work included four fertilizer plants (phosphoric acid plant, concentrated-phosphoric acid plant, single-superphosphate plant, and triple-superphosphate plant), which contained eight contaminant sources of hydrofluorosilicic acid (H_2SiF_6), sulfuric acid (H_2SO_4), and phosphorus pentoxide (P_2O_5). In the presented case study, some source streams were completely discharged as waste. Furthermore, several sinks were completely supplied for their water needs by freshwater; this consequently resulted in a higher consumption of freshwater and in an extra discharge of wastewater. Figure 2 shows the source and sink streams that were not integrated into the network in the current actual case study.

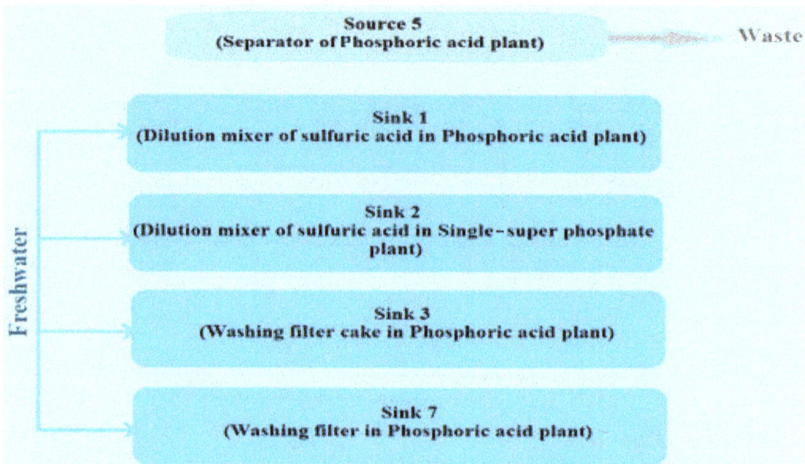

Figure 2. Source and sink streams were not integrated into the existing water network.

As shown in Figure 2, before applying the proposed mathematical model, source 5 (separator of phosphoric acid) discharged all its wastewater (20 m^3/h) to waste with its high concentration of H_2SiF_6. Four sinks (dilution mixer of sulfuric acid in the phosphoric acid plant, dilution mixer of sulfuric acid in the single-superphosphate plant, washing filter cake in the phosphoric acid plant, and washing filter in the phosphoric acid plant) were supplied only with freshwater with a flowrate of 270 m^3/h.

The flowrates and concentrations of the eight sources are shown in Table 1; the flowrate of source 1 (condenser of the phosphoric acid plant) was 280 m^3/h, and the source had normal concentrations of H_2SiF_6, H_2SO_4, and P_2O_5 (14, 0.3, and 4 wt% respectively); the flowrates of sources 2, 3, and 4 (reaction vacuum pump 1 of the phosphoric acid plant, filter vacuum pump 1 of the phosphoric acid plant, and filter vacuum pump 2 of the phosphoric acid plant) were 15, 18, and 18 m^3/h, respectively, and these sources had the same concentrations of H_2SiF_6, H_2SO_4, and P_2O_5 (8, 0.1, and 3 wt% respectively).

Source 5 (separator of the phosphoric acid plant) had a high concentration of H_2SiF_6 (30 wt%); so, its wastewater, with flowrate of 20 m^3/h, was discharged to waste. Source 6 (cooling water of the phosphoric acid plant) had a wastewater flowrate of 120 m^3/h and concentrations of H_2SiF_6, H_2SO_4, and P_2O_5 of 10, 0.6, and 5 wt% respectively.

Source 7 (cooling water of the single-superphosphate plant) had a flowrate of 160 m^3/h and concentrations of H_2SiF_6, H_2SO_4, and P_2O_5 of 12, 0.4, and 5 wt%, respectively, while source 8 (condenser of the concentrated unit in the phosphoric acid plant) had a high flowrate (250 m^3/h), a high concentration of H_2SiF_6 (20 wt%), and normal concentrations of H_2SO_4 and P_2O_5 (0.2, and 4 wt%, respectively).

Table 1. Data regarding the sources in the investigated fertilizer plants.

Sources	Flowrate (m^3/h)	H_2SiF_6 (wt%)	H_2SO_4 (wt%)	P_2O_5 (wt%)
Source 1 (condenser of phosphoric acid plant)	280	14	0.3	4
Source 2 (reaction vacuum pump 1 of phosphoric acid plant)	15	8	0.1	3
Source 3 (filter vacuum pump 1 of phosphoric acid plant)	18	8	0.1	3
Source 4 (filter vacuum pump 2 of phosphoric acid plant)	18	8	0.1	3
Source 5 (separator of phosphoric acid plant)	20	30	0.5	4
Source 6 (cooling water of phosphoric acid plant)	120	10	0.6	5
Source 7 (cooling water of single-superphosphate plant)	160	12	0.4	5
Source 8 (condenser of the concentrated unit in phosphoric acid plant)	250	20	0.2	4

The flowrate and contaminant concentration limits (H_2SiF_6, H_2SO_4, and P_2O_5) for the seven sinks are shown in Table 2; the flowrate of sink 1 (dilution mixer of sulfuric acid in the phosphoric acid plant) was 10 m^3/h, with concentration limits for H_2SiF_6, H_2SO_4, and P_2O_5 of 12, 0.5, and 4.5 wt%, respectively.

Table 2. Data regarding the sinks in the considered fertilizer plants.

Sinks	Flowrate (m^3/h)	H_2SiF_6 (wt%)	H_2SO_4 (wt%)	P_2O_5 (wt%)
Sink 1 (dilution mixer of sulfuric acid in phosphoric acid plant)	10	12	0.5	4.5
Sink 2 (dilution mixer of sulfuric acid in single-superphosphate plant)	140	12	0.8	3.69
Sink 3 (washing filter cake in phosphoric acid plant)	50	10	3	5
Sink 4 (gas scrubber in phosphoric acid plant)	280	14	3	5
Sink 5 (gas scrubber in single-superphosphate plant)	280	16	3	6
Sink 6 (gas scrubber in triple-superphosphate plant)	180	11	3	5
Sink 7 (washing filter in phosphoric acid plant)	70	14	5	4

The data for the three gas scrubbers in the three different plants are shown in Table 2; sink 4 (gas scrubber in the phosphoric acid plant) and sink 5 (gas scrubber in the single-superphosphate plant) had the same flowrate of 280 m^3/h, but the concentration limits of H_2SiF_6, H_2SO_4, and P_2O_5 in sink 4 and sink 5 were 14, 3, 5 wt% and 16, 3, 6 wt%, respectively. Sink 6 (gas scrubber in the triple-superphosphate plant) had a flowrate of 180 m^3/h with concentration limits of H_2SiF_6, H_2SO_4, and P_2O_5 of 11, 3, and 5 wt%, respectively.

The washing filter in the phosphoric acid plant (sink 7) had a flowrate of 70 m^3/h, with concentration limits of H_2SiF_6, H_2SO_4, and P_2O_5 of 14, 5, and 4 wt%, respectively.

The objective function of the model constructed for handling these fertilizer plants was to minimize freshwater consumption as well as wastewater discharge.

3.2. Case Study 2

The Brazilian petrochemical plant presented by Hansen et. al. [8] was the chosen second case study in the current research work. This case study included eight sources (clarified water, filtered water, cooling water 1, cooling water 2, cooling water 3, cooling water 4, bearing water 1, and bearing water 2) and six sinks (cooling water 1, cooling water 2, cooling water 3, cooling water 4, bearing water 1, and bearing water 2). We considered the COD parameter as the only indicator of contamination. The flowrate and COD limits in each plant stream are shown in Table 3.

Table 3. Data regarding the sources and sinks in the considered Brazilian petrochemical plant.

Sources	Flowrate (m^3/h)	COD (mg/L)	Sinks	Flowrate (m^3/h)	COD (mg/L)
Clarified water	1131	0	Cooling water 1	717	3.7
Filtered water	21	0	Cooling water 2	277	4.7
Cooling water 1	717	30	Cooling water 3	92	5.8
Cooling water 2	277	25	Cooling water 4	45	6.1
Cooling water 3	92	25	Bearing water 1	6	14.7
Cooling water 4	45	25			
Bearing water 1	6	20	Bearing water 2	15	3.5
Bearing water 2	15	20			

4. Results and Discussions

After applying the data regarding the sources and sinks of the two investigated case studies to the suggested mathematical model (using LINGO software), the obtained results showed the optimum flowrates from sources to sinks, the freshwater consumption in each sink, and the wastewater flowrate from each source. After that, the obtained results were analyzed simultaneously by the predesigned Excel software, automatically achieving the drawing of the optimal wastewater network. This approach aimed to minimize freshwater consumption and reduce wastewater discharge in the two case studies under investigation.

4.1. Results and Discussion for Case Study 1

The data regarding the fertilizer plants (case study 1) were introduced into the designed mathematical model and solved by LINGO optimization software. The obtained results showed the flowrates from sources to sinks ($G_{k\text{-}Q}$), the flowrates from sources to waste, and the freshwater flowrates to the sinks; these results are listed in Table 4.

Table 4. The optimum flowrates from sources to sinks and to waste of the studied fertilizer plants.

Stream	Flowrate (m^3/h)	Stream	Flowrate (m^3/h)	Stream	Flowrate (m^3/h)
F_W	129	$G_{4\text{–}6}$	9.0697	$G_{8\text{–}2}$	42.29
G_{Waste}	Zero	$G_{5\text{–}4}$	2.7824	$G_{8\text{–}3}$	19.84
$G_{1\text{–}2}$	25.966	$G_{5\text{–}5}$	17.217	$G_{8\text{–}4}$	59.69
$G_{1\text{–}4}$	93.598	$G_{6\text{–}4}$	68.7	$G_{8\text{–}5}$	78.04
$G_{1\text{–}5}$	65.056	$G_{6\text{–}5}$	37.34	$G_{8\text{–}6}$	44.15
$G_{1\text{–}6}$	25.378	$G_{6\text{–}6}$	13.97	F_{W1}	4.0201
$G_{1\text{–}7}$	70	$G_{7\text{–}2}$	23.15	F_{W2}	24.8
$G_{2\text{–}2}$	5.9302	$G_{7\text{–}3}$	8.425	F_{W3}	21.738
$G_{2\text{–}6}$	9.0697	$G_{7\text{–}4}$	41.05	F_{W4}	14.175
$G_{3\text{–}2}$	8.9302	$G_{7\text{–}5}$	55.66	F_{W5}	26.684
$G_{3\text{–}6}$	9.0697	$G_{7\text{–}6}$	31.71	F_{W6}	37.581
$G_{4\text{–}2}$	8.9302	$G_{8\text{–}1}$	5.98		

The obtained total freshwater flowrate (F_W) was 129 m^3/h; this freshwater was distributed to sink 1, sink 2, sink 3, sink 4, sink 5, and sink 6, with flowrates of 4.0201, 24.8, 21.738, 14.175, 26.684, and 37.581 m^3/h, respectively. The obtained results were then analyzed by Excel software to draw an optimal wastewater network for the investigated fertilizer plants, as shown in Figure 3.

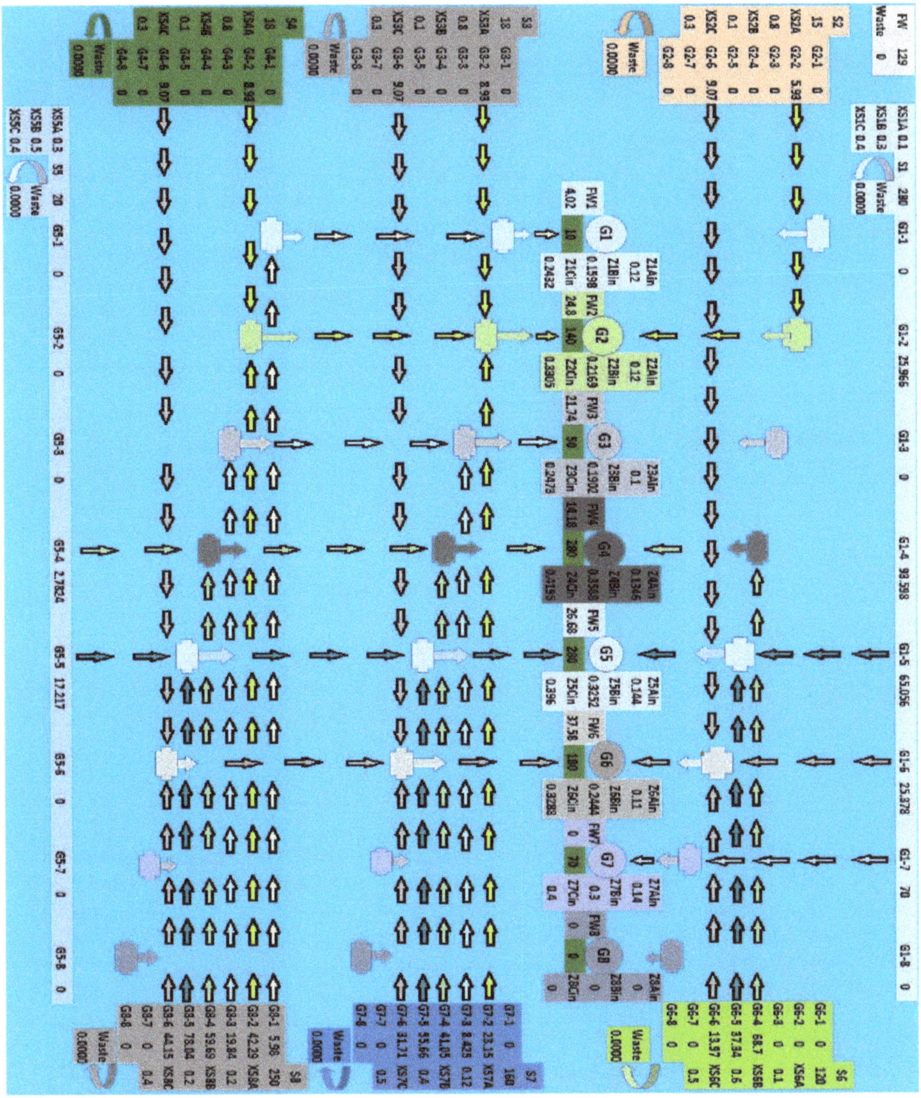

Figure 3. Design of water–wastewater network for the fertilizer plants studied.

Source 1 supplied sink 5, sink 6, and sink 7, with flowrates of 65.056, 25.378, and 70 m^3/h, respectively. Source 2 fed sink 2 and sink 6, with flowrates of 5.93 and 9.07 m^3/h, respectively; also source 3 supplied sink 2 and sink 6, with flowrates of 8.93 and 9.07 m^3/h, respectively. Source 4 fed sink 2, with a flowrate of 8.93 m^3/h, and sink 6, with a flowrate of 9.07 m^3/h. Source 5 supplied sink 4 and sink 5, with flowrates of 2.7824 and 17.217 m^3/h, respectively, while source 6 supplied sink 4, sink 5, and sink 6, with flowrates of 68.7, 37.34, and 13.97 m^3/h, respectively.

Source 7 fed sink 2, sink 3, sink 4, sink 5, and sink 6, with flowrates of 23.15, 8.425, 41.05, 55.66, and 31.71 m^3/h, respectively, while source 8 supplied sink 1, sink 2, sink 3, sink 4, sink 5, and sink 6, with flowrates of 5.98, 42.29, 19.84, 59.69, 78.04, and 44.15 m^3/h, respectively. Zero wastewater discharge was obtained in this case study.

By comparing the obtained results with the original case study data, freshwater consumption was reduced from 270 m^3/h to 129 m^3/h, with a reduction percentage of 52.2%, and the total flowrate of wastewater discharge decreased from 20 m^3/h to zero discharge.

4.2. Results and Discussion for Case Study 2

By feeding the data regarding the second case study (Brazilian petrochemical plant) to the proposed model, the optimum flowrates from sources to sinks and to waste could be determined. The obtained results, listed in Table 5, were used as feed data for the Excel software to obtain the automated drawing of the optimum water–wastewater network, shown in Figure 4.

Table 5. Flowrates from sources to sinks and to waste for the studied Brazilian petrochemical plant.

Sinks	Sources Flowrates (m^3/h) Distributed between Sinks and Waste							
	S1 (Clarified Water)	S2 (Filtered Water)	S3 (Cooling Water 1)	S4 (Cooling Water 2)	S5 (Cooling Water 3)	S6 (Cooling Water 4)	S7 (Bearing Water 1)	S8 (Bearing Water 2)
K1	623.57	0	78.43	0	0	0	0	15
K2	233.6	0	43.4	0	0	0	0	0
K3	57.213	15	13.79	0	0	0	6	0
K4	35.85	0	9.15	0	0	0	0	0
K5	3.06	0	2.94	0	0	0	0	0
K6	13.25	0	1.75	0	0	0	0	0
Waste	0	0	567.54	277	92	45	0	0
Source Total Flowrate	966.54	15	717	277	92	45	6	15

The clarified water source and filtered water source were considered as freshwater sources because the concentration of contaminants in these sources was zero. Regarding the results reported in Table 5, the clarified water flowrate of 966.54 m^3/h was distributed between sinks K1, K2, K3, K4, K5, and K6, with flowrates of 623.57, 233.6, 57.213, 35.85, 3.06, and 13.25 m^3/h, respectively. However, the filtered water source fed only the third sink (k3), with a flowrate of 15 m^3/h. The cooling water 1 source stream supplied six sinks (K1, K2, K3, K4, K5, and K6) and was in part discharged as waste, with flowrates of 78.43, 43.4, 13.79, 9.15, 2.94, 1.75, and 567.54 m^3/h, respectively. On the other hand, the cooling water 2, cooling water 3, and cooling water 4 sources discharged their water to waste, with flowrates of 277, 92, and 45 m^3/h, respectively. The bearing water 1 source fed only K3, with a flowrate 6 m^3/h, while the bearing water 2 source fed only sink 1, with a flowrate of 15 m^3/h. The total wastewater discharge, with a flowrate of 982 m^3/h, was collected from the S3, S4, S5, S6, S7, and S8 sources, with flowrates of 567.54, 277, 92, and 45 m^3/h, respectively.

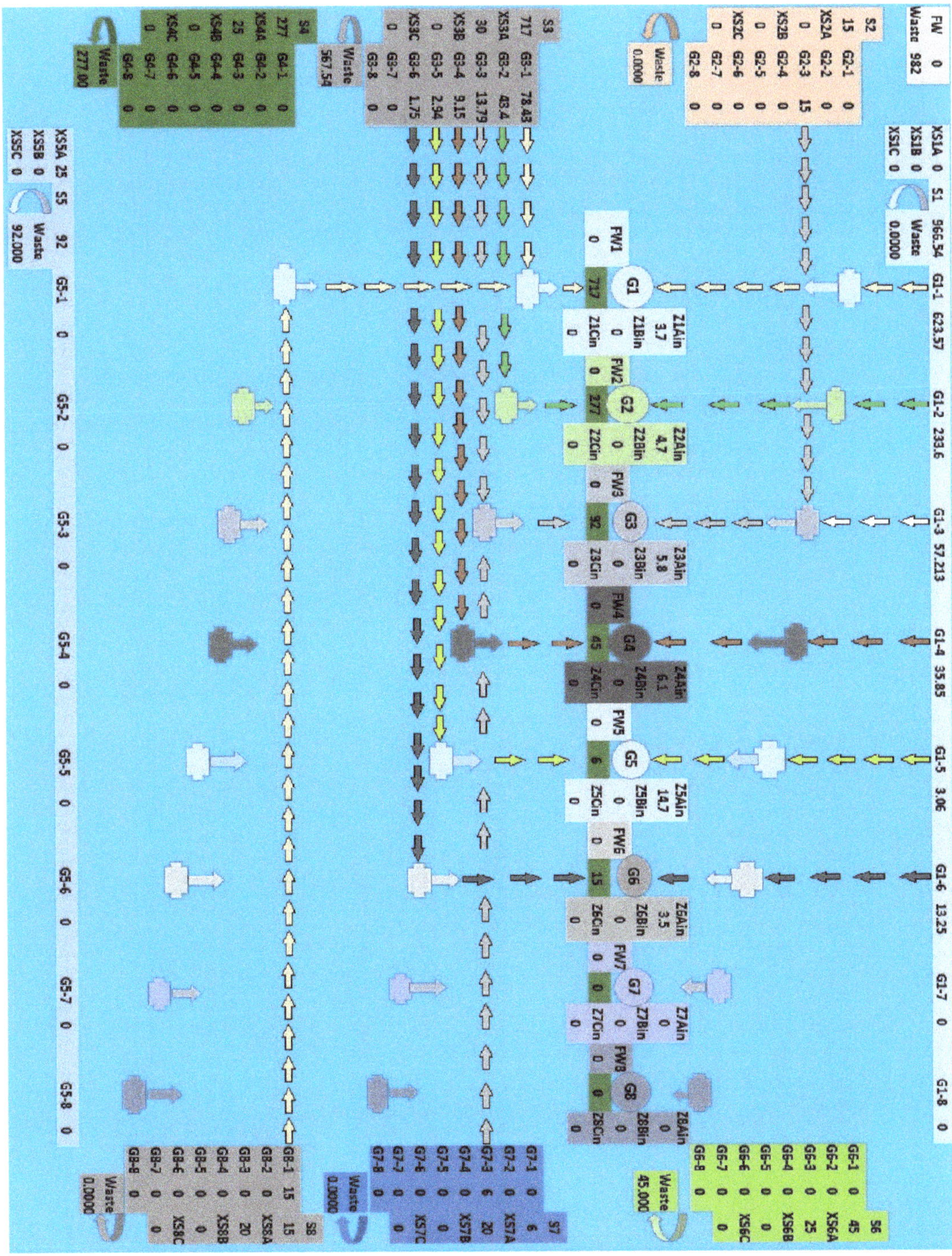

Figure 4. Design of a water–wastewater network for a Brazilian petrochemical plant (case study 2).

By comparing the results of the proposed mathematical model with those of the original case study, it was observed that the clarified water flowrate was reduced from 1102 m^3/h to 966.54 m^3/h, with a reduction percentage of 12.3%. It was also noticed that the filtered water flowrate was reduced from 18 m^3/h to 15 m^3/h, with a reduction percentage of 16.5%. Additionally, the wastewater discharge decreased by 4.5%, from 1029 m^3/h to 982 m^3/h.

Regarding the abovementioned results of the two investigated case studies, it is clear that the introduced approach is an effective and economical technique for designing an optimal water–wastewater network. These results showed that the achieved water–wastewater networks for the considered case studies are economically effective and more profitable compared to the current networks, minimizing freshwater consumption and reducing wastewater discharge. Moreover, the proposed technique can be applied to other case studies to optimize their existing water networks.

5. Conclusions

The water requirement of industrial processes in chemical plants is high; in the industry, water is used for washing, cooling, diluting, and processing in chemical plants. In this work, a mathematical model (LINGO optimization software, version 14.0) was proposed to design optimal mass-exchange networks; the presented model was formulated as a nonlinear program to minimize both freshwater consumption and wastewater discharge. Two sequential steps were established to design optimal water–wastewater networks; the first step consisted in applying the collected data into LINGO software; in the second step, the obtained results were simultaneously analyzed by the proposed Excel software, which was responsible for drawing an automated optimal water–wastewater network. Two case studies (fertilizer plants and a Brazilian petrochemical plant) were presented to show the validity of the proposed model. The obtained results for the abovementioned two case studies showed a reduction in freshwater consumption by 52.2 and 12.3%, respectively. Furthermore, the wastewater discharge was reduced to zero in the fertilizer plants studied, while the reduction percentage of wastewater discharge for the Brazilian petrochemical plant was 4.5%. It was also noticed that the obtained water-wastewater networks are more efficient and profitable than the present networks. This can be attributed to the reduction in freshwater consumption and wastewater discharge. Furthermore, the proposed approach is easy to use and can be applied to different industries, including fertilizer plants and petrochemical plants.

Author Contributions: Conceptualization, A.M.S., M.H.H. and A.A.B.; methodology, A.M.S. and M.H.H.; software, A.M.S. and A.A.A.; validation, A.A.A., A.M.S. and A.A.B.; formal analysis, A.A.A. and M.H.H.; investigation, A.A.A. and A.G.G.; resources, A.M.S. and A.A.B.; data curation, A.A.A. and A.G.G.; writing—original draft preparation, A.A.A.; writing—review and editing, A.M.S., A.G.G. and A.A.B.; visualization, A.M.S., M.H.H. and A.G.G.; supervision, A.M.S., M.H.H. and A.A.B.; funding acquisition, A.G.G. and A.A.B. All authors have read and agreed to the published version of the manuscript.

Funding: This work was supported and funded by the Deanship of Scientific Research at Imam Mohammad Ibn Saud Islamic University (IMSIU) (grant number IMSIU-RG23020).

Data Availability Statement: Data are available upon request through the corresponding author.

Conflicts of Interest: The authors declare no conflicts of interest.

References

1. Galan, B.; Grossmann, I.E. Optimal Design of Distributed Wastewater Treatment Networks. *Ind. Eng. Chem. Res.* **1998**, *37*, 4036–4048. [CrossRef]
2. Gabriel, B.; El-Halwagi, M. Simultaneous Synthesis of Waste Interception and Material Reuse Networks: Problem Reformulation for Global Optimization. *Am. Inst. Chem. Eng.* **2005**, *24*, 171–180. [CrossRef]
3. Faria, D.C.; Bagajewicz, M. A new approach for the design of multicomponent water/wastewater networks. *Comput. Aided Chem. Eng.* **2008**, *4*, 43–48. [CrossRef]

4. Galan, B.; Grossmann, I.E. Optimal Design of Real-World Industrial Wastewater Treatment Networks. *Comput. Aided Chem. Eng.* **2011**, *37*, 1251–1255. [CrossRef]
5. Buabeng, E.; Majozi, T. Effective Synthesis and Optimization Framework for Integrated Water and Membrane Networks: A Focus on Reverse Osmosis Membranes. *Ind. Eng. Chem. Res.* **2015**, *54*, 9394–9406. [CrossRef]
6. Padron, J.I.; Almaraz, S.; Martinez, A.R. Sustainable Wastewater Treatment Plants Design Through Multi objective Optimization. *Comput. Chem. Eng.* **2020**, *140*, 106–122. [CrossRef]
7. Nejad, S.M.; Ataei, A.; Reza, G.; Mehrdadi, N.; Ebadati, F.; Lotfi, F. Water Pinch Analysis for Water and Wastewater Minimization in Tehran Oil Refinery Considering Three Contaminants. *Environ. Monit. Assess.* **2012**, *184*, 2709–2728. [CrossRef] [PubMed]
8. Hansen, E.; Rodrigues, M.A.; Aragao, M.E.; Aquim, P.M. Water and Wastewater Minimization in a petrochemical Industry Through Mathematical Programming. *J. Clean. Prod.* **2017**, *12*, 959–981. [CrossRef]
9. Li, Z.; Majozi, T. Optimal Design of Batch Water Network with a Flexible Scheduling Framework. *Ind. Eng. Chem. Res.* **2019**, *58*, 9500–9511. [CrossRef]
10. Mohammad, A.; Ahmadi, A.; Mirzapour, S.M. Sustainable closed-loop Supply Chain Network for an Integrated Water Supply and Wastewater Collection System Under Uncertainly. *J. Environ. Manag.* **2020**, *275*, 111–128. [CrossRef]
11. Arola, K.; Manttari, M.; Kallioinen, M. Two-Stage Nanofiltration for Purification of Membrane Bioreactor Treated Municipal Wastewater—Minimization of Concentrate Volume and Simultaneous Recovery of Phosphorus. *Sep. Purif. Technol.* **2020**, *20*, 317–344. [CrossRef]
12. Chin, H.H.; Jia, X.; Sabev, P.; Klemes, J.; Liu, Z.Y. Internal and Total Site Water Network Design with Water Mains Using Pinch-Based and Optimization Approaches. *ACS Sustain. Chem. Eng.* **2021**, *9*, 6639–6658. [CrossRef]
13. Tuba, D.; Ozel, S.; Bulkan, S. Water and energy minimization in industrial processes through mathematical programming: A literature review. *J. Clean. Prod.* **2021**, *284*, 124752. [CrossRef]
14. Ho, L.Y.; Cheah, Y.T.; Chan, Y.J.; Chan, Y.J.; Wan, Y.K. Water Stress Reduction Using Superstructure-Based Mathematical Model: Synthesise a Wastewater Treatment Plant Associated with Nutrient Recovery for Palm Oil Industry. *Process Integr. Optim. Sustain.* **2024**, *8*, 393–405. [CrossRef]
15. Liang, Y.; Weijun, H.; Dagmawi, M.; Wan, Z.; Thomas, S.; Xia, W. A system dynamics simulation model for water conflicts in the Zhanghe River Basin, China. *Int. J. Water Resour. Dev.* **2023**, *39*, 1039–1055. [CrossRef]
16. Gajendra, K.; Ram, S.; Canxing, F.; Kuppusamy, S.; Fuchun, Z.; Tariq, M.; Karolina, K.; Xiang, G.; Xinghui, L.; Wesam, A. Development of optimum waste water using network. *Energy* **2024**, *297*, 131297. [CrossRef]
17. Irene, M.L.; Dominic, C.Y. Flowrate Targeting Algorithm for Interplant Resource Conservation Network. Part 2: Assisted Integration Scheme. *Ind. Eng. Chem. Res.* **2010**, *49*, 6456–6468. [CrossRef]
18. Sahu, G.C.; Garg, A.; Majozi, T.; Bandyopadhyay, S. Optimum Design of Waste Water Treatment Network. *Ind. Eng. Chem. Res.* **2013**, *52*, 5161–5171. [CrossRef]
19. Farrag, N.M.; Kamel, D.A.; Ghallab, A.O.; Gadalla, M.A.; Fouad, M.K. Graphical Design and Analysis of Mass Exchange Networks Using Composition Driving Forces. *South Afr. J. Chem. Eng.* **2021**, *36*, 94–104. [CrossRef]
20. Yamin, Y.; Yan, W.; Jie, Y.; Zhengguang, L.; Qi, L.; Bohong, W. Tech-economic modeling and analysis of agricultural photovoltaic-water systems for irrigation in arid areas. *J. Environ. Manag.* **2023**, *338*, 117858. [CrossRef]
21. Pranit, D.; Kul, V.; Vijendra, K.; Aneesh, M. Water distribution system modelling of GIS-remote sensing and EPANET for the integrated efficient design. *J. Hydroinformatics* **2024**, *26*, 567–588. [CrossRef]
22. Khezri, S.M.; Lotfi, F.; Tabibian, S.; Erfani, Z. Application of water pinch technology for water and wastewater minimization in aluminum anodizing industries. *Int. J. Environ. Sci. Tech.* **2010**, *7*, 281–290. [CrossRef]
23. Mughees, W.; Al-Ahmad, M. Application of water pinch technology in minimization of water consumption at a refinery. *Comput. Chem. Eng.* **2015**, *73*, 34–42. [CrossRef]
24. Liu, Z.Y.; Yang, Y.Z.; Zhang, Y. Determination the pinch point and calculating the freshwater target for water-using systems with single contaminant. *Chem. Eng. Res. Des.* **2007**, *85*, 1485–1490. [CrossRef]
25. Chin, H.H.; Liew, P.Y.; Varbanov, P.S.; Klemes, J.J. Extension of pinch analysis to targeting and synthesis of water recycling networks with multiple contaminants. *Chem. Eng. Sci.* **2022**, *248*, 117223. [CrossRef]
26. Kim, Y.; Lim, J.; Shim, J.Y.; Lee, H.; Cho, H.; Kim, J. Optimizing wastewater heat recovery systems in textile dyeing processes using pinch analysis. *Appl. Therm. Eng.* **2022**, *214*, 118880. [CrossRef]
27. Skouteris, G.; Ouki, S.; Foo, D.; Saroj, D.; Altini, M.; Melidis, P.; Cowley, B.; Ells, G.; Palmer, S.; Odell, S. Water footprint and water pinch analysis techniques for sustainable water management in the brick-manufacturing industry. *J. Clean. Prod.* **2018**, *172*, 786–794. [CrossRef]
28. Souifi, M.; Souissi, A. Simultaneous water and energy saving in cooling water networks using pinch approach. *Mater. Today Proc.* **2019**, *13*, 1115–1124. [CrossRef]
29. Zhang, Q.; Yang, M.; Liu, G.; Feng, X. Relative concentration based pinch analysis for targeting and design of hydrogen and water networks with single contaminant. *J. Clean. Prod.* **2016**, *112*, 4799–4814. [CrossRef]
30. Deng, C.; Feng, X.; Sum, D.K.; Foo, D.C.Y. Process-based Graphical approach for simultaneous targeting and design of water network. *AIChE J.* **2011**, *57*, 3085–3104. [CrossRef]
31. Mabitla, S.S.; Majozi, T. A hybrid method for synthesis of integrated water and regeneration networks with variable removal ratios. *J. Environ. Manag.* **2019**, *231*, 666–678. [CrossRef] [PubMed]

32. Quintero, V.; Quiroga, A.G.; Delgado, A.D.G. A hybrid methodology to minimize freshwater consumption during shrimp shell waste valorization combining multi-contaminant pinch analysis and superstructure optimization. *Polymers* **2021**, *13*, 1887. [CrossRef] [PubMed]
33. Karthick, R.; Kumaraprasad, G.; Sruti, B. Hyprid optimization approach for water allocation and mass exchange network. *Resour. Conserv. Recycl.* **2010**, *54*, 783–792. [CrossRef]
34. Cao, K.; Feng, X.; Ma, H. Pinch multi-agent genetic algorithm for optimizing water-using networks. *Comput. Chem. Eng.* **2007**, *31*, 1565–1575. [CrossRef]
35. Esmaeeli, A.; Sarrafzadeh, M.H. Reducing freshwater consumption in pulp and paper industries using pinch analysis and mathematical optimization. *J. Water Process Eng.* **2023**, *53*, 10364. [CrossRef]

water

MDPI

Article

Transport of Floating Plastics through the Fluvial Vector: The Impact of Riparian Zones

Manousos Valyrakis [1,*], Gordon Gilja [2,*], Da Liu [3,4] and Gaston Latessa [3]

1 Department of Civil Engineering, Aristotle University of Thessaloniki, 54124 Thessaloniki, Greece
2 Department of Hydroscience and Engineering, Faculty of Civil Engineering, University of Zagreb, 10000 Zagreb, Croatia
3 Infrastructure and Environment Research Division, School of Engineering, University of Glasgow, Glasgow G12 8QQ, UK; d.liu.2@research.gla.ac.uk (D.L.); g.latessa.1@research.gla.ac.uk (G.L.)
4 China Construction Science and Technology Group South China Co., Ltd., Shenzhen 518118, China; liuda6@cscec.com
* Correspondence: mvalyra@civil.auth.gr (M.V.); gordon.gilja@grad.unizg.hr (G.G.)

Abstract: This study presents results from an experimental campaign to explore how different riparian zone characteristics may facilitate the transport or capturing of plastics floating through the fluvial system. Specifically, following field observations for the transport of plastics through fluvial vectors, a substantial number of flume experiments has been designed to assess the effect of floating macro-plastics and riparian zone characteristics. The results from flume experiments were analyzed using particle tracking velocimetry techniques to derive transport metrics (such as transport velocities) of macro-plastics of different sizes and shapes, released at five locations across a wide channel with distinct distance from the vegetated riverbank. The findings are discussed while considering the trapping mechanisms along the vegetated riverbank, which include a range of vegetation densities and arrangements, aiming to identify and quantify the degree of impact of each of the control parameters on the transport of floating plastics. The flow velocimetry records obtained at locations near and within the riverbank correlate well with the transport velocities of the floating plastics. Macro-plastic litter carried downstream away from the riverbank can have up to nine times the transport velocity, compared to those found within the riverbank. The change from a low to a high average density can result in about three times decrease in the transport velocity of floating macro-plastic litter within the riparian zone. These outcomes can help inform better practices for the management of riparian vegetation to maximize the trapping efficiency of macro-plastics, adapted to different flow conditions and river morphologies.

Keywords: macro-plastic litter; water surface flow; turbulence; riparian zone; open channel flow; fluvial transport; particle tracking velocimetry

1. Introduction

Plastics are widely used in consumer products due to their low cost, low weight, and high durability compared to other types of materials [1]. The contamination of marine ecosystems due to plastics is a challenge that has received a lot of attention due to the significant risks it poses for the environment, marine mammals, and human health [2–4]. The abundance and continued accumulation of floating plastics in oceans across the globe have been growing challenges receiving increased recognition over the last decades [5,6]. It is estimated that the amount of plastic that flows into oceans each year is between 0.7 and 3.8 million tons [7]. This is a growing challenge for our consumer society, as disposable plastics that break up but do not break down are virtually used in our life on a daily basis [8]. It also signals the poor management of non-recyclable and non-recycled plastics, which eventually are not handled in a responsible manner and can be dispersed and introduced into the river network (from land via aeolian transport, as well as from untreated waste and storm water). Thus, plastics can find their way to the ocean via earth's river systems [9–11]. They may be temporarily trapped as floating litter or deposited sediment in these river systems, until mobilized again during high flow and flooding events (demonstrating a significant increase of up to 14 times the average in terms of the number of microplastics) [12]. In this manner, the fluvial network becomes one of the primary pathways for the transport of plastics into the ocean [13–16]. There has been little research considering the issue of transportation of floating plastic litter along the river system, which contributes to the global challenge of environmental pollution by such persistent contaminants [17]. Thus, obtaining a better understanding of the processes affecting their fate and transport is of outmost importance and urgency [18,19].

A study by Morritt et al. [20] identified the composition of litter in the upper Thames estuary as comprising plastic bags (more than 20%) and food wrappers/containers (another 20%), as well as sanitary towels, plastic cutlery, tobacco packing/wrappers, and general plastics, which is representative of the UK's coasts [21,22]. Gasperi et al. [23] assessed the quantity of floating litter in the river Seine. Their finding showed that plastic litter was 0.8–5.1% of the total litter in terms of weight, consisting mostly of one-off use plastic cutlery, food wrappers, and containers. Similar observations can be made for other rivers in the United Kingdom, as seen in Figure 1a,b (obtained during own field campaign). It can be clearly seen that the majority of floating plastics (food containers and wrapping) are trapped on the side of the stream, river, or canal, and within its typically vegetated banks [24,25]. Extruded Polystyrene Foam (XPS) is widely used due to its good thermal insulation properties as the primary material for food wrapping and has been studied in the past due to its potential to pollute the environment (e.g., with mercury and bacteria [26]). A study of small plastic litter abundance at Heungnam Beach (Korea), found that more than 90% comprised XPS spherules [27], which potentially originated from plastic containers that broke up during their transport from the nearby fluvial system.

Figure 1. Forth and Clyde Canal, showing (**a**) downstream view of the canal's cross-section including the main channel and vegetated riverbank and (**b**) top view of the canal's riparian zone retaining plastic litter (notice food containers and wrappers made of XPS plastic).

This study aims to advance the understanding of the processes affecting the fate of macro-plastic litter along the fluvial vector before they reach the coast. Specifically, this experimental study focuses on the initial to intermediate transport stages where macro-plastic litter have been introduced to the freshwater system and have only started to degrade and break up into smaller fragments as they are transported along the fluvial vectors [28].

The riparian zone has been found to control the flow hydrodynamics and subsequently, the flow's transport capacity at the vicinity and within the vegetated bank, as well as the main channel [29], having implications for the channel's stability. Previous research put emphasis on near-bed surface dynamics. Hence, the focus of this study is to assess how different vegetation densities and arrangements may influence the transport of floating plastics, both indirectly via modifying the water surface flow field, as well as directly via blocking their downstream movement.

The objective of this study is to assess the effect of the riparian zone on the downstream advection of floating plastic and obtain a better understanding of the processes involved via the use of flow diagnostics (acoustic Doppler velocimetry) and particle tracking velocimetry techniques. The study focuses on particles ranging from "small" macro-plastics [30]—onwards termed macro-plastic litter, up to 5 cm plastics, usually termed as "large" meso-plastics. The assessment will focus on extracting qualitative observations and conducting analysis to quantify the impact of the examined control parameters (such as particle size and shape) on their transport dynamics. To this goal, a series of purpose-specific physical experiments are carried out at one of the large research flumes of the Water Engineering Lab at the University of Glasgow to examine the mechanisms of dispersion of plastic litter along the water surface of a channel with simulated riparian vegetation. Field observation supports the development of experimental set ups and confirms the behavior of macro-plastics and its interaction with vegetated banks.

2. Materials and Methods

In the following sections the description of the experimental setup (including flow conditions and the flume and test section characteristics), the experimental matrix, and protocol for the specifically designed physical experiments aiming to examine the mechanisms of dispersion of floating macro-plastic litter along the water surface of a channel with simulated riparian vegetation are offered. Details of the supporting field campaigns along a selected water course are also included.

2.1. Experimental Design

The physical experiments described herein were conducted in one of the large (14 m long by 1.8 m wide) glass-walled water recirculating flume at the Water Engineering Lab of the University of Glasgow, which has flat and fixed bed and riverbank slopes to allow uniform flow conditions to develop at the test section, where the transport experiments take place. The flume has a simulated riparian zone comprising a large number of acrylic rods (of diameter, D = 6 mm), placed in a controlled manner (held in position by a suspended support panel) on the top of an inclining riverbank section (Figure 2). The stream bank was modeled by an inclined acrylic panel (of width W_v = 0.8 m), with an angle, $\theta = 17°$, running a streamwise length of 8 m, starting a short distance downstream of the inlet section of the flume. The main channel section (of width W_m = 1 m) is layered with coarse sand of nominal diameter of about 2 mm. Flow through the section of the flume with the riparian zone was smoothed using appropriate streamlining sections. Filter layers at the inlet and outlet of the vegetated channel, comprising coarse bed material, allowed the retaining of the finer sediment comprising about 8 mm fake sediment bed.

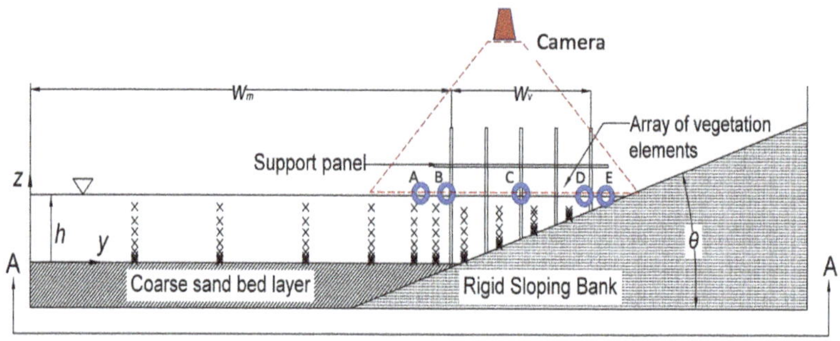

Figure 2. Upstream cross-section of the channel test section showing the stable main channel and inclined vegetated riverbank. Also shown is a subset of the measurement locations for the acoustic Doppler velocimetry (with symbol "x") and the locations where the plastic litter was initially released (blue circle). The flow direction is along the *X* axis, perpendicular to the *ZY* plane (into the page). The open blue circles (A to E) show the projection of the floating plastic elements release location to the illustrated test cross-section.

A uniform flow depth of about 122 mm was established along the test section by means of fixing the tailgate height to a certain level. A mean flow discharge (as assessed by means of electromagnetic flow meters placed at the inflow pipes downstream of the two pumps) of $Q = 8.5 \times 10^{-3}$ m^3/s was controlled via the pump inverters. Only one flow condition is tested herein, yet the flow field around and within the riparian zone, as well as at the main channel, demonstrated a strong variability, as assessed by the flow velocimetry, dependent on the riverbank vegetation density. The transect where the flow velocimetry measurements were taken was located 7.2 m from the start of the vegetated riverbank.

Flow hydraulics (flow discharge and channel characteristics such as channel width, flow depth, and bed surface roughness and slope), water surface hydrodynamics (defined by the three-dimensional character of the flow and especially riverbank vegetation), and riverbank vegetation features (such as vegetation type, stem, flexibility, and density, see for example [31]) are the main factors influencing the transport of the floating plastics (for a wide range of individual macro-plastic elements features, such as size, shape, and the material-dependent specific density). In making a choice to prioritize the design variables to be assessed in this study, the specific research objectives need be considered. Herein, the effects of riverbank vegetation densities on the transport of floating macro-plastics are considered to be important for a range of vegetation densities and macro-plastics features. In these experiments only the rod-like feature of the vegetation is seen by the flow, which may directly emulate the effects of vegetation stems or indirectly the trunk of scaled tree models, as done in [32]. The experimental matrix comprises the range of the control parameters, which are selected in such a way to allow the widest range of macro-plastics transport dynamics to be observed. The flow rate for which the experiments were conducted, was assessed from pilot lab experiments to introduce sufficient variability of the water surface dynamics across the test area of the channel cross-section near the riverbank for the range of vegetation densities assessed herein, thus introducing a wide range of macro-plastics transport dynamics. For example, pilot lab experiments (similar to [33]) helped optimally choose the flow discharge so that for the chosen smallest riverbank vegetation density and smallest size of macro-plastic elements tested herein, a measurable difference in the transport rate was still observed (higher flow discharge would result in virtually little differences of solids transport). Likewise, for the highest riverbank density and largest floating macro-plastic element, some finite amount of transport should still occur for the same flow discharge (lower flow discharge would mean all get trapped at the riverbank).

The test section ran for a length of 1 m, which was contained within the viewing area of the camera recording the transport of floating plastic litter. Five initial release locations for the plastic litter were chosen, as shown in Figure 2 (see locations A to E, denoted with symbols "o"). From left to right (looking downstream, Figure 2) the first and last release locations were at X_A = 975 mm and X_E = 1315 mm, while the second and fourth at the edges of the riparian zone (X_B = 985 mm and X_D = 1305 mm). The third location was found in the middle of the riparian zone (X_C = 1145 mm). The transect of these locations was about 5 m from the start of the vegetated riverbank, so as to ensure uniform and fully developed flow conditions (as assessed from flow velocity measurements at planes along consecutive downstream locations). Pilot experiments demonstrated that the floating litter obtain over relatively short distances that span up to five to ten times their characteristic dimension, a steady transport velocity (unless their transport is disrupted by the vegetation elements), dictated by the near-water-surface flow field [34,35]. Thus, it is observed that the length of the test section is sufficient to capture (via the use of a camera, Figure 2) the complete range of transport dynamics demonstrated by the floating particles.

2.2. Experimental Matrix

The type of material, size, and shape of floating plastic litter was determined according to the findings of an initial field campaign along certain rivers and canals in Scotland (including segments of rivers Kelvin and Clyde and the Forth and Clyde Canal, e.g., see Figure 1), during which the plastic litter captured in the riparian zone were sampled. There has been an increasing number of recently published studies on the methods of sampling floating macro- and meso-plastics, which range from passive to active, also using bespoke and elaborate or expensive equipment or just sourcing information via citizen science participation projects [36–42]. Given the plethora of sampling procedures, the optimal sampling may depend on the location and flow conditions that are characteristic of a specific river system. Even though our sampling and subsequent design choices may well reflect local conditions, it is important to note that the results presented below are considered to be representative of the transport dynamics of meso- and macro-plastics in water surface bodies, as affected by riverbank vegetation of different densities. These observations are of course limited to the range of size and shapes of the plastics, as well as the range of riverbank densities, but even in these cases, the same framework as presented herein can be used for extracting useful information for different ranges of interest. Based on this information and focusing on the floating macro-plastics, the matrix, comprising the control parameters, for this flume experiment was determined. Given that most of the sampled macro-plastics were made of XPS fragments, combinations of two shapes (circular and rectangular) and 3 sizes per shape, with their maximum dimension ranging from 2 to 4 cm, were selected, as shown in Figure 3.

Using the vegetation density per unit meter, defined as $\lambda = D/x_i^2$, where x_i is the distance between each rod and D is the simulated rigid vegetation element (acrylic rod, distributed by RS components©, London, UK) diameter [36], riparian vegetation was configured in 2 different riparian densities from low to high (λ = 0.9 to 1.9 m^{-1}, respectively). For each of these two average macroscopic densities, 3 different patterns for the placements of vegetation elements (one regular and two randomly spaced arrangements exhibiting low or high variability of the local density, as shown in Figure 4) were employed to aid in identifying the effect of the spatial variability on the transport mechanics of floating plastic litter. The regular configuration resulted in a linear arrangement for the low vegetation density case while a staggered arrangement was seen for the high-density case.

For each of the above six riverbank vegetation configurations, each of the six floating plastic elements was released from each of the five different locations upstream (A to E, as shown in Figure 2), and the experiment was repeated three times for each combination, resulting in a total of 540 individual experimental flume runs.

To allow conducting trajectory analysis for the downstream transport of the plastic litter, a camera (Sony FDR-X1000V, with ZEISS Tessar lens, manufactured in Wuxi City, Jiangsu Province, China) recording at 120 fps offering a wide-angle top view of the test section was appropriately placed above the flume. The camera was looking down at the center of the test section along axis Z and perpendicular to the plane of the riverbed (*XY*).

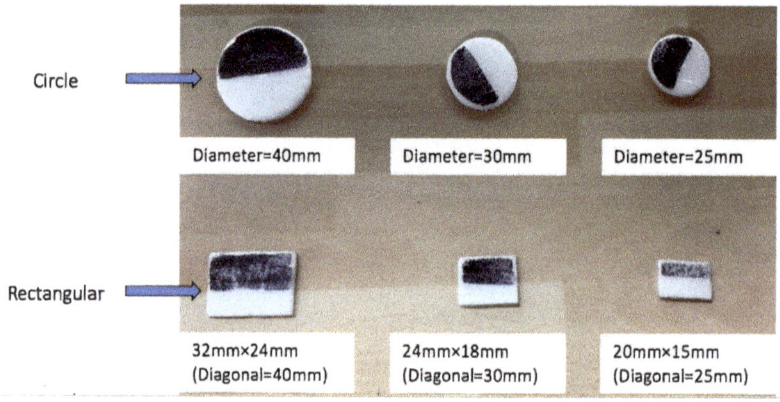

Figure 3. Comparison of the size and shape of the 6 floating macro-plastic fragments used in the experiments. Each half of the plastic fragments surface is white or dark colored, to allow assessing their rate of rotation as they transverse along the riverbank.

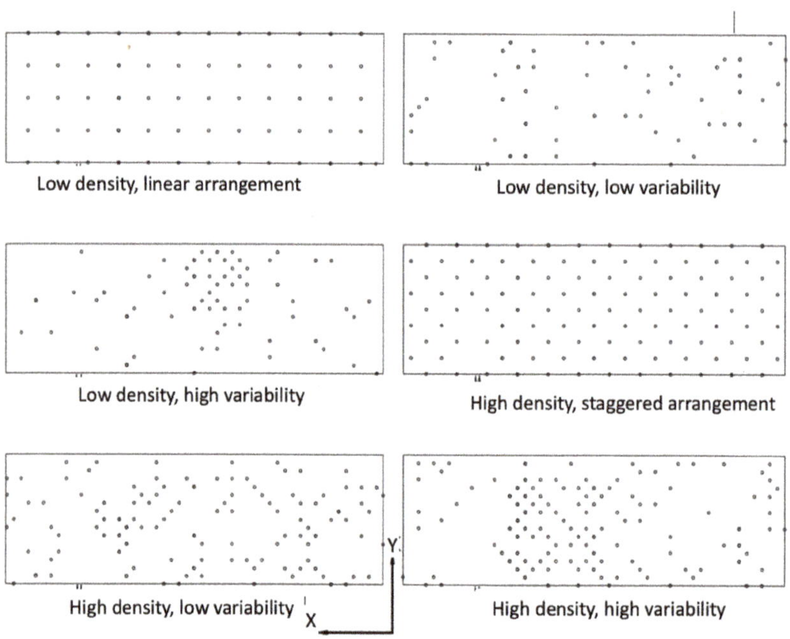

Figure 4. Demonstration of the regular and random (low and high variability of layout non-uniformity) riverbank vegetation arrangements for the assessed mean vegetation densities (linear low-density arrangement of $\lambda = 0.9\ m^{-1}$ and staggered high-density configuration of $\lambda = 1.9\ m^{-1}$). The flow direction is along the X axis. The open circles pinpoint the location of the rigid riverbank vegetation elements, to better comprehend the vegetation density non-uniformity.

2.2.1. Trajectory Analysis of Floating Macro-Plastics

At first, the videos were assessed qualitatively to identify a wide range of dynamics and any potential issues with the analysis proposed below. At this stage, focus was given on assessing mechanisms enhancing the mobilization or trapping of the floating plastics along or within the riverbank. A collage of overlaid pictures taken at equidistant temporal instances, focusing on the first stage of transport, and outlining the location of one of the circular floating plastics, is shown in Figure 5. For example, it may be observed (as is particularly usual for plastics released within the riverbank) that the instantaneous transport velocity of floating plastics along the longitudinal as well as lateral direction, may change (Figure 5). This may be because of the influence of three-dimensional flow structures upwelling onto the water surface and carrying the plastic in this direction. Also, for denser (locally or on average) configurations, the flow transporting the floating plastic can interact with the patch of vegetation, or in other cases the plastic can directly interact with individual vegetation elements (rods) within the riparian zone. Such observations allow a better comprehension of the dynamics of transport of floating plastics and facilitate the interpretation of the results obtained.

Figure 5. Demonstration of the procedure to obtain part of the trajectory of a downstream-advected plastic litter, in the form of a composite picture comprising subsequent overlaid frames, to demonstrate the change in location of the transported plastic element (outlined with the light blue filled circle).

For the follow-up stage of quantitative analysis, a number of accessible software exist (such as freeware ImageJ) that allow us to conduct analysis of the trajectories of the floating plastic litter. However, there exists no readily automated routine for conducting this analysis, as there are a lot of time-consuming video and image post-processing procedures that are required.

For the present study, each video was first split into its constituent frames, which were then imported into appropriate image processing software (here Adobe Photoshop© version 19 was used), for correcting the parallax effect (removing geometric distortion) of the wide-angle lens. Once this was done, the range of selected images was fed manually into the particle tracking velocimetry software (ImageJ version 1.52i) for extracting the trajectories and from these, the mean transport distances and velocities. In this manner, a variety of transport metrics were obtained (including maximum local transport velocities along the longitudinal and lateral directions, $u_{max,X}$ and $u_{max,Y}$, and time to trapping) as well as the trapping location (to identify if there is any persistence), allowing us to extract the dependence of these features to the control parameters (primarily the size of the litter and the riparian vegetation density and configuration). Considering the total time for each trial, capped to a maximum duration for the cases the particle was entrapped within the vegetation patch, the bulk transport velocities along the longitudinal and lateral direction, U_X and U_Y, were estimated.

2.2.2. Acoustic Doppler Velocimetry

The flow field near the water surface is the primary forcing mechanism for the downstream advection of the macro-plastic litter. In order to understand how the changing arrangements of riparian zones affect the flow field, a flow diagnostics campaign was pursued along a representative cross-section of the flume. To that goal, acoustic Doppler velocimetry was used across a very dense measurement grid comprising a few thousands of measurement locations (seen as white dots in Figure 6), capturing the whole flow field along the ZY plane. While there is no obvious mean flow effect for the various arrangements (linear or staggered versus low and high-density variability), it can be clearly seen that the bulk flow at the main channel is increasing with vegetation density (compare low and high-density cases in Figure 6). Likewise, the flow through the riparian zone is clearly reduced for the high density compared to the case of low vegetation density.

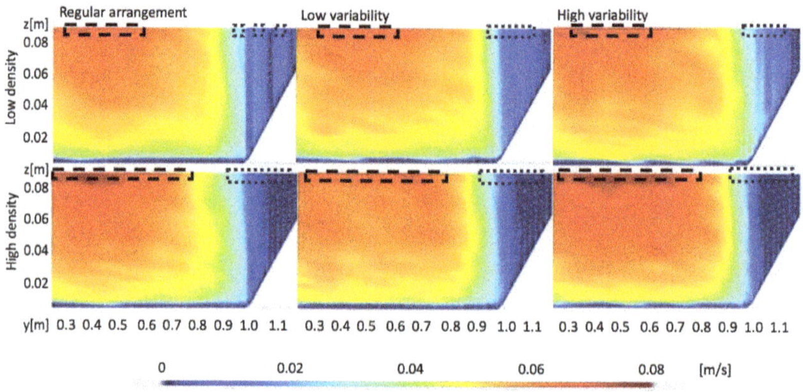

Figure 6. Contour plots of the mean flow velocity field along a cross-section of the flume at the test section (including measurements both at the main channel and the riverbank) for the range of riparian vegetation densities and configurations examined. The high density of riparian vegetation results in a demonstrably higher fast-flowing cross-sectional area at the main channel (see section outlined with the dashed rectangle), as well as a more slow flow and transport at the riverbank through the riparian vegetation (see sections outlined with the dotted rectangles).

This may result in a higher transport efficiency of floating plastics near location A (closer to the main channel) for the case of high density. On the contrary, for high densities, plastic litter found within the riparian zone are expected to be transported at a slower rate, while also having a greater probability of getting trapped (until potentially a higher flow event, e.g., during a flood, can again mobilize them).

2.3. Field Campaigns

A series of field campaigns were conducted to collect real-world data related to plastic pollution in water courses. These campaigns permitted the categorization of the presence of different plastic sources and the analysis of how the vegetation patches located on the riverbanks acted as retention areas. Field surveys were carried out along the River Kelvin and the Forth and Clyde Canal in Glasgow, UK. These surveys permitted to identify the presence of multiple plastic sources, varying in size and density. These macro-plastics were found to be buoyant given its relatively low density and therefore were categorized under the floating plastics source. The site surveys along the Canal (Figure 1) highlighted the interaction of riverbank vegetation with these floating elements and confirmed its importance in trapping and retaining plastics in water courses. Field trips at the River Kelvin also presented a different source of macro-plastics in the water course that is not related to food containers and wrappers but to existing outdoors infrastructure (Figure 7). These plastics were also found to be floatable and of similar size to the plastics used in the flume experiments.

(a) (b) (c) (d)

Figure 7. (**a**) Plastic elements from worn-out paint of lighting poles, (**b**) these poles sit next to River Kelvin, at Kelvingrove Park, Glasgow, UK, and can potentially contribute to the plastic discharge in the water course, front (**c**) and reverse (**d**) of a sample of a maximum length of 40 mm.

3. Results and Discussion

3.1. Qualitative Description of Transport Mechanics

A qualitative description of the transport mechanics of macro-plastic litter by means of direct observation of the visual material obtained will aid in obtaining a better comprehension of the wide range of litter–flow–vegetation dynamic interactions. Firstly, there exist the cases where the plastics are fully transported downstream, following a straight or slightly curved path. These are in general the cases where the plastics are released near the main channel (e.g., location A), where the downstream flow velocity at the water surface is much greater compared to the riverbank (Figure 8). On the contrary, the trajectories of plastic litter, released within the vegetated riverbank (locations B, D, and particularly C) may exhibit a wide range of interactions directly bouncing off or getting trapped by individual vegetation elements or indirectly via the modification of the flow field in the vicinity of the vegetation elements or their patch. Thus, it may be expected that the rate of rotation as well as the lateral advection may be greater for the cases of litter transported at the main channel compared to those within the vegetated riverbank.

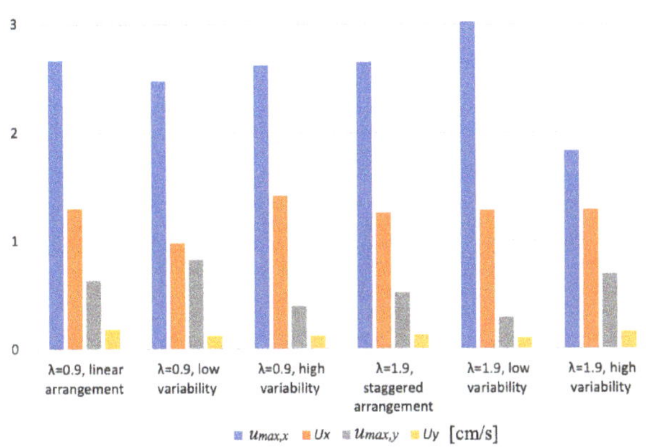

Figure 8. The effect of riparian vegetation densities and arrangements on floating macro-plastic transport velocities.

In certain cases, the litter are seen to be momentarily captured at the upstream face of the vegetation element and twitching until an approaching flow structure may result in its episodic sweep downstream (similar to transport near the bed surface [43,44]). For smaller litter and high vegetation patch densities, it may be observed that the slowly transported plastic is captured downstream the vegetation element (e.g., at the recirculation region where the mean flow velocity is almost zero). It is expected that this effect will be greater as the size of the recirculation region, which scales with the vegetation element, is of the same order as the size of the plastic litter. There is the case where plastic litter could also be stuck between two or more vegetation elements if the distance between the vegetation elements is smaller than the maximum dimension of the plastic litter.

3.2. Transport Velocity

Herein, the rate of transport of floating macro-plastics of various characteristic shapes and sizes along a riverbank vegetated at different representative densities and arrangements is analyzed and quantified. In addition to improving our understanding of transport processes, such information can be used to calibrate numerical models towards improving their performance, e.g., [45].

3.2.1. Effect of Riparian Vegetation Density and Arrangement

The analysis of trajectory data from all experiments indicates that increased riverbank vegetation density does not affect the efficiency of transport of floating plastics with any statistical significance. For example, the maximum longitudinal and lateral velocity of the litter, $u_{max,X}$ and $u_{max,Y}$ which remain about 2.6 cm/s and 0.56 cm/s, respectively, do not demonstrate any consistent variability for the different arrangements (Figure 8). The same holds true for U_X, U_Y with even smaller variability, at 1.2 cm/s and 0.1 cm/sec (Figure 8).

3.2.2. Effect of Size and Shape

Overall, size and shape did not appear to have a significant effect on the transport velocities. Specifically, when comparing the results for circular and rectangular shape, the local and mean streamwise and lateral transport speed was relatively invariant at $u_{max,X}$ = 2.5 cm/s, U_X = 1.3 cm/s and U_Y = 0.14 cm/s (Figure 9a). However, for the lateral local transport velocities, with an overall mean value of $u_{max,Y}$ = 0.6 cm/s, a reduction of about 30% and 13% was observed for the circular- and rectangular-shaped litter, as the vegetation density increases from λ = 0.9 m^{-1} to λ = 1.9 m^{-1}, respectively (Figure 9a). Figure 9b illustrates that the size of floating macro-plastics has no significant impact on the fate of transport (for the assessed sizes and ratio of litter maximum dimension to the average spacing of vegetation elements comprising the riparian zone).

3.2.3. Effect of Release Location

As floating litter are advected downstream with the flow, they can be found at different locations across the open channel (considering the three-dimensional nature of macroscopic flow structures motions, due to which litter will experience a level of spatio-temporally varying lateral hydrodynamic forcing). As it is expected that at the main channel the floating particle will act as a tracer to the water surface flow, this study focused on the transport of litter in the vicinity or within the riparian zone.

Indeed, past field studies have shown that floating particles (such as eco-foam chips [46]) can be used for assessing the water surface flow field and two-dimensional flow structures (e.g., via large scale particle image velocimetry), under the assumption they act as tracers to the flow [47–49].

Figure 10 showcases the various transport velocities the floating litter has for the different release locations and vegetation densities. Comparing the velocities of the floating plastics with the near-water-surface flow velocities (Figure 6), for the corresponding locations, clearly demonstrates that the assumption that the macro-plastics acts as tracers to the local flow is valid. As the water surface velocity is location dependent (Figure 6), it is reasonable to consider that the release location will directly control the velocity with which the plastic litter is transported. For example, for both the low and high vegetation density, the transport velocity of the plastics will reduce significantly as the location changes from the shear region near the main channel, with U_X about 3.1 and 3.5 cm/s (for location A), to about 0.3 and 0.1 cm/s (for locations B to D), respectively. Particularly, for locations B to D within the riverbank, for the high-density vegetation, streamwise flow velocity at the near-surface region is about 50% smaller compared to the low vegetation density. This results in the plastics' velocity getting up to two times higher for the case of low vegetation density, compared to the high vegetation density.

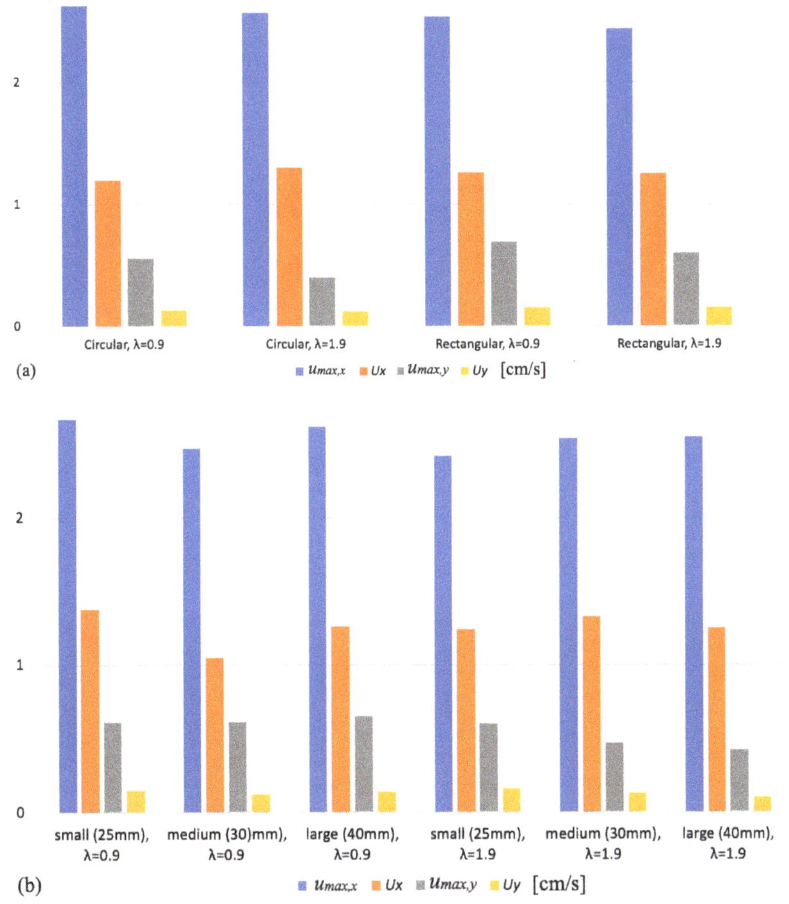

Figure 9. Floating macro-plastic transport velocities for the assessed riparian vegetation densities and different (**a**) shapes and (**b**) sizes.

While the plastics are slowing down within the vegetated riverbank for increasing vegetation densities (λ = 0.9 m^{-1} to 1.9 m^{-1}), the opposite happens for the release location (A) near the main channel, where plastic transport mean streamwise velocity can increase by about 13% (following the same trend, as shown in Figure 6), while the local variation can decrease by about 5%. The plastic litter that are carried downstream away from the riverbank (location A) can have a manyfold (9-fold for λ = 0.9 m^{-1}, or 30 times for λ = 1.9 m^{-1}) increase on the transport velocity, compared to those found within the riverbank. The change from a low to a high average density can result in about three times decrease in the transport velocity of floating litter within the riparian zone.

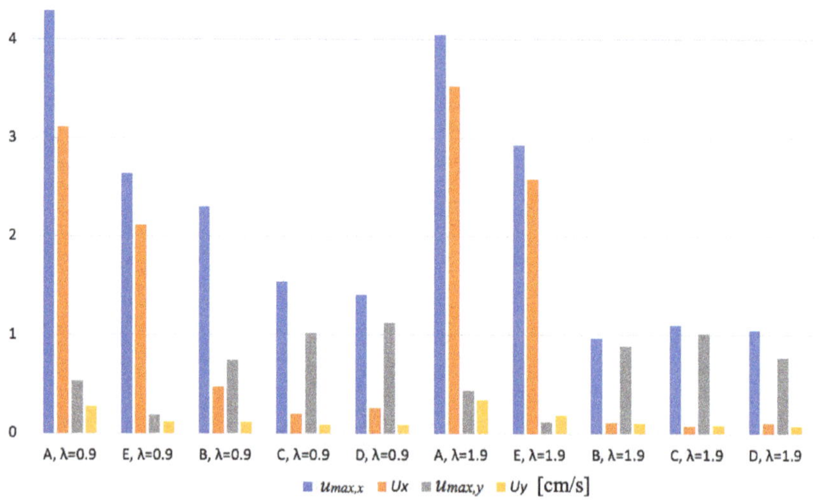

Figure 10. The effect of release location (from A to E) on floating macro-plastic transport velocities for the assessed riparian vegetation densities.

3.3. Rate of Rotation

Vegetation density and the size and shape of plastic litter influences its rotation rate for the assessed distance travelled (largely clockwise on average). On average, for the low vegetation densities (λ = 0.9 m^{-1}) the rotation was 41 degrees for circular-shaped litter, while it was about half (22 degrees) for rectangular. For the high vegetation density (λ = 1.9 m^{-1}) the degree of rotation dropped fourfold for the circular litter (10 degrees) and at about 16 degrees for the rectangular. For the high density, the litter size mattered more, as the rotation increased from 5 degrees for small (2.5 cm) to 14 degrees for medium size (3 cm) and 22 degrees for the large litter (4 cm). The same increasing trend for the degree of rotation with size was observed for the low vegetation density but with a much smaller variation, from about 30 to 33 degrees. The degree of rotation is greater for the higher vegetation density (90 compared to 72 degrees) if the plastic litter is close to the main channel (location A).

3.4. Implications for Fluvial Storage

Most of the plastic litter will be eventually captured (more than about 90%) if found at the vicinity or within the riparian zone for λ = 0.9 m^{-1}. It is certain that all plastic litter (for the cases examined herein) will get captured for the vegetation density of λ = 1.9 m^{-1}. Thus, it is apparent that a vegetation density of λ > 1 m^{-1}, may be considered as a threshold value for the effective capturing of macro-plastic litter (2.5 cm < D < 5 cm). The probability that plastic litter found at a distance 2–3 times their size away from the riparian zone will get captured within the riparian zone is reduced from 35% to 26% as the density increases from λ = 0.9 m^{-1} to 1.9 m^{-1}. Overall, the probability that plastic litter is captured within the riparian zone is the same for both densities at about 69%. The identification of macro-plastic litter capturing location can be used as a basis for the deployment of river-cleaning measures such as rotating screen drums [50] to improve their efficiency.

Changing the flow discharge rate would generate an even greater range of variability, which for small changes is expected to scale with plastics transport rates. A study dedicated to the effects of substantial changes in flow discharge can be pursued in the near future but has to address complexities with the three-way interaction of the flow field, vegetation elements, and floating plastics. For combinations of flows and types of vegetation that are flexible, additional mechanisms of trapping and releasing plastics, that have been discussed here can be observed and must be accounted for.

Even though the direct observations of this idealized experimental study may be limited by the range of the parameters employed herein (such as shape and size of plastics, steady flow conditions, as well as riverbank vegetation density), these are still useful in capturing the full range of plastic transport dynamics. Flume experiments, while controlled and replicable, often simplify the conditions found in natural river systems. To bridge this gap, it is crucial to scale up the findings by incorporating factors such as varying flow rates (including unsteady flow conditions, and those representing high flooding conditions), river morphology, and the presence of natural and anthropogenic obstacles that can affect plastic transport and accumulation. There exists strong interest in expanding the range of these observations both experimentally and in the field, by adopting the presented framework to such studies, towards a more holistic approach, obtaining a more comprehensive understanding of the involved processes and eventually developing models that more accurately reflect the retention and release mechanisms of plastics in diverse riverine environments, by incorporating the role of turbulence and their interaction with vegetation. Additionally, these models should be validated with field data from different types of river systems to ensure their applicability and broader capacity to represent a range of such river systems. By integrating detailed flume experiment results with robust modeling and field validation, researchers can develop a comprehensive understanding of floating plastic transport dynamics that can be applied to manage and mitigate plastic pollution in riverine environments across the earth's surface.

4. Conclusions

Plastic litter, continuously accumulating on our coasts, estuaries, seas, and oceans, is persistent and has an extended lifespan, threatening both the natural environment and human health. This study aims to shed light on the mechanics of transport or entrapment of floating plastic litter in the receiving fluvial systems and assess the role the riparian zones may play in controlling these. To this goal, a series of detailed flow diagnostic measurements were acquired across a flume with a vegetated riverbank and videos of floating plastics transport experiments were analyzed with particle tracking velocimetry, for a range of densities and configurations, representative of natural riparian zones. This experimental campaign resulted in the generation of a first experimental data set to help provide initial insights into the transport mechanics by which floating macro-plastics are transported into the oceans, as well as a novel method for systematically studying such transport processes for a wider range of control parameters. Additionally, a series of field campaigns were conducted to assess the behavior of floating plastics in selected water courses, with particular focus on vegetated banks and to support the description of plastic materials commonly found in water courses in the UK.

The qualitative description of the range of motions and dynamic interactions between the plastics–flow–riparian zone can help better understand the processes affecting the spatio-temporal variation of the transport (including capture and release processes for a range of flow events). Via conducting appropriate laboratory scale experiments under various well controlled conditions, representative of the processes present for studying the fluvial transport of plastic litter, the impacts of their characteristics (e.g., size, shape, and density) as well as the features of the riparian zone were assessed. The major findings demonstrate that there exists no variation for the shape and size of the particles with any strong statistical significance. The same holds true for the different arrangements of vegetation (with low or high local densities), which implies that for the conditions tested here, the controlling parameter is the mean (macroscopic) vegetation density. This is a useful observation, as traditionally physical models of riparian vegetation comprise uniformly arranged vegetation in the environmental hydraulics literature. The control parameter with the most significant impact in this study is the location where the litter is transported downstream. This parameter is important as it defines the region where the litter is found in the water surface during their transport and directly relates to the local flow field, which eventually controls the velocity with which the litter (acting like a tracer to the flow) are advected downstream. It is seen that the unobstructed release locations, particularly near the main channel, have higher flow velocities compared to the locations where flow is slowed down due to the riverbank vegetation.

It is also found that a riparian zone with average density $\lambda > 1\ m^{-1}$ may allow the capture of all macro-plastic litter. Thus, riverbank restoration and revegetation may help stabilize the riverbank against erosion, but also can act as a means for storage of plastics along the fluvial vector, which may have profound implications for ecological restoration. However, more studies are needed to identify the controls for transport or capture of a wider range of prevalent floating plastics, and to assess the degree to which the riparian zone contributes to exporting or fluvial storage, under a wider range of flow conditions and considering a wider range for the ratio of average spacing of vegetation elements to the plastic litter size. Studying and obtaining a better understanding of the mechanisms contributing to the fate of plastic litter is therefore important in helping to inform environmental risk assessment, policymaking, improved treatment, and proactively devise methods to reduce their abundance and inform new designs towards their entrapment to prevent them from entering the ocean through fluvial vectors.

Author Contributions: Conceptualization, M.V., G.G. and D.L.; methodology, M.V.; validation, G.G. and G.L. formal analysis, M.V., D.L. and G.G.; investigation, M.V., D.L. and G.L.; resources, M.V. and D.L.; data curation, D.L.; writing—original draft preparation, M.V. and D.L.; writing—review and editing, M.V., G.G. and G.L.; visualization, M.V. and D.L.; supervision, M.V.; project administration, M.V.; funding acquisition, M.V. and G.G. All authors have read and agreed to the published version of the manuscript.

Funding: This work was supported by the Newton Fund and the British Council (E046-2017-01) and the Croatian Science Foundation (R3PEAT, UIP-2019-04-4046).

Data Availability Statement: The original contributions presented in the study are included in the article, further inquiries can be directed to the corresponding author/s.

Acknowledgments: The authors want to acknowledge the help and support of the University of Glasgow Water Engineering laboratory technician, Timothy Montgomery, as well as postgraduate research students Tsz Hin Aldus Ian, Wassim Hage Chehade, and Jokha Khalid Al Hinai.

Conflicts of Interest: Author Da Liu was employed by the company China Construction Science and Technology Group South China. The remaining authors declare that the research was conducted in the absence of any commercial or financial relationships that could be construed as a potential conflict of interest.

References

1. Zulkernain, N.H.; Gani, P.; Chuck Chuan, N.; Uvarajan, T. Utilisation of plastic waste as aggregate in construction materials: A review. *Constr. Build. Mater.* **2021**, *296*, 123669. [CrossRef]
2. Nelms, S.E.; Galloway, T.S.; Godley, B.J.; Jarvis, D.S.; Lindeque, P.K. Investigating microplastic trophic transfer in marine top predators. *Environ. Pollut.* **2018**, *238*, 999–1007. [CrossRef] [PubMed]
3. Cole, M.; Lindeque, P.K.; Halsband, C.; Galloway, T.S. Microplastics as contaminants in the marine environment: A review. *Mar. Pollut. Bull.* **2011**, *62*, 2588–2597. [CrossRef] [PubMed]
4. Carney Almroth, B.M.; Åström, L.; Roslund, S.; Petersson, H.; Johansson, M.; Persson, N.-K. Quantifying shedding of synthetic fibers from textiles; a source of microplastics released into the environment. *Environ. Sci. Pollut. Res.* **2018**, *25*, 1191–1199. [CrossRef] [PubMed]
5. Blair, R.M.; Waldron, S.; Phoenix, V.; Gauchotte-Lindsay, C. Micro- and Nanoplastic Pollution of Freshwater and Wastewater Treatment Systems. *Springer Sci. Rev.* **2017**, *5*, 19–30. [CrossRef]
6. Schmaltz, E.; Melvin, E.C.; Diana, Z.; Gunady, E.F.; Rittschof, D.; Somarelli, J.A.; Virdin, J.; Dunphy-Daly, M.M. Plastic pollution solutions: Emerging technologies to prevent and collect marine plastic pollution. *Environ. Int.* **2020**, *144*, 106067. [CrossRef]
7. Zhang, Y.; Wu, P.; Xu, R.; Wang, X.; Lei, L.; Schartup, A.T.; Peng, Y.; Pang, Q.; Wang, X.; Mai, L.; et al. Plastic waste discharge to the global ocean constrained by seawater observations. *Nat. Commun.* **2023**, *14*, 1372. [CrossRef] [PubMed]
8. Talbot, R.; Chang, H. Microplastics in freshwater: A global review of factors affecting spatial and temporal variations. *Environ. Pollut.* **2022**, *292*, 118393. [CrossRef]
9. Suaria, G.; Avio, C.G.; Mineo, A.; Lattin, G.L.; Magaldi, M.G.; Belmonte, G.; Moore, C.J.; Regoli, F.; Aliani, S. The Mediterranean Plastic Soup: Synthetic polymers in Mediterranean surface waters. *Sci. Rep.* **2016**, *6*, 37551. [CrossRef]

10. Lebreton, L.; van der Zwet, J.; Damsteeg, J.-W.; Slat, B.; Andrady, A.; Reisser, J. River plastic emissions to the world's oceans. *Nat. Commun.* **2017**, *8*, 15611. [CrossRef]
11. Range, D.; Scherer, C.; Stock, F.; Ternes, T.A.; Hoffmann, T.O. Hydro-geomorphic perspectives on microplastic distribution in freshwater river systems: A critical review. *Water Res.* **2023**, *245*, 120567. [CrossRef] [PubMed]
12. Gündoğdu, S.; Çevik, C.; Ayat, B.; Aydoğan, B.; Karaca, S. How microplastics quantities increase with flood events? An example from Mersin Bay NE Levantine coast of Turkey. *Environ. Pollut.* **2018**, *239*, 342–350. [CrossRef] [PubMed]
13. Meijer, L.J.J.; van Emmerik, T.; van der Ent, R.; Schmidt, C.; Lebreton, L. More than 1000 rivers account for 80% of global riverine plastic emissions into the ocean. *Sci. Adv.* **2021**, *7*, eaaz5803. [CrossRef] [PubMed]
14. Schmidt, C.; Krauth,, T.; Wagner, S. Export of Plastic Debris by Rivers into the Sea. *Environ. Sci. Technol.* **2017**, *51*, 12246–12253. [CrossRef] [PubMed]
15. Valero, D.; Belay, B.S.; Moreno-Rodenas, A.; Kramer, M.; Franca, M.J. The key role of surface tension in the transport and quantification of plastic pollution in rivers. *Water Res.* **2022**, *226*, 119078. [CrossRef] [PubMed]
16. Lofty, J.; Valero, D.; Wilson, C.A.M.E.; Franca, M.J.; Ouro, P. Microplastic and natural sediment in bed load saltation: Material does not dictate the fate. *Water Res.* **2023**, *243*, 120329. [CrossRef] [PubMed]
17. van Emmerik, T.; Strady, E.; Kieu-Le, T.-C.; Nguyen, L.; Gratiot, N. Seasonality of riverine macroplastic transport. *Sci. Rep.* **2019**, *9*, 13549. [CrossRef] [PubMed]
18. Horton, A.A.; Walton, A.; Spurgeon, D.J.; Lahive, E.; Svendsen, C. Microplastics in freshwater and terrestrial environments: Evaluating the current understanding to identify the knowledge gaps and future research priorities. *Sci. Total Environ.* **2017**, *586*, 127–141. [CrossRef]
19. Sutherland, B.R.; DiBenedetto, M.; Kaminski, A.; van den Bremer, T. Fluid dynamics challenges in predicting plastic pollution transport in the ocean: A perspective. *Phys. Rev. Fluids* **2023**, *8*, 070701. [CrossRef]
20. Morritt, D.; Stefanoudis, P.V.; Pearce, D.; Crimmen, O.A.; Clark, P.F. Plastic in the Thames: A river runs through it. *Mar. Pollut. Bull.* **2014**, *78*, 196–200. [CrossRef] [PubMed]
21. Nelms, S.E.; Coombes, C.; Foster, L.C.; Galloway, T.S.; Godley, B.J.; Lindeque, P.K.; Witt, M.J. Marine anthropogenic litter on British beaches: A 10-year nationwide assessment using citizen science data. *Sci. Total Environ.* **2017**, *579*, 1399–1409. [CrossRef]
22. Horton, A.A.; Svendsen, C.; Williams, R.J.; Spurgeon, D.J.; Lahive, E. Large microplastic particles in sediments of tributaries of the River Thames, UK—Abundance, sources and methods for effective quantification. *Mar. Pollut. Bull.* **2017**, *114*, 218–226. [CrossRef]
23. Gasperi, J.; Dris, R.; Bonin, T.; Rocher, V.; Tassin, B. Assessment of floating plastic debris in surface water along the Seine River. *Environ. Pollut.* **2014**, *195*, 163–166. [CrossRef] [PubMed]
24. van Emmerik, T.; Roebroek, C.; De Winter, W.; Vriend, P.; Boonstra, M.; Hougee, M. Riverbank macrolitter in the Dutch Rhine–Meuse delta. *Environ. Res. Lett.* **2020**, *15*, 104087. [CrossRef]
25. van Emmerik, T.; Vriend, P.; Roebroek, J. *An Evaluation of the River-OSPAR Method for Quantifying Macrolitter on Dutch Riverbanks*; Wageningen University: Wageningen, The Netherlands, 2008; p. 86. [CrossRef]
26. Graca, B.; Bełdowska, M.; Wrzesień, P.; Zgrundo, A. Styrofoam debris as a potential carrier of mercury within ecosystems. *Environ. Sci. Pollut. Res.* **2014**, *21*, 2263–2271. [CrossRef] [PubMed]
27. Heo, N.W.; Hong, S.H.; Han, G.M.; Hong, S.; Lee, J.; Song, Y.K.; Jang, M.; Shim, W.J. Distribution of small plastic debris in cross-section and high strandline on Heungnam beach, South Korea. *Ocean Sci. J.* **2013**, *48*, 225–233. [CrossRef]
28. Maes, T.; Van der Meulen, M.D.; Devriese, L.I.; Leslie, H.A.; Huvet, A.; Frère, L.; Robbens, J.; Vethaak, A.D. Microplastics Baseline Surveys at the Water Surface and in Sediments of the North-East Atlantic. *Front. Mar. Sci.* **2017**, *4*, 135. [CrossRef]
29. Liu, D.; Valyrakis, M.; Williams, R.J. Flow Hydrodynamics across Open Channel Flows with Riparian Zones: Implications for Riverbank Stability. *Water* **2017**, *9*, 720. [CrossRef]
30. Eriksen, M.; Lebreton, L.; Carson, H.S.; Thiel, M.; Moore, C.J.; Borerro, J.C.; Galgani, F.; Ryan, P.G.; Reisser, J. Plastic Pollution in the World's Oceans: More than 5 Trillion Plastic Pieces Weighing over 250,000 Tons Afloat at Sea. *PLoS ONE* **2014**, *9*, e111913. [CrossRef]
31. Yagci, O.; Celik, F.; Kitsikoudis, V.; Ozgur Kirca, V.S.; Hodoglu, C.; Valyrakis, M.; Duran, Z.; Kaya, S. Scour patterns around individual vegetation elements. *Adv. Water Resour.* **2016**, *97*, 251–265. [CrossRef]
32. Valyrakis, M.; Liu, D.; Turker, U.; Yagci, O. The role of increasing riverbank vegetation density on flow dynamics across an asymmetrical channel. *Environ. Fluid Mech.* **2021**, *21*, 643–666. [CrossRef]
33. Xu, Y.; Valyrakis, M.; Gilja, G.; Michalis, P.; Yagci, O.; Przyborowski, L. Assessing riverbed surface destabilization risk downstream isolated vegetation elements. *Water* **2022**, *14*, 2880. [CrossRef]
34. Valyrakis, M.; Al Hinai, J.K.; Liu, D. Transport of floating plastics along a channel with a vegetated riverbank. In Proceedings of the 12th International Symposium on Ecohydraulics ISE 2018, Tokyo, Japan, 19–24 August 2018.
35. Nepf, H.M. Drag, turbulence, and diffusion in flow through emergent vegetation. *Water Resour. Res.* **1999**, *35*, 479–489. [CrossRef]
36. Abeynayaka, A.; Kojima, F.; Miwa, Y.; Ito, N.; Nihei, Y.; Fukunaga, Y.; Yashima, Y.; Itsubo, N. Rapid Sampling of Suspended and Floating Microplastics in Challenging Riverine and Coastal Water Environments in Japan. *Water* **2020**, *12*, 1903. [CrossRef]
37. Pasquier, G.; Doyen, P.; Carlesi, N.; Amara, R. An innovative approach for microplastic sampling in all surface water bodies using an aquatic drone. *Heliyon* **2022**, *8*, e11662. [CrossRef] [PubMed]

38. Campanale, C.; Savino, I.; Pojar, I.; Massarelli, C.; Uricchio, V.F. A Practical Overview of Methodologies for Sampling and Analysis of Microplastics in Riverine Environments. *Sustainability* **2020**, *12*, 6755. [CrossRef]
39. van Emmerik, T.; Schwarz, A. Plastic debris in rivers. *WIREs Water* **2020**, *7*, e1398. [CrossRef]
40. de Lange, S.I.; Mellink, Y.; Vriend, P.; Tasseron, P.F.; Begemann, F.; Hauk, R.; Aalderink, H.; Hamers, E.; Jansson, P.; Joosse, N.; et al. Sample size requirements for riverbank macrolitter characterization. *Front. Water* **2023**, *4*, 1085285. [CrossRef]
41. Chowdhury, Z.; Yokota, K.; Nguyen, M.N.; Inoue, T. Bank Accumulation of Macroplastic Samples in a Farmland River in Japan. *Water Air Soil Pollut.* **2023**, *12*, 234. [CrossRef]
42. Cowger, W.; Markley, L.A.T.; Moore, S.; Gray, A.B.; Upadhyay, K.; Koelmans, A.A. How many microplastics do you need to (sub)sample? *Ecotoxicol. Environ. Saf.* **2024**, *275*, 116243. [CrossRef]
43. Valyrakis, M.; Diplas, P.; Dancey, C.L. Entrainment of coarse grains in turbulent flows: An extreme value theory approach. *Water Resour. Res.* **2011**, *47*, W09512. [CrossRef]
44. Valyrakis, M.; Diplas, P.; Dancey, C.L. Entrainment of coarse particles in turbulent flows: An energy approach. *J. Geophys. Res. Earth Surf.* **2013**, *118*, 42–53. [CrossRef]
45. Courtene-Jones, W.; Quinn, B.; Gary, S.F.; Mogg, A.O.M.; Narayanaswamy, B.E. Microplastic pollution identified in deep-sea water and ingested by benthic invertebrates in the Rockall Trough, North Atlantic Ocean. *Environ. Pollut.* **2017**, *231*, 271–280. [CrossRef]
46. Jodeau, M.; Hauet, A.; Paquier, A.; Le Coz, J.; Dramais, G. Application and evaluation of LS-PIV technique for the monitoring of river surface velocities in high flow conditions. *Flow Meas. Instrum.* **2008**, *19*, 117–127. [CrossRef]
47. Gunawan, B.; Sun, X.; Sterling, M.; Shiono, K.; Tsubaki, R.; Rameshwaran, P.; Knight, D.W.; Chandler, J.H.; Tang, X.; Fujita, I. The application of LS-PIV to a small irregular river for inbank and overbank flows. *Flow Meas. Instrum.* **2012**, *24*, 1–12. [CrossRef]
48. Lewis, Q.W.; Rhoads, B.L. Resolving two-dimensional flow structure in rivers using large-scale particle image velocimetry: An example from a stream confluence. *Water Resour. Res.* **2015**, *51*, 7977–7994. [CrossRef]
49. Laxague, N.J.M.; Özgökmen, T.M.; Haus, B.K.; Novelli, G.; Shcherbina, A.; Sutherland, P.; Guigand, C.M.; Lund, B.; Mehta, S.; Alday, M.; et al. Observations of Near-Surface Current Shear Help Describe Oceanic Oil and Plastic Transport. *Geophys. Res. Lett.* **2018**, *45*, 245–249. [CrossRef]
50. Fuchs, Y.; Scherbaum, S.; Huber, R.; Rüther, N.; Hartlieb, A. Removing Plastic Waste from Rivers: A Prototype-Scale Experimental Study on a Novel River-Cleaning Concept. *Water* **2024**, *16*, 248. [CrossRef]

water

MDPI

Article

Spatial Variations in Microplastics in the Largest Shallow Lake of Central Europe and Its Protecting Wetland Area

Bence Prikler [1,2,†], Réka Svigruha [3,†], Judit Háhn [1,*], Péter Harkai [1], István Fodor [3], Edit Kaszab [1], Balázs Kriszt [1], Gergő Tóth [1], István Szabó [1], Zsolt Csenki [1], Gábor Bordós [2] Adrienn Micsinai [2], Brigitta Nyírő-Fekete [2], Zoltán Palotai [2], Zsófia Lovász [4], Zsolt Pirger [3,‡] and Sándor Szoboszlay [1,‡]

1 Institute of Aquaculture and Environmental Safety, Hungarian University of Agriculture and Life Sciences, 2100 Gödöllő, Hungary; prikler.bence@laboratorium.hu (B.P.); harkai.peter@uni-mate.hu (P.H.); kaszab.edit@uni-mate.hu (E.K.); kriszt.balazs@uni-mate.hu (B.K.); toth.gergo@uni-mate.hu (G.T.); szabo.istvan@uni-mate.hu (I.S.); csenki-bakos.zsolt.imre@uni-mate.hu (Z.C.); szoboszlay.sandor@uni-mate.hu (S.S.)

2 Eurofins Analytical Services Hungary Ltd., 1045 Budapest, Hungary; bordos.gabor@laboratorium.hu (G.B.); micsinai.adrienn@laboratorium.hu (A.M.); fekete.brigitta@laboratorium.hu (B.N.-F.); palotai.zoltan@laboratorium.hu (Z.P.)

3 Ecophysiological and Environmental Toxicological Research Group, HUN-REN Balaton Limnological Research Institute, 8237 Tihany, Hungary; svigruha.reka@blki.hu (R.S.); fodor.istvan@blki.hu (I.F.); pirger.zsolt@blki.hu (Z.P.)

4 Western Transdanubia Water Directorate, 9700 Szombathely, Hungary; lovasz.zsofia@nyuduvizig.hu

* Correspondence: hahn.judit@uni-mate.hu

† These authors contributed equally to this work.

‡ These authors contributed equally to this work.

Abstract: The concentration of microplastics (MPs) in two important Hungarian freshwater habitats was determined in the size range of 50 μm–1 mm. Lake Balaton (LB) is the largest shallow lake in Central Europe, with a significant role in recreation and tourism. Its main inflow, the Zala River, enters the lake through an artificially constructed wetland, the Kis-Balaton Water Protection System and its catchment area (KB), which helps preserve the water quality of the lake. From these two areas, 15 samples were taken with an in situ filtration sampling method. After preparation, the samples were analyzed automatically by FT-IR microscopy. All samples, from both areas, contained MPs; the dominant microplastic (MP) shape was the fragment, while the most frequently polymer types were polyethylene, polypropylene and alkyd. Small MPs were dominant in both areas; around 90% of the MPs were smaller than 500 μm. On average, LB contained more MPs (21.0 ± 12.5 MPs/m^3) compared to the KB, which presented an average concentration of MPs of 7.8 ± 5.9 MPs/m^3. In the examined areas, two potential MP sources were determined, i.e., treated wastewater and road traffic. The importance of tourism should also be further investigated.

Keywords: FTIR; MPs; types of polymers; fiber; fragment; filtration; freshwater; Lake Balaton; Kis-Balaton Water Protection System

1. Introduction

In recent years, the issue of microplastics (MPs), synthetic polymers within the size range of 1 μm–5 mm, has become an increasing environmental concern [1]. According to the ISO standard, two sub-categories of MPs can be differentiated: small MPs (1–1000 μm) and large MPs (1–5 mm) [2]. The establishment of these sub-categories was necessary because, in this wide range of sizes, it is not possible to use only one sample preparation and detection method, because this can lead to errors during quantification [3].

Studies focusing on the identification and quantification of MPs started in the marine environment, but in recent years, freshwater ecosystems have also come to the fore [4]. The presence of MPs has since been confirmed in various water bodies, such as rivers and lakes, across Asia, Europe and North America. Nonetheless, there are still less data available about the abundance of MPs in freshwater ecosystems than in the marine environment [1,5]. In the case of the freshwater environment, some direct microplastic (MP) sources can be identified, and the majority of these sources can be linked to anthropogenic factors. Consequently, MPs are present in higher concentrations in regions that are highly populated, have more wastewater treatment plants (WWTPs), have higher levels of urbanization or have intensive tourist activity [6]. MPs have become omnipresent pollutants that are observed not only near their sources but also at large distances. According to their material types, the most frequently detected MPs in freshwater are polyethylene (PE), polypropylene (PP), polystyrene (PS) and polyethylene terephthalate (PET), which correlates with the production and utilization data of plastics [5–7]. MPs, especially in the small size range, pose an environmental risk, as they can be consumed by aquatic organisms [8] and can enter the food chain. Furthermore, MPs can adsorb various hazardous substances (e.g., organic pollutants and heavy metals) and can act as abiotic surfaces for biofilm formation, which facilitates the spread of pathogenic and antibiotic-resistant microorganisms in the environment [9].

The largest shallow lake of Central Europe, Lake Balaton (LB), is an important summer holiday destination visited by millions of tourists every year [10,11]. Therefore, the preservation of the good water quality and biodiversity of the lake is of strategic importance from both economic and ecological aspects. Based on the intricate findings of our current survey, our research team previously published an article focusing on the concentration of MPs in LB. However, that publication exclusively presented the concentration results for MPs within the range of 50–100 µm, as this size range is crucial for conducting toxicology tests on *Daphnia* spp. [8]. A larger-scale study investigating MPs in a wider size range and involving the catchment area is still needed to obtain a more complete picture of the possible threats caused by MPs to the ecosystem of LB. Keeping this in mind, the present study aimed to gain comprehensive knowledge of the presence of and spatial variations in MPs in the size range of 50–1000 µm in two connected freshwater ecosystems: (1) LB; (2) the Kis-Balaton Water Protection System and its catchment area (KB), an artificially constructed wetland that helps preserve the water quality of LB.

2. Material and Methods

2.1. Study Area and Sampling

With a surface area of more than 590 km^2, a mean depth of 3.25 m, and a volume of 1.8 km^3, LB is the largest freshwater shallow lake in Central Europe [12,13]. LB has 20 permanent water inflows (Figure 1), with its main inflow, the Zala river, having a considerable impact on the lake's water quality. The Zala river reaches the lake through the KB (Figure 1) [14], which is an extensive, artificially constructed wetland region that acts as a natural filter of organic pollutants to improve the water quality of LB [12,15]. The KB was built in two phases and plays a crucial role in the water management system of the region. During phase one, the Hídvégi Pond was created, which increased the water retention time by 30 days. During phase two, the Fenéki Pond was constructed, which additionally increased the water retention time by another 90 days. From the Zala river, the first pond retains 60% of the suspended solids, while the second pond retrains 75% of the remaining solids before they enter the lake [16]. Even though KB is a built environment, it is a valuable, protected wetland ecosystem known for its biodiversity and listed as a Wetland of International Importance under the Ramsar Convention (https://rsis.ramsar.org/ris/185, accessed on 27 March 2024).

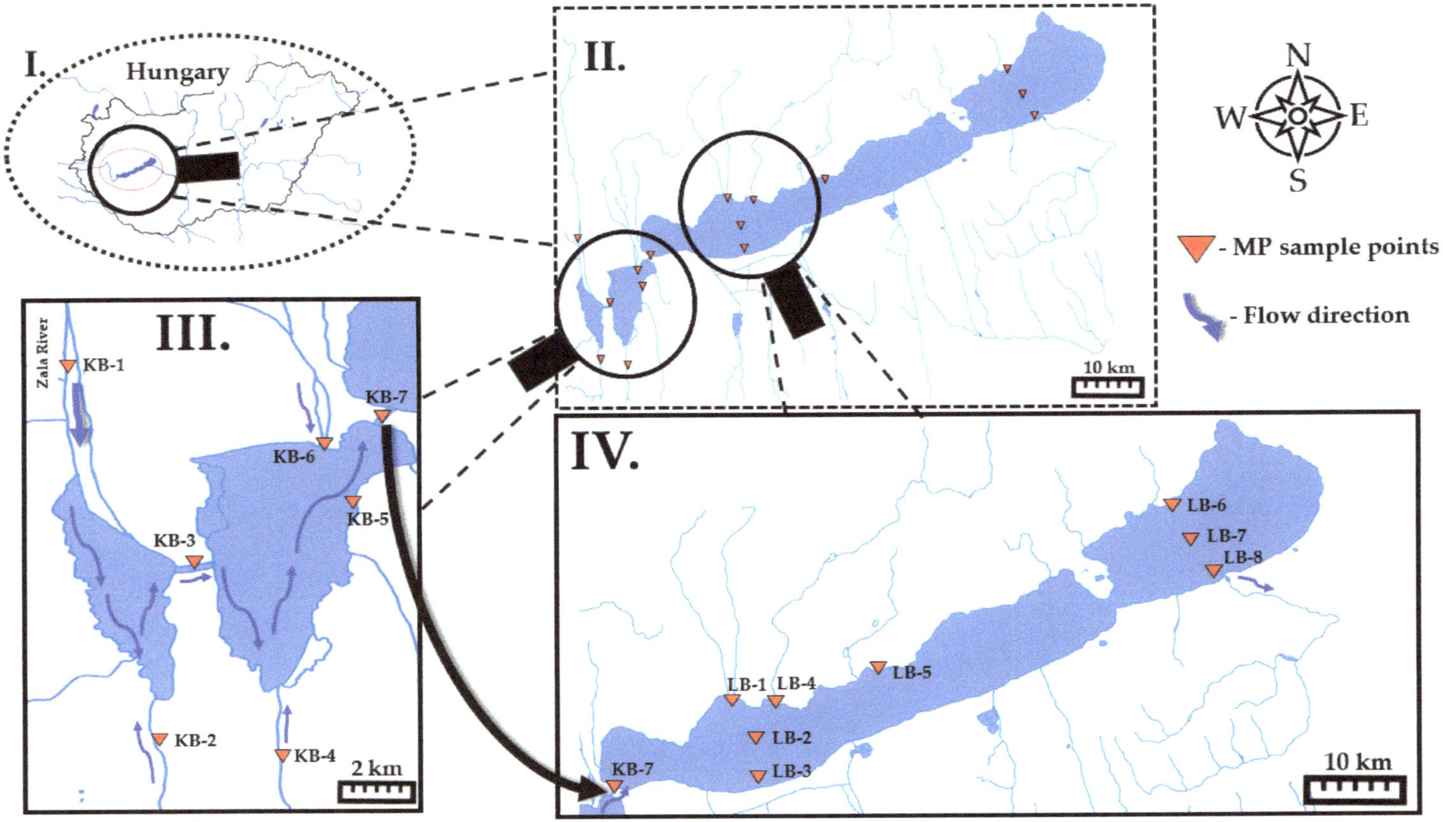

Figure 1. Subfigure (**I**) shows the geographical location of the studied area in Hungary, and subfigure (**II**) indicates the magnified area with the sampling points. The location and designation of the MP sample points in the Kis-Balaton Water Protection System and its catchment area (KB) are presented in subfigure (**III**), and the Lake Balaton's (LB) sample points are figured in Subfigure (**IV**) with the main water flow directions.

Nowadays, more than 40 WWTPs with varying capacities (2–50,000 m^3/day) are operating in the catchment area of LB (Figure 1) [10]; the largest one is located in the city of Zalaegerszeg. To reduce the direct discharge of treated sewage into the lake, a sewage transfer duct system was implemented in the southern and eastern coastal regions. This system effectively gathers and diverts a significant portion of treated communal sewage from the catchment area. However, the WWTPs located away from the lake release their effluent into the tributaries of LB.

To reveal MP concentrations in the KB and LB, 15 sampling points were marked out: seven sampling locations were chosen in the KB, while eight sampling locations were designated in LB, with the aid of the Western Transdanubia Water Directorate. The designated monitoring points represent locations where substantial anthropogenic impact presumably affects the water quality (shown in Figure 1). In KB, sampling was carried out from 15 to 20 June 2022, while LB was sampled from 27 to 29 July 2022. Sampling was performed with a previously described in situ filtration method [17]. The applied sampling device starts the filtration process with a pre-filter (pore size 1 mm), which prevents larger particles from entering the device. Filtration was driven by a jet pump through three stainless-steel in-line filters (pore sizes: 300, 100, 50 μm) (Infiltec GmbH, Speyer, Germany). Thus, the tested particles fell into the range from 50 μm to 1 mm. This technique allowed us to acquire a representative amount of water (one to two cubic meters per sample).

2.2. Microplastic Extraction and Analysis

The samples were prepared and examined as described previously [8]. Briefly, to remove the organic matter from the samples, two oxidation steps were applied: a hydrogen peroxide-based pre-oxidation step and a Fenton reaction to eliminate any residual organic materials. After oxidation, a density separation step was carried out. The samples were put into small-volume glass separators (SVGSs) filled with a zinc chloride brine solution (1.7 g/cm^3) to separate the microplastics. The sample preparation method with SVGSs was previously described and validated by our research group [18]. Anodisc filters (Ø = 25 mm; pore = 0.2 μm) were used for the final filtration, and the concentrated samples on the filter were directly placed under a Fourier Transform Infrared imaging microscope (FTIR) (NicoletTM In10 MX; Thermo Fisher Scientific, Waltham, MA, USA).

The complete surface of the filter underwent scanning using the FTIR microscope, outfitted with a linear array detector capable of achieving a spatial resolution of 25 μm in transmission mode, along with a spectral resolution of 8 cm^{-1}, and employing 4 scans. The collected data were then analyzed automatically with the siMPle software (https://simple-plastics.eu/, accessed on 27 March 2024), developed by Primpke et al., 2020, and applied by other research groups extensively [19,20], as demonstrated in Supplementary Figure S1. In order to accept the collected spectra as representing MPs, in comparison to the database, a hit quality index of 70% was determined. With this method, the smallest identified particle size was 50 μm, as the software is programmed to require at least two positive neighboring pixels to produce a credible result during MP identification [8]. The estimated mass concentration of MPs was also determined using the same software based on the predicted volume of each particle, assuming an ellipsoid shape and considering the specific density of its polymer type [21]. As the toxicity of MPs also depends on their shape [22], the distribution of fragments or fibers was also determined in the samples. MPs with a calculated length-to-width ratio greater than 3:1 (based on the length and width dimensions of the identified MPs measured by the software) were considered fibers [23,24].

2.3. Contamination Prevention and Quality Control

During sample preparation, precautions were taken to prevent sample contamination. Beakers and any glassware were rinsed before usage with ultra-pure water (UPW), which was produced as deionized water and was filtered through a 5 μm metal filter and a 0.7 μm glass fiber filter. During sample preparation, beakers and any glassware were covered with aluminum foil, and the laboratory staff wore cotton lab coats. Most sample preparation steps were conducted under a laminar flow hood, but due to work safety reasons, the oxidation steps were carried out in a fume hood.

To increase the reliability of MP quantification, it was necessary to run process blank samples (i.e., conduct background measurement) alongside the environmental samples to calculate the limit of detection (LOD) and the limit of quantification (LOQ) values. To accomplish this task, four blank samples were generated using UPW. The blank samples were prepared with the previously described procedure (for both preparation and analysis) [8]. From the results of the blanks, the LOD and LOQ values were calculated as follows: the LOD was determined as 3.3 times the standard deviation plus the mean values of the blanks; similarly, the LOQ was established as 10 times the standard deviation of the blanks plus their mean values. Although these values were different between the polymer types, they were calculated for the total quantity of MPs in accordance with the guidance provided by previous research articles [19,25]. Of note, following the suggestion of a previous research article [26], we did not adjust our results using the mean of the blanks.

For every sample, the concentrations of MPs were computed per cubic meter using the recorded sample volume during the sampling process. When determining the mean concentrations of MPs for KB and LB, outlier data points were excluded from the calculations.

3. Results

3.1. LOD and LOQ Values

To guarantee the reliability of our results about the MP contamination of KB and LB surface water samples, the background contamination level was assessed, and the LOD and LOQ values were calculated in all cases. In four of the process blanks, altogether eight MP particles were found that were distributed among four polymer groups as follows: three of them were PE; other three were PP; one of them was PS; and one was cellulose acetate (CA). The defined LOD value was 7.22, while the LOQ was 17.81.

3.2. MP Abundance in KB and Lake Balaton

Figure 2 shows the determined abundance of MPs in KB and LB. In LB, the MP concentrations fell into the range from 1.50 to 106.84 MPs/m^3, and the average abundance of MPs was 21.0 $\pm$ 12.5 MPs/m^3. In KB, the concentrations of MPs were between 1.51 and 196.85 MPs/m^3, averagely 7.8 $\pm$ 5.9 MPs/m^3. Of the 15 examined surface water samples, two were not above the LOD value. Moreover, although another two samples reached the LOD value, they did not reach the LOQ value (shown in Supplementary Table S1). The estimated mass concentration of MPs in LB was in the range from 4.89 to 99.46 µg/m^3, with an average value of 44.4 $\pm$ 27.2 µg/m^3. In contrast, the calculated mass concentrations in KB were between 0.10 and 5140.98 µg/m^3 and, averagely, 128.2 $\pm$ 144.4 µg/m^3.

3.3. Polymer Types of the Identified MPs

In terms of average particle numbers, the MPs detected in LB fell into 12 polymer groups, while the MPs in KB were classified in only 5 polymer groups. In both areas, PE was the most abundant (in the case of LB, 33.29% of MPs were PE; in the case of KB, 55.72% of MPs were PE), followed by PP (LB: 31.22%; KB: 30.66%) and alkyd (LB: 15.46%; KB: 10.36%). In KB, other polymer types (polyester (POS) and PS) represented only 3.26% of the total MPs (Figure 3). In contrast, in LB, other polymer types such as POS, PS, CA, epoxy, acrylic, polyvinyl chloride (PVC), polyurethane (PU), polyamide (PA) and acrylonitrile butadiene styrene (ABS) represented 20.00% (Figure 3).

3.4. MP Distribution according to Their Shapes and Sizes

For size determination, major size dimensions of the particles were used. In LB, the most frequent size category was the smallest one, with particles between 50 and 100 µm; 26.27% of the total MPs belonged to this category. In KB, this smallest MP category was just the third most abundant (11.70% of all MPs), while the most common category was that containing particles between 100 and 150 µm, representing 17.80% of all MPs. In general, our results suggest that the smaller-size category represented the greater number of the identified MPs. In both sampling areas, around 90% of all MPs were under 500 μm (shown in Figure 4).

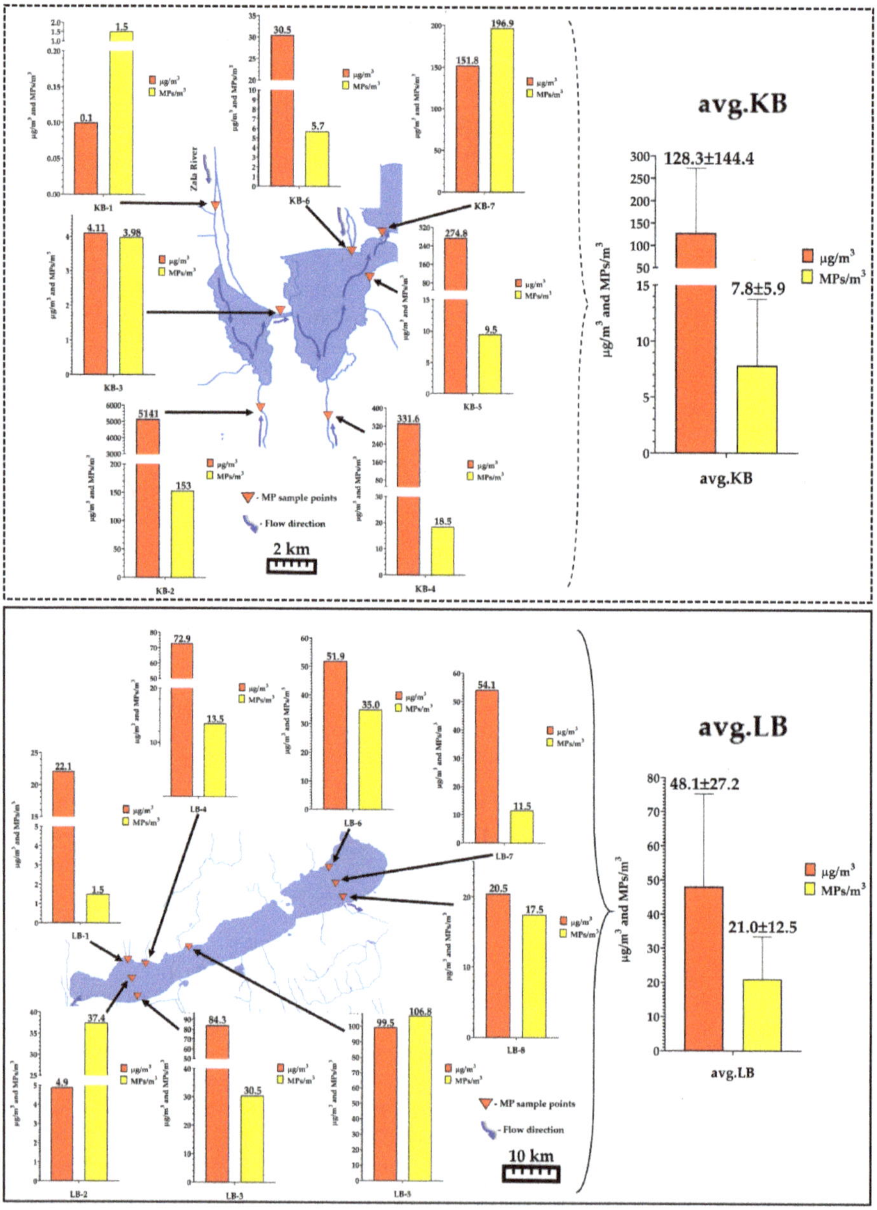

Figure 2. The determined abundance of MPs and their estimated mass in the areas of Kis-Balaton and its catchment area and Lake Balaton, Hungary. Bar graphs indicate average MP concentrations in MPs/m^3 and estimated mass concentrations in μg/m^3 in the two examined regions. Error bars indicate the standard deviation between the sampling points per region. (LB: Lake Balaton; KB: Kis-Balaton Water Protection System and its catchment area; bold straight arrows link the bar charts to their respective sampling points).

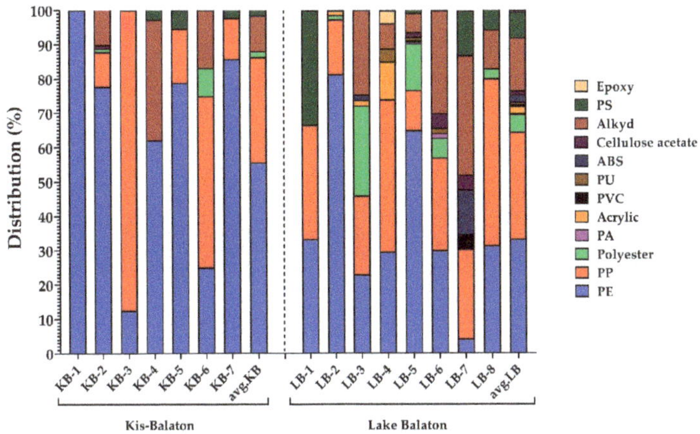

Figure 3. The abundance of different polymer types in Kis-Balaton and its catchment area and Lake Balaton water samples. Average microplastic distribution by polymer type in the two examined regions in percentage. (PE: polyethylene; PP: polypropylene; PS: polystyrene; ABS: acrylonitrile butadiene styrene; PA: polyamide; PU: polyurethane; PVC: polyvinyl chloride; LB: Lake Balaton; KB: Kis-Balaton Water Protection System and its catchment area).

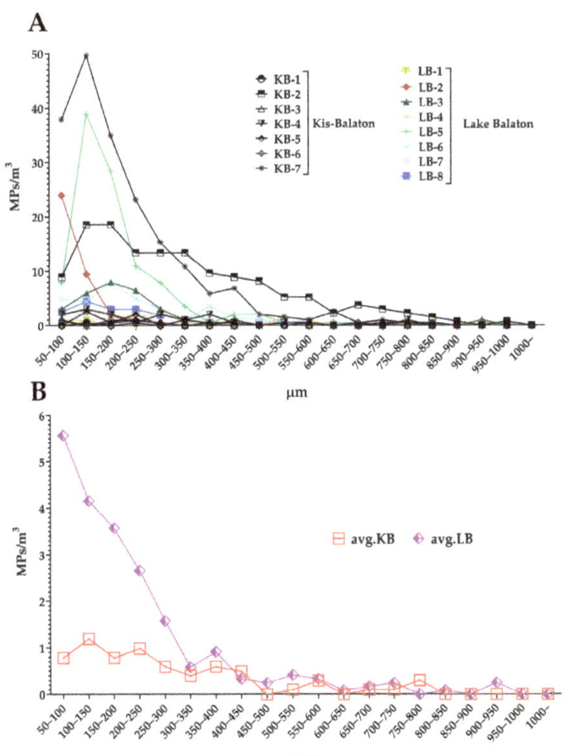

Figure 4. MP distribution in the area of Kis-Balaton and its catchment area (KB) and Lake Balaton (LB). (**A**) size distribution of MPs in the 15 sampling points, (**B**) average size distribution of MPs in the two examined regions.

Regarding the shape of the identified MPs (Figure 5), the fragment shape proved to be dominant in both sampling sites: in LB, 86.60% of the MPs were identified as fragments, while their abundance was 62.46% in KB.

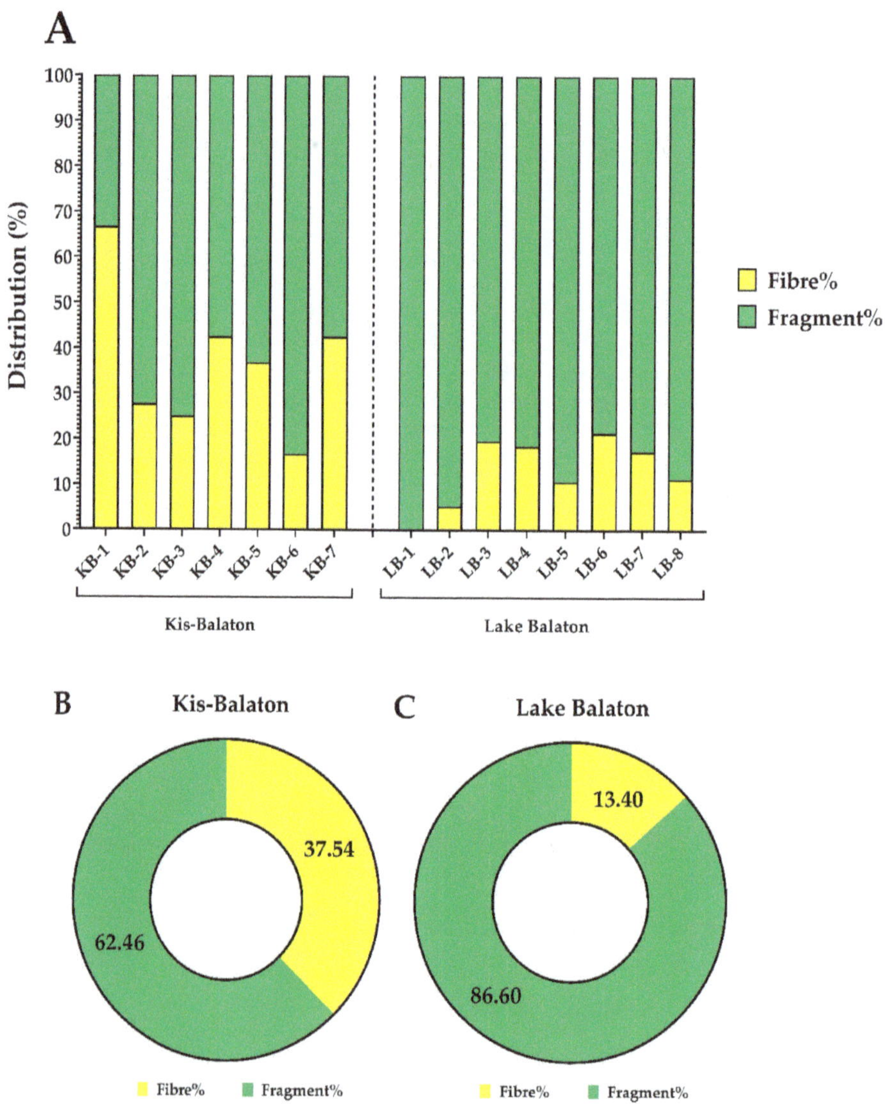

Figure 5. (**A**) Percentages of MP fibers and fragments in the area of Kis-Balaton and its catchment area (KB) and Lake Balaton (LB) in individual sampling points, (**B**,**C**) proportions of fibers and fragments within the entire set of examined samples from KB and LB.

4. Discussion

At the time of our research, only one study had been conducted in LB dealing with the concentration and composition of MPs and investigating their potential top–down effects on the water flea (*Daphnia magna*) [8], and no study was available about the KB region. In that study, MPs were only investigated within a narrow size range (50–100 μm), as specimens of *D. magna* are capable of ingesting MPs in this size range due to potential food confusion. To extend our current understanding, the present study offers the first comprehensive determination of MP concentrations for both regions, which gives a basis for the evaluation of MP spread and distribution in the area. In this study, the concentration of MPs in LB ranged from 1.50 to 106.84 MPs/m^3, and the average MP abundance was 21.0 ± 12.5 MPs/m^3. In KB, the MP concentrations were between 1.51 and 196.85 MPs/m^3, averagely, 7.8 ± 5.9 MPs/m^3. According to these results, on average, KB contained a smaller concentration of MPs than LB; however, two KB samples had the highest concentration among all examined samples. Our results show marginally elevated MP concentrations in contrast to the previously determined values for 67 European lakes, which indicated an average of 0–7.3 particles/m^3 (median = 0.28) in the size range of 310–5000 μm [27]. This disparity can potentially be attributed to divergent methodological approaches.

It is known that the smaller the size range one can investigate, the more particles can be found [28]. As we could operate within smaller size ranges than those considered in previous studies, it was anticipated that more particles would be identified. Furthermore, the automation of MP identification contributed to a reduction in human bias, and the imaging μFTIR technique is more likely to perform better at analyzing smaller particles. The sampling technique, sample preparation protocol and identification method are known to influence the final reported MP concentrations; hence, it is worth comparing new research findings with those derived by analogous techniques [29]. Compared to another study that was conducted in Finland [30], an in situ pump filtration method similar to our sampling device was applied. Our results are close to their findings, determining the number of MPs as 12 ± 17 MPs/m^3 in a size range of 100–300 μm. The outcomes of our current study exhibit a significant resemblance to the results of our prior research conducted in Hungarian aquatic environments: the concentration of MPs in the previous study was 13.8 ± 9.3 MPs/m^3. Nevertheless, it is important to note that the MP size range examined in the prior study was between 100 μm and 2 mm, and the MP identification method was different from the one we used in our present study [31]. MP pollution is less studied in wetlands compared to other aquatic ecosystems. Significant differences can be observed among the existing studies regarding MP concentrations, ranging from 1.44 to 101,60 MPs/m^3. As indicated before, the comparison of MP abundances from different studies is difficult due to the variation in sampling methods, sample preparation and identification practices [32]. As expected, our results fall into the wide range indicated by previous studies.

The estimated mass concentration was determined in LB as 44.4 ± 27.2 μg/m^3 and in KB as 128.2 ± 144.4 μg/m^3. To the best of our knowledge, there is only one study dealing with the estimated MP mass in freshwater lakes in Poland. In that case, the highest value was determined as 4.7 μg/m^2, which indicated about 25 g of MPs in the studied lake to a depth of 0.2 m [33]. If we extrapolate our results, assuming that our pump sucked water from the upper 0.2 m layer of water and calculating the surface area (590 km^2) of the lake, LB contains 5.23 kg of MPs. This is a much higher value, which can also be explained by the different sampling and detection techniques. It is also obvious that a direct connection cannot be established between the mass and the particle number concentration of MPs: a small number of larger MPs can significantly contribute to the total mass, while having a limited impact on the particle number, and conversely, a multitude of small MPs will influence the particle number, while having a low impact on the total mass. In our study, the highest mass was measured at the KB-2 point (5140.98 μg/m^3), while the highest particle number was recorded at the KB-7 point (196.85 MPs/m^3). Also, in light of the current knowledge on risk assessment of MPs, it is recommended that the concentrations in toxicity tests be calculated as both mass and particle number [34]; this is why we reported both values.

In both areas, the most commonly found polymer types were PE and PP. Their abundance is not surprising, because these materials possess low-density characteristics (density below 1 g/cm^3), which results in their floating on the water surface. Also, these two polymer types cover 60% of the global plastic production [35]. Alkyd, which was also a commonly found polymer type, is a commonly used painting material, including in coatings for roads and buildings as well as boat painting [36,37]. Alkyd is also commonly found in WWTP effluent water [38]. Our sampling period coincided with the main peak of the tourist season, with increased recreational activity (e.g., swimming, fishing, sailing), which could have influenced our results in the case of the lake [11]. Furthermore, the abundant presence of other polymer types, among which PS (7.44%) and POS (5.16%) were the two dominant ones, supports the hypothesis that tourism influences the MP load in the lake. The presence of both polymer types can clearly be connected with the increased human activity during the summer period. PS is commonly used in microbeads in personal care products and could enter the ecosystem, for example, through WWTPs [39], while POS is a commonly used synthetic fiber (e.g., in swimming suits) [40]. Also, the polymer type distribution indicate the possibility that most of the MP load in the lake did not originate from the KB area. Not only the polymer type distribution supports this theory, but also the MP concentration data. Our results revealed that, though the lake receives water from the KB, the average MP concentration in the lake was around three times higher than the KB average value. It is therefore conceivable that some MPs present in the lake did not enter the lake in conjunction with the KB water. Thus, a portion of the MPs present in the lake can likely be associated with tourism, entering the lake either directly from tourists or indirectly through wastewater treatment plants due to the increased population around the lake [11,40,41].

This is the first study to report the MP size distribution in these areas. As previously mentioned, our study also supports the notion that if we examine a smaller size range, more particles can be found. In our investigation, the prevailing size categories of MPs were found to be below 150 μm, indicating that they are susceptible to ingestion by aquatic organisms, including various zooplankton species such as *Daphnia* spp. [8]. As *Daphnia* spp. are prey for other aquatic organisms, MPs can enter and accumulate alongside the food chain [9]. The determination of the MP shapes is important, as the shape can influence the particles' toxicity [22]. In both examined areas, fragment-shaped MPs were dominant: of the total MPs identified, they represented 86.60% in LB and 62.46% in KB. A higher prevalence of fragments among smaller MPs can be anticipated, as fragments can originate from both the degradation of larger particles and the disintegration of larger fibers [42].

According to the analysis of the MP concentrations at the sampling points, outlier values were determined in several cases: the number of MP particles in the KB-2, KB-7 and LB-5 samples was significantly higher compared to the average values. These sampling locations were marked out to further investigate the possible origin of MPs. At the LB-5 sampling point, the concentration of MPs was determined to be 106.84 MPs/m^3; this value was approximately five times higher than the average concentration of MPs in the lake. Notably, this site is situated close to a treated wastewater inlet. In line with the literature, our result also supports the hypothesis that treated wastewater can be considered the major source of MPs in the case of freshwater ecosystems [43]. In the case of the other sampling points from the KB territory (KB-2, KB-7), KB-7 showed 196.85 MPs/m^3, which is around 25 times higher than the average value for this region (7.8 $\pm$ 5.9 MPs/m^3). As an explanation, KB-7 was situated under a newly built highway overpass; hence, an increased concentration of MPs at this sampling point was most probably due to road traffic, as supported by the scientific literature [44,45]. KB-2 was also situated close to a heavily congested road; furthermore, near this site, a highway rainwater drainage channel is connected to the location of the sampled surface water.

5. Conclusions

The present study, for the first time, dealt with the concentration of MPs in two unique aquatic ecosystems (Lake Balaton and the Kis-Balaton Water Protection System and its catchment area). The concentrations of MPs were expressed not only in particle number but also in estimated mass (the average particle number was 21.0 $\pm$ 12.5 MPs/m^3 in LB and 7.8 $\pm$ 5.9 MPs/m^3 in KB, while the average mass concentration was 44.4 $\pm$ 27.2 μg/m^3

in LB and was 128.2 ± 144.4 μg/m^3 in KB). Given the current scarcity of data concerning the presence of MPs in European lakes and an even more limited dataset regarding the MP levels within wetlands, our results definitely mitigate a void in knowledge. Our findings revealed that the presence of MPs could be detected at all sampling points, and the most abundant polymer types were PE, PP and alkyd. The MPs were mostly fragments in shape. In the lower size ranges, more MPs could be detected, and most of the MPs (≈90%) were below 500 μm. In our study, we defined two MP sources, i.e., WWTPs and traffic. Furthermore, the lake apparently receives most of the MPs not from its biggest water supplier but from different sources. Tourism is likely to be an additional source of MPs in the case of the lake; however, further (seasonal) investigations are necessary to prove this idea. Our study can contribute to future research that aims to determine the origin and fate of MPs in freshwater ecosystems.

Supplementary Materials: The following supporting information can be downloaded at: https://www.mdpi.com/article/10.3390/w16071014/s1, Figure S1: (**A**) The digital picture from the LB-4 sample presented by Dino-Lite Edge AM4115TL; (**B**) The collected spectra intensity of LB-4 are represented in the false color intensity map generated by siMPle software (Version 1.1.β), after the data collection by the FTIR microscope (NicoletTM In10 MX; Thermo Fisher Scientific, USA); (**C**) The data evaluation for LB-4 sample was done automatically by the by siMPle software (Version 1.1.β), where all individual spectra were compared to library spectrum in the software, orange spectrum represents, the spectrum coming from the sample, while blue spectrum represents, spectrum coming from the library; (**D**) A false color microplastic map of LB-4 sample created by the software, where each color represents other kind of polymers. Table S1: (A) The ascertainable limits of detection (LOD) and quantification (LOQ) are determinable through the examination of blank samples, enabling the computation of these values concerning the overall amount of microplastics (MPs); (B) Total microplastic concentrations of samples and sampling volumes can be seen; also, colors indicate how samples relate to the LOD and LOQ values (PE: polyethylene; PP: polypropylene; PS: polystyrene; ABS: acrylonitrile butadiene styrene; PA: polyamide; PU: polyurethane; PVC: polyvinyl chloride; LB: Lake Balaton; KB: Kis-Balaton Water Protection System and its catchment area).

Author Contributions: Conceptualization, B.P. and S.S.; Methodology, B.P., G.B., Z.L. and S.S.; Formal analysis, R.S.; Investigation, B.P., R.S., J.H., P.H., I.F., G.B., B.N.-F., Z.L. and S.S.; Resources, B.K., I.S., A.M., Z.P. (Zoltán Palotai) and S.S.; Writing—original draft, B.P., R.S., G.T. and S.S.; Writing—review & editing, J.H., E.K., G.B. and Z.P. (Zsolt Pirger); Visualization, B.P., J.H., G.T. and S.S.; Supervision, E.K., Z.C., A.M., Z.P. (Zoltán Palotai), Z.P. (Zsolt Pirger) and S.S.; Project administration, B.K., I.S., A.M., B.N.-F., Z.P. (Zoltán Palotai), Z.L., Z.P. (Zsolt Pirger) and S.S.; Funding acquisition, B.K. and I.S. All authors have read and agreed to the published version of the manuscript.

Funding: Project no. 2020-1.1.2-PIACI-KFI-2021-00239 was implemented with the support provided by the Ministry of Innovation and Technology of Hungary with the National Research, Development, and Innovation Fund, financed under the PIACI KFI funding scheme, the National Laboratory for Water Science and Water Security (#RRF 2.3.1-21-2022-00008, HUN-REN BLRI) and the Sustainable Development and Technologies National Program of the Hungarian Academy of Sciences (#NP2022-II3/2022, HUN-REN BLRI). This research was also supported by the Ministry of Innovation and Technology within the framework of the Thematic Excellence Program 2021, National Defense and Security sub-program (TKP2021-NVA-22, Hungarian University of Agriculture and Life Science). Project NO. C2270802 was implemented with the support provided by the Ministry of Culture and Innovation of Hungary with the National Research, Development and Innovation Fund, financed under the KDP-2023. This project was supported by the Research Excellence Program 2024 of the Hungarian University of Agriculture and Life Sciences (Z.C.). The APC was funded by the Hungarian University of Agriculture and Life Sciences.

Data Availability Statement: Further data available on request from the corresponding author.

Conflicts of Interest: Authors Bence Prikler, Gábor Bordós, Adrienn Micsinai, Brigitta Nyírő-Fekete, Zoltán Palotai were employed by the company Eurofins Analytical Services Hungary Ltd. The remaining authors declare that the research was conducted in the absence of any commercial or financial relationships that could be construed as a potential conflict of interest.

References

1. Eerkes-Medrano, D.; Thompson, R.C.; Aldridge, D.C. Microplastics in Freshwater Systems: A Review of the Emerging Threats, Identification of Knowledge Gaps and Prioritisation of Research Needs. *Water Res.* **2015**, *75*, 63–82. [CrossRef] [PubMed]
2. Plastics—Environmental Aspects—State of Knowledge and Methodologies. Available online: https://cdn.standards.iteh.ai/samples/72300/65b73cb6030f442aa1bd7eaa97386292/ISO-TR-21960-2020.pdf (accessed on 27 March 2024).
3. Adhikari, S.; Kelkar, V.; Kumar, R.; Halden, R.U. Methods and Challenges in the Detection of Microplastics and Nanoplastics: A Mini-Review. *Polym. Int.* **2022**, *71*, 543–551. [CrossRef]
4. Scherer, C.; Weber, A.; Stock, F.; Vurusic, S.; Egerci, H.; Kochleus, C.; Arendt, N.; Foeldi, C.; Dierkes, G.; Wagner, M.; et al. Comparative Assessment of Microplastics in Water and Sediment of a Large European River. *Sci. Total Environ.* **2020**, *738*, 139866. [CrossRef] [PubMed]
5. Li, C.; Busquets, R.; Campos, L.C. Assessment of Microplastics in Freshwater Systems: A Review. *Sci. Total Environ.* **2020**, *707*, 135578. [CrossRef] [PubMed]
6. Talbot, R.; Chang, H. Microplastics in Freshwater: A Global Review of Factors Affecting Spatial and Temporal Variations. *Environ. Pollut.* **2022**, *292*, 118393. [CrossRef] [PubMed]
7. Bhagat, J.; Zang, L.; Nishimura, N.; Shimada, Y. Zebrafish: An Emerging Model to Study Microplastic and Nanoplastic Toxicity. *Sci. Total Environ.* **2020**, *728*, 138707. [CrossRef] [PubMed]
8. Svigruha, R.; Prikler, B.; Farkas, A.; Ács, A.; Fodor, I.; Tapolczai, K.; Schmidt, J.; Bordós, G.; Háhn, J.; Harkai, P.; et al. Presence, Variation, and Potential Ecological Impact of Microplastics in the Largest Shallow Lake of Central Europe. *Sci. Total Environ.* **2023**, *883*, 163537. [CrossRef] [PubMed]
9. Bowley, J.; Baker-Austin, C.; Porter, A.; Hartnell, R.; Lewis, C. Oceanic Hitchhikers—Assessing Pathogen Risks from Marine Microplastic. *Trends Microbiol.* **2021**, *29*, 107–116. [CrossRef] [PubMed]
10. Maasz, G.; Mayer, M.; Zrinyi, Z.; Molnar, E.; Kuzma, M.; Fodor, I.; Pirger, Z.; Takács, P. Spatiotemporal Variations of Pharmacologically Active Compounds in Surface Waters of a Summer Holiday Destination. *Sci. Total Environ.* **2019**, *677*, 545–555. [CrossRef]
11. Molnar, E.; Maasz, G.; Pirger, Z. Environmental Risk Assessment of Pharmaceuticals at a Seasonal Holiday Destination in the Largest Freshwater Shallow Lake in Central Europe. *Environ. Sci. Pollut. Res.* **2021**, *28*, 59233–59243. [CrossRef]
12. Kovács, J.; Hatvani, I.G.; Korponai, J.; Kovács, I.S. Morlet Wavelet and Autocorrelation Analysis of Long-Term Data Series of the Kis-Balaton Water Protection System (KBWPS). *Ecol. Eng.* **2010**, *36*, 1469–1477. [CrossRef]
13. Istvánovics, V.; Clement, A.; Somlyódy, L.; Specziár, A.; G.-Tóth, L.; Padisák, J. Updating Water Quality Targets for Shallow Lake Balaton (Hungary), Recovering from Eutrophication. *Hydrobiologia* **2007**, *581*, 305–318. [CrossRef]
14. Farkas, M.; Szoboszlay, S.; Vörös, L.; Lovász, Z.; Méhes, N.; Mátyás, K.; Sebők, R.; Kaszab, E.; Háhn, J.; Tóth, G.; et al. Bacterial Community Dynamics along a River-Wetland-Lake System. *Water* **2022**, *14*, 3519. [CrossRef]
15. Dömötörfy, Z.; Reeder, D.; Pomogyi, P. Changes in the Macro-Vegetation of the Kis-Balaton Wetlands over the Last Two Centuries: A GIS Perspective. *Hydrobiologia* **2003**, *506*, 671–679. [CrossRef]
16. Marinović, Z.; Tokodi, N.; Backović, D.D.; Šćekić, I.; Kitanović, N.; Simić, S.B.; Đorđević, N.B.; Ferincz, Á.; Staszny, Á.; Dulić, T.; et al. Does the Kis-balaton Water Protection System (Kbwps) Effectively Safeguard Lake Balaton from Toxic Cyanobacterial Blooms? *Microorganisms* **2021**, *9*, 960. [CrossRef] [PubMed]
17. Bordós, G.; Gergely, S.; Háhn, J.; Palotai, Z.; Szabó, É.; Besenyő, G.; Salgó, A.; Harkai, P.; Kriszt, B.; Szoboszlay, S. Validation of Pressurized Fractionated Filtration Microplastic Sampling in Controlled Test Environment. *Water Res.* **2021**, *189*, 116572. [CrossRef] [PubMed]
18. Mári, Á.; Bordós, G.; Gergely, S.; Büki, M.; Háhn, J.; Palotai, Z.; Besenyő, G.; Szabó, É.; Salgó, A.; Kriszt, B.; et al. Validation of Microplastic Sample Preparation Method for Freshwater Samples. *Water Res.* **2021**, *202*, 117409. [CrossRef] [PubMed]
19. Liu, Y.; Lorenz, C.; Vianello, A.; Syberg, K.; Nielsen, A.H.; Nielsen, T.G.; Vollertsen, J. Exploration of Occurrence and Sources of Microplastics (>10 Mm) in Danish Marine Waters. *Sci. Total Environ.* **2023**, *865*, 161255. [CrossRef]
20. Primpke, S.; Cross, R.K.; Mintenig, S.M.; Simon, M.; Vianello, A.; Gerdts, G.; Vollertsen, J. Toward the Systematic Identification of Microplastics in the Environment: Evaluation of a New Independent Software Tool (SiMPle) for Spectroscopic Analysis. *Appl. Spectrosc.* **2020**, *74*, 1127–1138. [CrossRef]
21. Simon, M.; van Alst, N.; Vollertsen, J. Quantification of Microplastic Mass and Removal Rates at Wastewater Treatment Plants Applying Focal Plane Array (FPA)-Based Fourier Transform Infrared (FT-IR) Imaging. *Water Res.* **2018**, *142*, 1–9. [CrossRef]
22. Bostan, N.; Ilyas, N.; Akhtar, N.; Mehmood, S.; Saman, R.U.; Sayyed, R.Z.; Shatid, A.A.; Alfaifi, M.Y.; Elbehairi, S.E.I.; Pandiaraj, S. Toxicity Assessment of Microplastic (MPs); a Threat to the Ecosystem. *Environ. Res.* **2023**, *234*, 116523. [CrossRef] [PubMed]
23. Vianello, A.; Jensen, R.L.; Liu, L.; Vollertsen, J. Simulating Human Exposure to Indoor Airborne Microplastics Using a Breathing Thermal Manikin. *Sci. Rep.* **2019**, *9*, 8670. [CrossRef] [PubMed]
24. Cole, M. A Novel Method for Preparing Microplastic Fibers. *Sci. Rep.* **2016**, *6*, 34519. [CrossRef] [PubMed]
25. Horton, A.A.; Cross, R.K.; Read, D.S.; Jürgens, M.D.; Ball, H.L.; Svendsen, C.; Vollertsen, J.; Johnson, A.C. Semi-Automated Analysis of Microplastics in Complex Wastewater Samples. *Environ. Pollut.* **2021**, *268*, 115841. [CrossRef] [PubMed]
26. Miller, E.; Sedlak, M.; Lin, D.; Box, C.; Holleman, C.; Rochman, C.M.; Sutton, R. Recommended Best Practices for Collecting, Analyzing, and Reporting Microplastics in Environmental Media: Lessons Learned from Comprehensive Monitoring of San Francisco Bay. *J. Hazard. Mater.* **2021**, *409*, 124770. [CrossRef] [PubMed]

27. Tanentzap, A.J.; Cottingham, S.; Fonvielle, J.; Riley, I.; Walker, L.M.; Woodman, S.G.; Kontou, D.; Pichler, C.M.; Reisner, E.; Lebreton, L. Microplastics and Anthropogenic Fibre Concentrations in Lakes Reflect Surrounding Land Use. *PLoS Biol.* **2021**, *19*, e3001389. [CrossRef] [PubMed]
28. Zheng, Y.; Li, J.; Sun, C.; Cao, W.; Wang, M.; Jiang, F.; Ju, P. Comparative Study of Three Sampling Methods for Microplastics Analysis in Seawater. *Sci. Total Environ.* **2021**, *765*, 144495. [CrossRef] [PubMed]
29. Liu, Y.; Prikler, B.; Bordós, G.; Lorenz, C.; Vollertsen, J. Does Microplastic Analysis Method Affect Our Understanding of Microplastics in the Environment? *Sci. Total Environ.* **2023**, *902*, 166513. [CrossRef] [PubMed]
30. Uurasjärvi, E.; Hartikainen, S.; Setälä, O.; Lehtiniemi, M.; Koistinen, A. Microplastic Concentrations, Size Distribution, and Polymer Types in the Surface Waters of a Northern European Lake. *Water Environ. Res.* **2020**, *92*, 149–156. [CrossRef]
31. Bordós, G.; Urbányi, B.; Micsinai, A.; Kriszt, B.; Palotai, Z.; Szabó, I.; Hantosi, Z.; Szoboszlay, S. Identification of Microplastics in Fish Ponds and Natural Freshwater Environments of the Carpathian Basin, Europe. *Chemosphere* **2019**, *216*, 110–116. [CrossRef]
32. Dalvand, M.; Hamidian, A.H. Occurrence and Distribution of Microplastics in Wetlands. *Sci. Total Environ.* **2023**, *862*, 160740. [CrossRef] [PubMed]
33. Kaliszewicz, A.; Winczek, M.; Karaban, K.; Kurzydłowski, D.; Górska, M.; Koselak, W.; Romanowski, J. The Contamination of Inland Waters by Microplastic Fibres under Different Anthropogenic Pressure: Preliminary Study in Central Europe (Poland). *Waste Manag. Res.* **2020**, *38*, 1231–1238. [CrossRef] [PubMed]
34. Thornton Hampton, L.M.; Brander, S.M.; Coffin, S.; Cole, M.; Hermabessiere, L.; Koelmans, A.A.; Rochman, C.M. Characterizing Microplastic Hazards: Which Concentration Metrics and Particle Characteristics Are Most Informative for Understanding Toxicity in Aquatic Organisms? *Microplastics Nanoplastics* **2022**, *2*, 20. [CrossRef]
35. Kye, H.; Kim, J.; Ju, S.; Lee, J.; Lim, C.; Yoon, Y. Microplastics in Water Systems: A Review of Their Impacts on the Environment and Their Potential Hazards. *Heliyon* **2023**, *9*, e14359. [CrossRef]
36. Gaylarde, C.C.; Neto, J.A.B.; da Fonseca, E.M. Paint Fragments as Polluting Microplastics: A Brief Review. *Mar. Pollut. Bull.* **2021**, *162*, 111847. [CrossRef]
37. Turner, A. Paint Particles in the Marine Environment: An Overlooked Component of Microplastics. *Water Res. X* **2021**, *12*, 100110. [CrossRef]
38. Kameda, Y.; Yamada, N.; Fujita, E. Source- and Polymer-Specific Size Distributions of Fine Microplastics in Surface Water in an Urban River. *Environ. Pollut.* **2021**, *284*, 117516. [CrossRef] [PubMed]
39. Habib, R.Z.; Al Kendi, R.; Thiemann, T. The Effect of Wastewater Treatment Plants on Retainment of Plastic Microparticles to Enhance Water Quality—A Review. *J. Environ. Prot.* **2021**, *12*, 161–195. [CrossRef]
40. Ronda, A.C.; Menéndez, M.C.; Tombesi, N.; Álvarez, M.; Tomba, J.P.; Silva, L.I.; Arias, A.H. Microplastic Levels on Sandy Beaches: Are the Effects of Tourism and Coastal Recreation Really Important? *Chemosphere* **2023**, *316*, 137842. [CrossRef]
41. Franco, A.A.; Iglesias-Arroyo, D.; Egea-Corbacho, Á.; Martín-García, A.P.; Quiroga, J.M.; Coello, M.D. Influence of Tourism on Microplastic Contamination at Wastewater Treatment Plants in the Coastal Municipality of Chiclana de La Frontera. *Sci. Total Environ.* **2023**, *900*, 165573. [CrossRef]
42. Li, C.; Wang, X.; Liu, K.; Zhu, L.; Wei, N.; Zong, C.; Li, D. Pelagic Microplastics in Surface Water of the Eastern Indian Ocean during Monsoon Transition Period: Abundance, Distribution, and Characteristics. *Sci. Total Environ.* **2021**, *755*, 142629. [CrossRef] [PubMed]
43. Wu, X.; Zhao, X.; Chen, R.; Liu, P.; Liang, W.; Wang, J.; Teng, M.; Wang, X.; Gao, S. Wastewater Treatment Plants Act as Essential Sources of Microplastic Formation in Aquatic Environments: A Critical Review. *Water Res.* **2022**, *221*, 118825. [CrossRef] [PubMed]
44. Chen, G.; Feng, Q.; Wang, J. Mini-Review of Microplastics in the Atmosphere and Their Risks to Humans. *Sci. Total Environ.* **2020**, *703*, 135504. [CrossRef] [PubMed]
45. Järlskog, I.; Strömvall, A.M.; Magnusson, K.; Galfi, H.; Björklund, K.; Polukarova, M.; Garção, R.; Markiewicz, A.; Aronsson, M.; Gustafsson, M.; et al. Traffic-Related Microplastic Particles, Metals, and Organic Pollutants in an Urban Area under Reconstruction. *Sci. Total Environ.* **2021**, *774*, 145503. [CrossRef]

water

MDPI

Article

Potential Pollution Loads of the Cikembar Sub-Watershed to the Cicatih River, West Java, Indonesia

Iwan Juwana [1], Ahyahudin Sodri [2,*], Nitin Muttil [3], Rafid Risandri Hikmat [4], Adryan Lukman Indira [4] and Arief Dhany Sutadian [5]

1 Environmental Engineering Department, Institut Teknologi Nasional Bandung, Bandung 40124, Indonesia; juwana@itenas.ac.id
2 Graduate Program in Environmental Science, University of Indonesia, Jakarta 10430, Indonesia
3 Institute for Sustainable Industries & Liveable Cities, Victoria University, Melbourne, VIC 8001, Australia; nitin.muttil@vu.edu.au
4 Environmental Engineering Department, Institut Teknologi Bandung, Bandung 40132, Indonesia; rafidrisandri@gmail.com (R.R.H.); adryan.lukman@gmail.com (A.L.I.)
5 West Java Research and Development Agency, Bandung 40286, Indonesia; ariefdhany@jabarprov.go.id
* Correspondence: ahyahudin.sodri@ui.ac.id

Abstract: The Cicatih River in the Cikembar sub-watershed, located in Sukabumi Regency, flows along the upstream of the Cikembar sub-watershed and Palabuhan Ratu Bay. For decades, the river has served as a vital water resource, catering to various needs such as domestic use, agriculture, livestock, fisheries, industry, mining, tourism, and hydroelectric power. Recently, the Sukabumi Regency Government made the pollution of coastal and river areas one of the priority issues on sustainable development initiatives for the Sukabumi Regency. This highlights the importance of managing the Cicatih River, one of the main rivers in the regency. This research aims to analyze the potential pollution loads in the Cikembar sub-watershed to the Cicatih River by assessing its water quality status and potential pollution loads from four main sectors: domestic, agricultural, livestock, and industrial. The water quality was calculated based on the Pollution Index method. The potential pollution load from different sources was analyzed using various emission factor calculations. For each pollutant source, the following parameters were analyzed: total suspended solids (TSS), biochemical oxygen demand (BOD), chemical oxygen demand (COD), Total-N (nitrogen), and Total-P (phosphorus). Results showed that the condition of the Cicatih River is slightly polluted. The parameter exceeding its maximum threshold is BOD. The largest contributor to the potential pollution load of BOD is the domestic sector (53%), while the agricultural sector is the least (2%). In the domestic sector, Bojongkembar Village is the largest contributor to the potential pollution load. The results of this analysis can be used to develop strategies for managing the overall Cikembar sub-watershed.

Keywords: Cikembar sub-watershed; water quality status; pollution load; Cicatih River

1. Introduction

Water pollution has been one of the global challenges for many countries [1]. The decline of river water quality, including in developing countries, is primarily because of urbanization, inadequate waste management policies, or basic infrastructure; this situation has repercussions for human health and for the ecosystem structure and function [2,3]. On the island of Java, Indonesia, particularly in the western part, pollutants entering the watershed and its rivers have been well documented [4]. However, in the southern part, where it is occurring at a rapid pace, development is often unavailable or has not been studied yet. Such a condition has brought attention to investigating and quantifying how urbanization, leading to land use change, affects streams.

One of the important watersheds in the southern part of the West Java Province is the Cikembar sub-watershed. Its significance lies in its role as a primary contributor to the Cicatih River, a vital water body that sustains local communities of Sukabumi Regency for agricultural activities and diverse ecosystems downstream [5]. The increasing pressures of human activities and rapid development within the Cikembar sub-watershed have sparked concerns regarding potential pollution loads that could significantly degrade the water quality of the Cicatih River. Understanding the pollution potential of human activities and their impact on water quality in this watershed is imperative for devising sustainable management strategies to preserve the ecological balance and safeguard human well-being.

The sources of pollution within the Cikembar sub-watershed encompass a wide spectrum of anthropogenic activities. For example, agriculture, particularly its traditional practices, plays a pivotal yet contentious role, contributing substantial amounts of pesticides, fertilizers, and sediment runoff into the waterways, subsequently polluting coastal waters [6]. The evolving landscape due to urbanization and industrialization introduces a complex array of pollutants, including domestic sewage, industrial effluents, and improper waste disposal practices. The combined effects of these activities necessitate a comprehensive investigation to delineate the sources and quantify the potential pollution loads that find their way into the Cicatih River, thereby discerning the dynamics influencing its water quality.

A recent Sukabumi Regency Strategic Environmental Study shows two strategic sustainable developments related to watersheds in Sukabumi Regency—West Java, which is located in the southern part of Java Island: the pollution of coastal and river areas and land use changes [6]. The growing changes in land use in Sukabumi Regency catchments have increased the risk of river pollution. In the last decade, increased areas for the industrial sector (45.50%), settlements (27.71%), and sand mining (64.50%) were the most significant land use changes in Sukabumi Regency. Land use changes occur in the watershed in Sukabumi Regency due to population and density increases, including urbanization [5]. Furthermore, water quality management practices in the Cikembar sub-watershed are currently fragmented and insufficient to address the multifaceted challenges posed by diverse pollution sources. While localized regulations and community-driven initiatives exist, they often lack the cohesive approach required for effective management. The absence of integrated strategies and inadequate monitoring exacerbates the situation, posing risks to the sustained health of the Cicatih River [5].

The use of water quality indices as a tool to evaluate the status of water quality in rivers has been introduced in the last four decades [4,7,8]. For instance, it has been successfully used to identify the water quality status in the Citarum River, Indonesia [9], or the Erdao Songhua River, China [10]. In this study, the Pollution Index (PI) specifically developed by the Ministry of Forestry and Environment of Indonesia was used to analyze water quality status as it has been formally applied across the country [9]. Similar to water quality indices, there are many studies on the non-point source pollution load modeling [11]. The calculation of the pollution loads was performed to investigate the amount, sources, and impact of Total-P (phosphorus) entering the Dongting Lake in China Field [12] and asses Total-N (nitrogen) and Total-P pollution loads in the upper Yellow River [13].

Considering the above-mentioned conditions, the principal objective of this study has two folds namely these were (1) to analyze the water quality status using Indonesia's Pollution Index and (2) to identify potential pollution loads from five main sectors, such as domestic, agriculture, livestock, industry, and commercial, in the Cikembar sub-watershed to the Cicatih River. The analysis was conducted on five key parameters: total suspended solids (TSS), biochemical oxygen demand (BOD), chemical oxygen demand (COD), Total-N, and Total-P. Prior to this study, information on the potential pollution load in the catchment from various pollutant sources was not available. This information is needed by the local government of Sukabumi Regency to determine priority locations and programs to improve the Cicatih River water quality. Therefore, this research contributes to scholarship on water quality management in developing countries by examining water quality status and pollution loads in a rapidly growing area in the southern part of the West Java Province.

2. Materials and Methods

2.1. Study Area Description

The Cikembar sub-watershed, located in Sukabumi Regency, West Java, is part of the Cimandiri watershed with a catchment area covering 84.6 km^2, with an elevation range of 150–700 m msl. The largest river in Sukabumi Regency, the Cimandiri River, flows along the Palabuhan Ratu Bay and the south of West Java. For decades, the Cimandiri River has been a source of domestic/clean water, agriculture/irrigation, livestock, fisheries, industry, sand mining, tourism, and hydroelectric power. The boundary of the Cikembar sub-watershed, and its monitoring locations are shown in Figure 1.

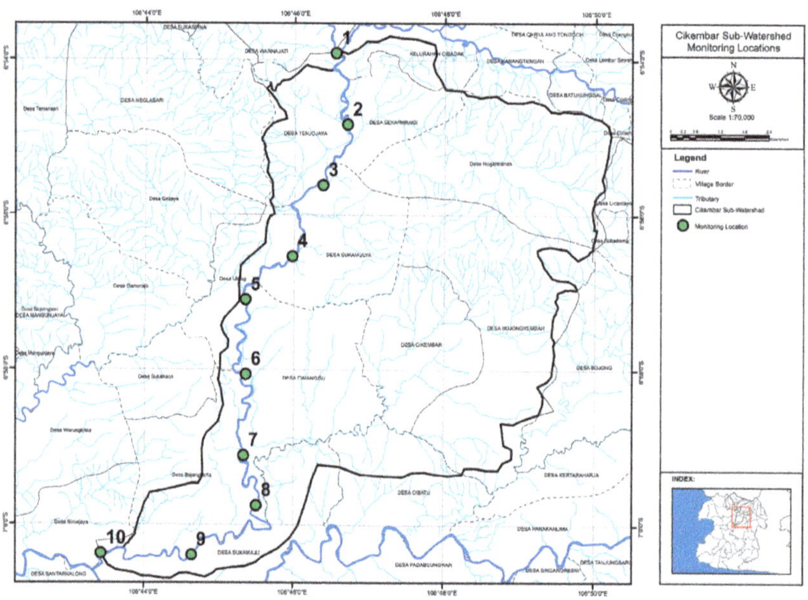

Figure 1. Cikembar Sub-Watershed Monitoring Locations.

Table 1 presents the characteristics of the population, land area, and density in the Cikembar Sub Watershed. Administratively, there are twelve villages with a total population of about 136,874 inhabitants. With regard to populations, Bojongkembar and Hegarmanah had the largest populations with 30,576 persons and 22,779 persons. Meanwhile, Cibatu and Ubrug had the lowest populations, with 2241 persons and 814 persons. Similar to the total population, in terms of total area, Bojongkembar and Hegarmanah are the two biggest villages, spanning 1522.00 ha and 1328.00 ha, respectively, and Cibatu and Ubrug are the two smallest villages, spanning total areas of 215.00 ha and 269.00 ha. Figure 1 also presents the characteristic population density wherein Cibadak (43.51 persons/ha) and Sekarwangi (24.52 persons/ha) are the most densely populated villages, while Tenonjaya and Cibatu are the least densely populated villages in the study area, with 9.38 persons/ha and 3.03 persons/ha [6].

In terms of the average annual rainfall, the Cikembar Sub Watershed was grouped into three categories, ranging from 2950 to 3200 mm/year for the 375.48 ha, between 3200 and 3450 mm/year for the 6580.79 ha, and between 3450 and 3700 mm/year for the 1508.17 ha, as presented in Table 2 below [6].

Table 1. Population, Land Area, and Density of Villages in Cikembar Sub Watershed.

No.	Villages	Number of Population (Persons)	Area (ha)	Population Density (Persons/ha)
1	Hegarmanah	22,779	1328.00	17.15
2	Ubrug	2241	215.00	10.42
3	Bojongkerta	8374	637.00	13.15
4	Cibadak	7266	167.00	43.51
5	Sekarwangi	11,232	458.00	24.52
6	Tenjojaya	4653	496.00	9.38
7	Sukamulya	10,023	795.00	12.61
8	Bojongkembar	30,576	1522.00	20.09
9	Cikembar	16,262	795.00	20.46
10	Cimanggu	15,736	1094.00	14.38
11	Cibatu	814	269.00	3.03
12	Sukamaju	6918	515.00	13.43
	Total	136,874	8291.00	16.84

Table 2. Characteristic of Rainfall in Cikembar Sub Watershed [6].

No.	Categories	Area (ha)
1	2950–3200 mm/year	375.48
2	3200–3450 mm/year	6580.79
3	3450–3700 mm/year	1508.17

Table 3 presents land use in the study area on ten different classifications, namely forest, industry, gardens/plantations, settlement area, grass/empty land, irrigation rice fields, rainfed rice fields, shrubs, moorland/field, and body of water. Overall, the land use of Cikembar Sub Watershed was dominated by rainfed rice fields with 695.378 ha of the total area, followed by the second and third largest: gardens/plantations and moorland with 3891.972 ha and 1631.324 ha, respectively. Meanwhile, industry and forest were the two smallest proportions of land use in the study area, with 31.292 ha and 58.591 ha. The third smallest proportion was body of water, which was 90.461 ha of the total land use area [6].

Table 3. The Characteristics of Land Use in Cikembar Sub Watershed.

No.	Categories	Area (ha)
1	Forest	58.591
2	Industry	31.292
3	Gardens/Plantations	3891.972
4	Settlement Area	1459.163
5	Grass/Empty Land	89.714
6	Irrigation Rice Fields	314.786
7	Rainfed Rice Fields	695.378
8	Shrubs	201.752
9	Moorland/Field	1631.324
10	Body of water	90.461

2.2. Methodology

2.2.1. Water Quality Status

In Indonesia, water quality status refers to the Decree of the State Minister of the Environment on the Guidelines for Determining Water Quality Status. In this decree, river quality status in Indonesia should be determined using the PI method as described in Section 1. Therefore, in this study, the PI method was used [14].

This method assesses water quality by evaluating various parameters such as pH, dissolved oxygen (DO), BOD, COD, TSS, and specific pollutants like heavy metals and nutrients.

Each parameter is assigned a score based on its derivation from set standards or thresholds. These scores are then aggregated to calculate an overall PI for the water body. The PI categorizes water quality into different uses like drinking, aquatic life, or irrigation.

In this study, the water quality status of the Cicatih River is categorized as Class II, as it is generally considered suitable for specific purposes such as irrigation, industrial cooling, or controlled waste disposal. This classification indicates that the water, despite some level of pollution, can be used for certain non-drinking purposes after appropriate treatment or management.

The PI is calculated using the following equation:

$$PI = \sqrt{\frac{\left(\left(\frac{C_i}{L_i}\right)^2_{maximum} + \left(\frac{C_i}{L_i}\right)^2_{average}\right)}{2}} \quad (1)$$

where:

C_i = quality of specific parameter (i) based on water quality sampling
L_i = maximum threshold value for parameter (i)
PI = Pollution Index
i = parameter 1, 2, 3...

Once the value of PI is obtained, the categories of water quality status is analyzed based on the following: $0 \leq PI \leq 1.0$: meet the quality standards (good condition); $1.0 < PI \leq 5.0$: slightly polluted; $5.0 < PI \leq 10$: moderately polluted; and $PI > 10$: heavily polluted.

2.2.2. Potential Pollution Load

Potential pollution load refers to the number of pollutants that could be introduced into an environment, including rivers. By estimating potential pollution loads, policymakers and environmental scientists can develop strategies to mitigate and control pollution effectively. Accurately assessing potential pollution loads is essential for maintaining the health and sustainability of our ecosystems and ensuring the well-being of human populations.

Assessing potential pollution loads involves emission factors, land-use planning, and pollution dispersion models. In this research, the potential pollution load of various sectors, namely domestic, agriculture, livestock, and industrial, will be calculated using available emission factors.

Domestic Sector

The potential pollution load of the domestic sector in a watershed is influenced by the population, emission factors, city equivalent ratio, and load transfer coefficient. The potential pollution load for the domestic sector is calculated using the following formula [6]:

$$\text{The Domestic Sector Pollution Load} = \text{Total Population} \times \text{Emission Factor} \times \text{City Equivalent Ratio} \times \text{Load Transfer Coefficient} \quad (2)$$

The total population in the Cikembar sub-watershed is calculated using the percentage of the settlement area of each village in the watershed area to the total area of settlements in Sukabumi Regency. The percentage is then multiplied by the total population of Sukabumi Regency.

The potential pollution load in the domestic sector is calculated on five key parameters: TSS, BOD, COD, Total-N, and Total-P. These five parameters are water quality parameters with a probability of $\geq$80% causing water pollution [6,15]. Emission factors of these five parameters from household activities are shown in Table 4 below. These emission factors were calculated based on the study by Iskandar [16], taking into account multiple measurements along the Ciliwung River:

1. Assessment of river water quality across 35 locations for various parameters.
2. Analysis of river flow at 25 locations along the Ciliwung River.
3. Collection of sludge samples from 10 locations in the Ciliwung River for DO.

Table 4. The Domestic Sector Emission Factor [6].

Parameter [1]	Emission Factor (g/Person/Day)
TSS	38
BOD	40
COD	55
Total-N	1.95
Total-P	0.21

Note: [1] In Indonesia, these are the five parameters with their emission factors available for domestic sector.

After this, water quality modeling was employed to calculate emission factors by establishing correlations between water quality and pollution sources (domestic, livestock, farming, and industries) along the river within their respective catchment areas. The calibration results for the parameter of BOD show a correlation value of 0.898.

The findings of this study conducted on the Ciliwung River hold particular significance, especially concerning the emission factors of various water quality parameters. These emission factors are used in estimating potential pollution loads in the Cicatih River, given the similarity in catchment characteristics between both rivers. Moreover, this study on the Ciliwung River stands as a pivotal resource, being the sole investigation available that determines emission factors for domestic, agricultural, and livestock activities in Indonesia.

The value of the city equivalent ratio is determined by applying villages to certain categories based on city type. Three categories, namely urban, rural, and inland, show their respective equivalent ratios in Table 5 below.

Table 5. City Equivalent Ratio [6].

Area	Value
Urban	1
Rural	0.8125
Inland	0.6250

The value of the load transfer coefficient for each village is determined by calculating the percentage of settlements within the range of 0–100 m, 100–500 m, and >500 m from the Cicatih River in the Cikembar sub-watershed. The percentage is then multiplied by the value of the load transfer coefficient (α), as is shown in Table 6.

Table 6. Load Transfer Coefficient (α) [6].

Value of α	Distance to the River (m)	Sanitary Pattern
1	0–100	Direct disposal to the river
0.85	100–500	Open canal
0.3	>500	Septic tank

When employing the city equivalent ratio, urban areas are delineated as regions inhabited by over 100,000 people, while rural zones encompass populations ranging between 20,000 and 100,000 inhabitants. Inland areas, in this context, refer to locations with populations of fewer than 20,000 inhabitants [14].

Similar to the emission factor calculation, the determination of α (load transfer coefficient) involved water quality modeling, utilizing multiple measurements along the Ciliwung River on river water quality, river flow, and DO measurements [16].

Subsequently, the model was employed to derive the load transfer coefficient by establishing correlations between water quality and pollution sources (domestic, livestock, farming, and industries) within their respective catchment areas [16].

Agricultural Sector

The type of land, area, and harvest period comprise the potential pollution load of the agricultural sector. The potential pollution load for the agricultural sector is calculated using the following formula [16]:

$$\text{The Agricultural Sector Pollution Load} = \frac{\text{Emission Factor} \times \text{Land Area (ha)} \times 10\%}{\text{Harvest Period (day)}} \quad (3)$$

The 10% in the formula indicated the percentage of agricultural waste entering the river.

The emission factor of agricultural waste is shown in Table 7 below.

Table 7. Agricultural Waste Emission Factor [16].

Type of Agriculture	Parameters of Agricultural Waste [1]				
	TSS	BOD	COD	Total-N	Total-P
	Kg/ha/Planting Season				
Rice field	0.4	225	337.5	20	10
Secondary crops	2.2	125	187.5	10	5
Other agricultures	0.6	32.5	48.75	3	1.5

Note: [1] In Indonesia, these are the five parameters with their emission factors available for agricultural sector.

Livestock Sector

The potential pollution load of livestock sector comes from livestock activities, such as animal manure. The following is the formula to calculate potential pollution load for the livestock sector [16]:

$$\text{The Livestock Sector Pollution Load} = \text{Number of Farm Animals} \times \text{Emission Factor} \times 20\% \quad (4)$$

The 20% in the formula indicated the percentage of livestock waste entering the river.

The emission factor of livestock waste is shown in Table 8 below.

Table 8. Livestock Waste Emission Factor [16].

Type of Livestock	Parameters of Agricultural Waste [1]			
	BOD	COD	Total-N	Total-P
	Kg/ha/Planting Season			
Cattle	292	717	0.933	0.153
Sheep	55.7	136	0.278	0.063
Chicken	2.36	5.59	0.002	0.003
Duck	0.88	2.22	0.001	0.005
Buffalo	207	530	2.6	0.39
Horse	226	558	38.083	0.306
Goat	34.1	92.9	1.624	0.115

Note: [1] In Indonesia, these are the five parameters with their emission factors available for livestock sector.

Industrial and Commercial Sector

There are four ways to calculate the pollutant load from point sources [17,18]:

1. Using concentration and discharge of wastewater data to monitor results;
2. If concentration data is available, but discharge data is unavailable from the monitoring results, then the maximum wastewater discharge in the permit is used;
3. If both concentration and discharge wastewater from the monitoring results is unavailable, then pollution load unit (PLU) or Emission Factor (EF) is used;
4. If (1) to (3) is unavailable, the pollution load of the similar sector(s) is used.

The available emission factor of the industrial sector based on a number of employees is shown in Table 9 below.

Table 9. Industrial Sector Emission Factor [17].

Industrial Sector	BOD (g/Day/Employee)
Dyeing	79.1
Other foods	37.9
Metal	10.3
Paper	17.9
Polyester fiber	47.1
Textile	219.2
Laundry	96.4
Machine	4.7
Plastic goods	57.3
Car & motorcycle parts	13.5
Ceramic and tile	2.0
Tannery	144.4
Soap and detergent	50.4
Chemical	1898.2
Metal goods	0.2
Printing	0.6
Glass	0.3
Hospital	123
Hotel	55
Restaurant	17

The potential pollution load for the industrial sector was calculated using the following formula [16]:

$$\text{The Industrial Sector Pollution Load}(\text{kg}/\text{day}) = \beta \times \gamma \times \delta \times \text{Qk} \times \text{Ei} \quad (5)$$

with:

β = unit conversion (0.0864)
γ = distance load transfer coefficient (0.3–1.0)
δ = load transfer coefficient discharge ratio (0.1–1.0)
Q_k = discharge (L/s)
E_i = pollutant emission (mg/L).

In applying the equation, the distance-based load transfer coefficient (gamma) is utilized, calculated from the industrial site's proximity to the river. For practical purposes in this study, a gamma value of 0.3 is employed for industries situated over 500 m away, 0.5 for distances between 100 and 500 m, and a gamma value of 1 for industries within 100 m of the river.

Regarding the discharge ratio-based load transfer coefficient (alpha), its value is determined by the industry's discharge recycling practices. A higher recycle ratio correlates with a lower alpha value. In this particular study, identified industries aren't engaged in discharge recycling, resulting in an alpha value of 1 across all industries.

The potential pollution load for the industrial sector could be calculated using another formula [17]:

$$I,i = Ci \times V \times OpHrs/1000000 \tag{6}$$

with:

I,i = load/pollutant emission i (kg/year)
Ci = industrial waste concentration (mg/L)
V = wastewater discharge flow rate (L/h)
OpHrs = number of operating hours per year (h/year)
1000000 = conversion factor (mg/kg).

3. Results and Discussion

3.1. Water Quality Status

As mentioned earlier, the water quality status of the Cicatih River is determined using the PI method, which was calculated based on the water quality sampling in 10 locations along the Cicatih River. The results of the samplings for various water quality parameters and their comparisons with respective threshold values are presented in Figure 2.

According to the Ministry of the Environment of Indonesia, water quality status juxtaposes the level of water quality conditions in a water body at a certain time with respective water quality standards [16]. This study calculates the water quality status using the Pollution Index (PI value) method and data derived from the water quality measurements of 10 different sampling locations. The results are shown in Table 10.

The results show that the condition of the river in the Cikembar sub-watershed is slightly polluted. Even though the river is still in the early category of pollution, this is an alarming condition due to the growing housings and industries in the catchment. In most rivers in Indonesia, river pollutions are mainly caused by domestic and industrial activities. For example, in Cimahi Catchment, the sub-catchment of Citarum, domestic activities have contributed 43% of the total river pollution in its respective catchment [16], and in the overall Citarum Catchment, domestic activities have contributed 85% of the total river pollution [19]. Therefore, the government of Sukabumi Regency should regularly monitor the water quality of the river and implement necessary policies to ensure that housings and industrial development in the catchment area will not increase its pollution level.

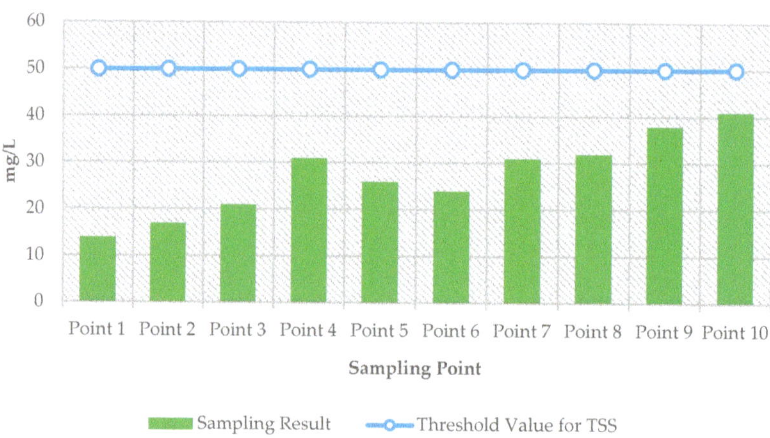

(a)

Figure 2. *Cont.*

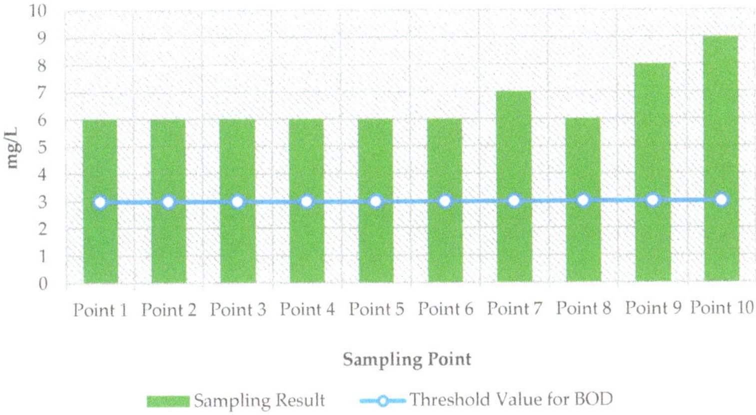

(b)

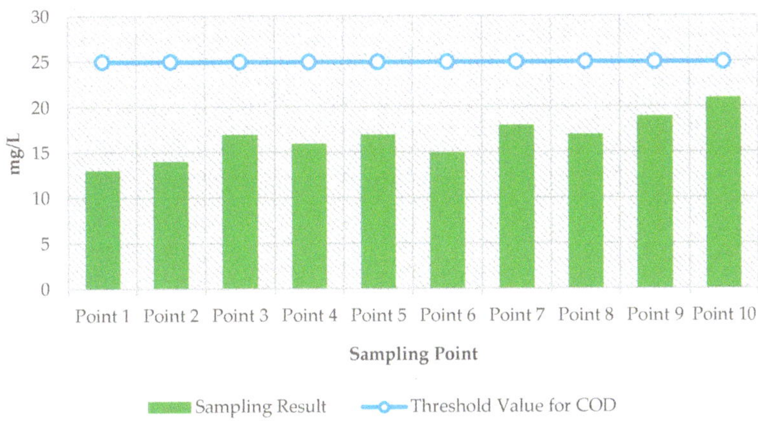

(c)

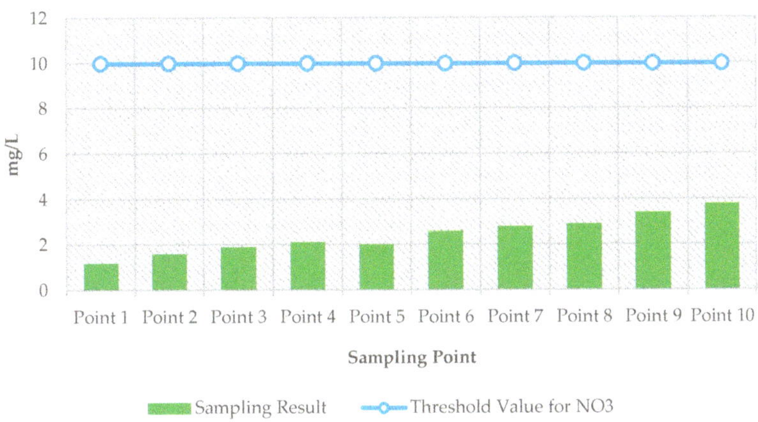

(d)

Figure 2. *Cont.*

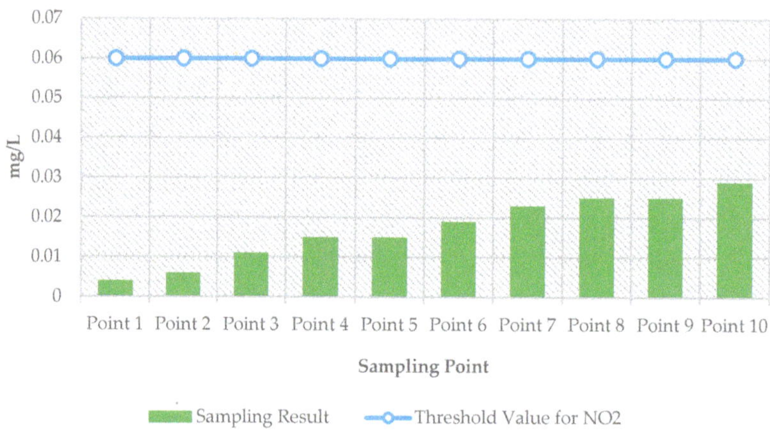

(e)

Figure 2. Water quality of Cicatih Rivers for various parameters (**a**) TSS; (**b**) BOD; (**c**) COD (**d**) Parameter NO_3; (**e**) NO_2.

Table 10. Water Quality Status of the Rivers in Cikembar Sub-watershed.

Monitoring Location	PI Value	Description
Cicatih River Point 1	1.82	Slightly polluted
Cicatih River Point 2	1.81	Slightly polluted
Cicatih River Point 3	1.82	Slightly polluted
Cicatih River Point 4	1.82	Slightly polluted
Cicatih River Point 5	1.82	Slightly polluted
Cicatih River Point 6	1.82	Slightly polluted
Cicatih River Point 7	2.06	Slightly polluted
Cicatih River Point 8	1.83	Slightly polluted
Cicatih River Point 9	2.28	Slightly polluted
Cicatih River Point 10	2.47	Slightly polluted

3.2. *Potential Pollution Load*

3.2.1. Domestic Sector

As indicated in the methodology section, the potential pollution load of the domestic sector was based on the total population in the Cikembar sub-watershed area. Figure 3 shows the settlement areas in the Cikembar sub-watershed, marked in light-brown color. The total area of settlements in Sukabumi Regency is 28,663.45 ha, with a population of 2,725,450. Based on calculation results, the largest population came from Bojongkembar Village (30,576 inhabitants), with 136,874 inhabitants in the Cikembar sub-watershed.

The Cikembar sub-watershed comprises two area categories: urban and rural areas. There are three villages falling into the category of urban areas: Cibadak, Sekarwangi, and Tenjojaya. Other remaining villages are included in the rural category. The calculation of the value of the load transfer coefficient begins with calculating the area of settlements within 0–100, 100–500, and >500 m ranges from the Cicatih River in the Cikembar sub-watershed. Based on the calculation results, most settlement areas are beyond 500 m distance.

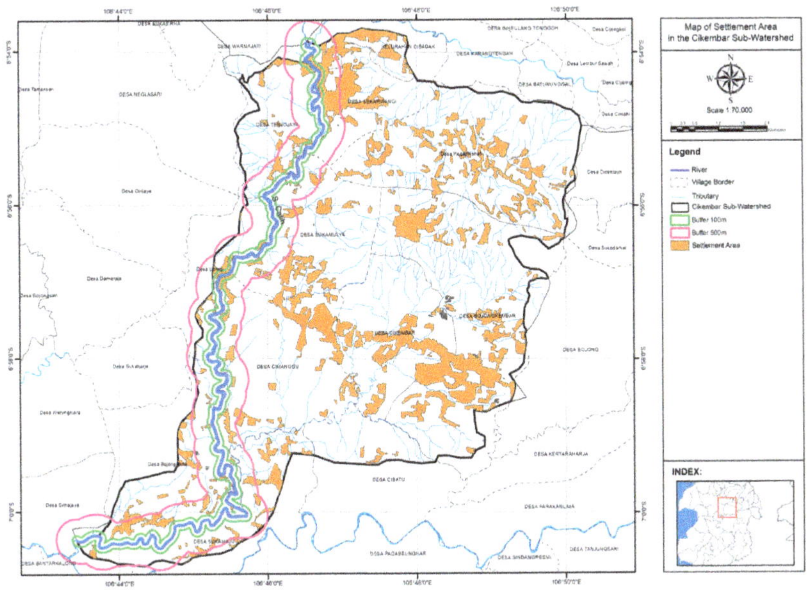

Figure 3. Settlement Area in the Cikembar Sub-Watershed.

The potential pollution load from the domestic sector is directly proportional to the number of populations. The value of the emission factor, city equivalent ratio, and the load transfer coefficient also affect the value of the potential pollution load from the domestic sector. High area category (urban) and close settlement to the river lead to the domestic sector's high potential pollution load discharged into the river [16,20]. The results of the calculation of the potential pollution load of the domestic sector in the Cikembar sub-watershed are shown in Table 11, and the contribution of each village to the potential pollution load of the domestic sector in the Cikembar sub-sub-watershed is shown in Figures 4 and 5.

Table 11. Potential Pollution Load of the Domestic Sector in the Cikembar Sub-watershed.

District	Village	Potential Pollution Load (kg/Day)				
		TSS	BOD	COD	Total-N	Total-P
Cicantayan	Hegarmanah	210.99	222.10	305.38	10.83	1.17
Warung Kiara	Ubrug	61.18	64.40	88.55	3.14	0.34
	Bojongkerta	167.16	175.96	241.95	8.58	0.92
Cibadak	Cibadak	130.87	137.76	189.42	6.72	0.72
	Sekarwangi	210.78	221.87	305.08	10.82	1.16
	Tenjojaya	121.13	127.51	175.32	6.22	0.67
Cikembar	Sukamulya	129.41	136.22	187.31	6.64	0.72
	Bojongkembar	283.21	298.12	409.91	14.53	1.57
	Cikembar	150.63	158.56	218.02	7.73	0.83
	Cimanggu	222.05	233.73	321.39	11.39	1.23
	Cibatu	7.54	7.93	10.91	0.39	0.04
	Sukamaju	107.96	113.65	156.26	5.54	0.60
Total		1802.93	1897.82	2609.50	92.52	9.96

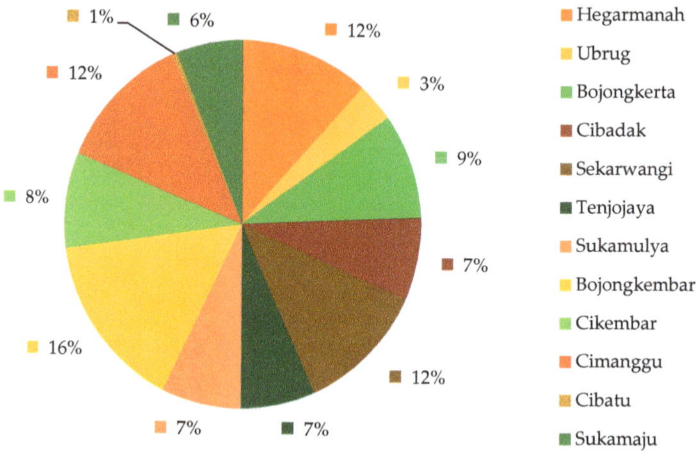

Figure 4. The Contribution of Each Village to the Potential Pollution Load of the Domestic Sector in The Cikembar Sub-Watershed.

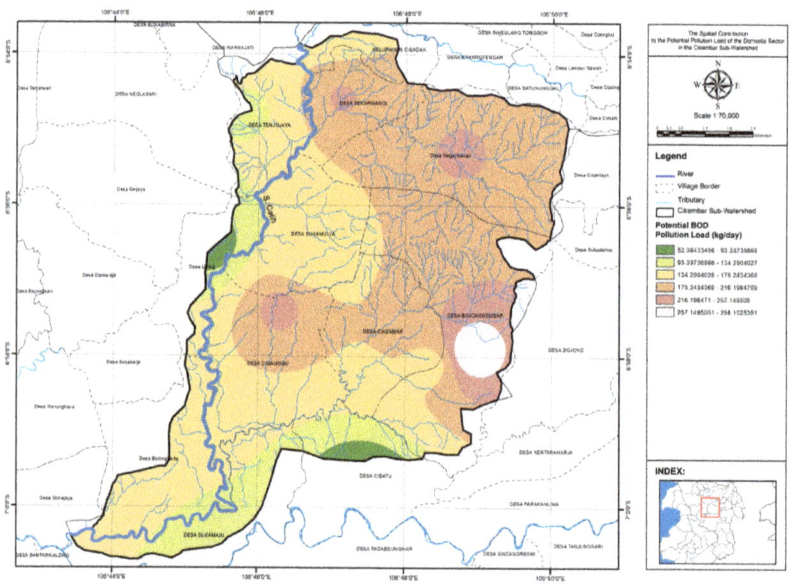

Figure 5. The Spatial Contribution to the Potential Pollution Load of the Domestic Sector in the Cikembar Sub-Watershed.

Based on the calculation, the potential pollution load for the domestic sector in the Cikembar sub-watershed for the BOD parameter is 1897.82 kg/day. This value indicates the potential pollution load from the domestic sector for the BOD parameter entering the Cicatih River daily in the Cikembar sub-sub-watershed.

The same explanation applies to other parameters. Based on the calculation results, Bojongkembar Village is the largest contributor to the potential pollution load of the domestic sector in the Cikembar sub-watershed, with 16% of the total pollution potential, followed by Hegarmanah, Sekarwangi, and Cimanggu Villages, with the same contribution of 12%. Meanwhile, the smallest contributor was Cibatu Village at 1%.

Yusuf [16] presents that the main contributors to the high level of BOD in rivers are domestic and industrial activities. In most of Indonesia's densely populated urban and peri-urban areas, untreated household sewage and wastewater are often discharged directly into the nearby rivers and streams. The same situation applies in most villages in Cikembar sub-watershed, where outlets of wastewater discharge either go directly to nearby river or to water channel leading to nearby rivers. These discharges can contain organic matter, detergents, and other pollutants, contributing to the BOD levels in the river. The organic matter in sewage becomes a food source for bacteria, which consume oxygen during decomposition [21]. Consequently, high BOD levels deplete dissolved oxygen in the water, harming aquatic life, including fish and other organisms. Further, oxygen-deprived water can lead to fish kills and negatively impact the health of the river's ecosystem [21]. In addition, elevated BOD levels can lead to algal bloom. These blooms can further deplete oxygen levels and release toxins harmful to aquatic organisms and human health [22].

3.2.2. Agricultural Sector

The agricultural sector in the Cikembar sub-watershed comprises rice fields and farms. The potential pollution load of the agricultural sector is calculated based on land areas, emission factors, and the number of planting days.

The results of the calculation of the potential pollution load of the agricultural sector in the Cikembar sub-watershed are presented in Table 12, and the contribution of each village to the potential pollution load of the agricultural sector in the Cikembar sub-watershed is shown in Figure 6. The agricultural area in the study area is presented in Figure 7. Rice fields and farm areas are marked in green and red, respectively.

Table 12. Potential Pollution Load of the Agricultural Sector in the Cikembar Sub-watershed.

District	Village	Potential Pollution Load (kg/Day)				
		TSS	BOD	COD	Total-N	Total-P
Cicantayan	Hegarmanah	0.02	4.66	7.00	0.42	0.21
Warung Kiara	Ubrug	0.01	3.45	5.17	0.31	0.15
	Bojongkerta	0.04	20.03	30.05	1.78	0.89
Cibadak	Cibadak	0.00	1.77	2.65	0.16	0.08
	Sekarwangi	0.01	1.09	1.64	0.10	0.05
	Tenjojaya	0.01	0.80	1.20	0.07	0.04
Cikembar	Sukamulya	0.02	5.35	8.02	0.48	0.24
	Bojongkembar	0.03	11.28	16.91	1.01	0.50
	Cikembar	0.02	4.02	6.03	0.36	0.18
	Cimanggu	0.03	9.31	13.97	0.83	0.41
	Cibatu	0.00	0.25	0.37	0.02	0.01
	Sukamaju	0.01	4.92	7.39	0.44	0.22
Total		0.20	66.94	100.41	5.97	2.98

The area of agricultural land is directly proportional to the amount of potential pollution load from the agricultural sector. In addition, agricultural land type, emission factor, multiplier factor, and the length of the growing season also affect the potential pollution load from this sector. Based on the calculation results, Bojongkerta Village has the largest contribution to the pollution load of the agricultural sector produced in the Cikembar sub-watershed, as they have the largest area of agriculture in the catchment.

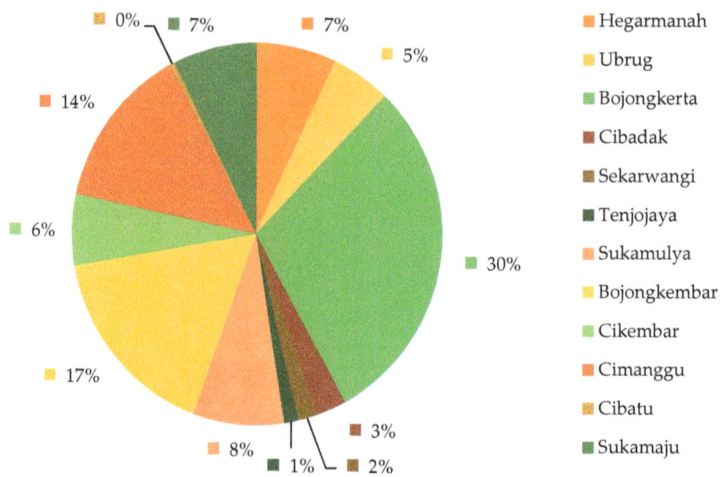

Figure 6. The Contribution of Each Village to the Potential Pollution Load of the Agricultural Sector in the Cikembar Sub-Watershed.

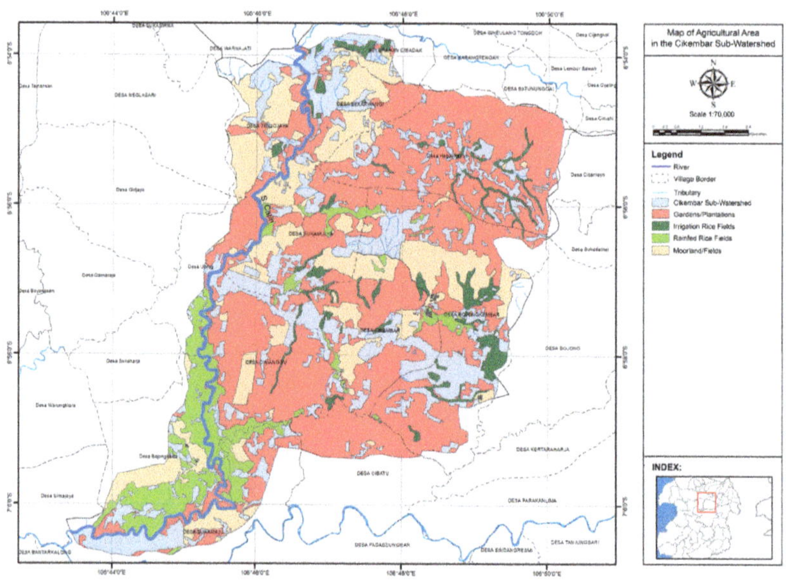

Figure 7. Agricultural Area in the Cikembar Sub-Watershed.

In Sukabumi Regency, farmers use fertilizers, pesticides, and herbicides to increase crop yields. When mismanaged or applied excessively, these chemicals wash into nearby rivers through surface runoff or leach into groundwater, eventually reaching the rivers. This leads to elevated levels of nutrients, such as nitrogen and phosphorus, and various harmful chemicals in the water, which eventually harm aquatic ecosystems and human health [23].

Furthermore, inadequate land management practices, such as inappropriate land clearing and logging, significantly exacerbate river pollution within the Cikembar watershed by intensifying soil erosion. The eroded soil, called sediment, gets carried into rivers during rainfall. This sediment smothers riverbeds and negatively impacts aquatic habitats by reducing the amount of light and oxygen that can reach aquatic organisms [1].

3.2.3. Livestock Sector

The livestock sector in the Cikembar sub-watershed comprises seven types of livestock: cattle, sheep, chicken, duck, buffalo, horse, and goat. The calculation is conducted based on the number of livestock and the emission factor of each type of livestock. The results of the calculation of the potential pollution load of the livestock sector in the Cikembar sub-watershed are shown in Table 13, and the contribution of each village to the potential pollution load and their spatial mapping of the livestock sector in the Cikembar sub-watershed are shown in Figures 8 and 9 below.

Table 13. Potential Pollution Load of the Livestock Sector in the Cikembar Sub-watershed.

District	Village	Potential Pollution Load (kg/Day)			
		BOD	COD	Total-N	Total-P
Cicantayan	Hegarmanah	26.668	64.513	0.134	0.035
Warung Kiara	Ubrug	7.072	16.998	0.030	0.010
	Bojongkerta	3.154	7.841	0.038	0.005
Cibadak	Cibadak	0	0	0	0
	Sekarwangi	0	0	0	0
	Tenjojaya	0	0	0	0
Cikembar	Sukamulya	0	0	0	0
	Bojongkembar	11.596	29.568	0.268	0.030
	Cikembar	256.408	608.766	0.350	0.329
	Cimanggu	0.935	2.428	0.031	0.002
	Cibatu	11.051	27.167	0.091	0.015
	Sukamaju	131.370	313.807	0.317	0.169
Total		448.253	1071.089	1.259	0.595

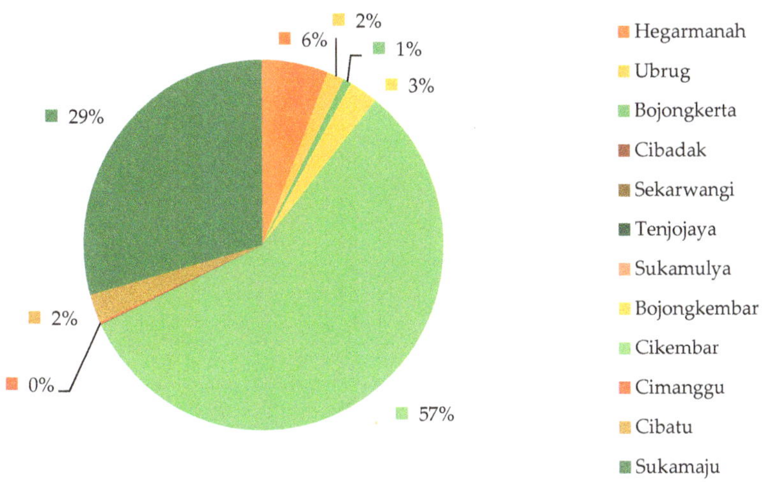

Figure 8. The Contribution of Each Village to the Potential Pollution Load of the Livestock Sector in the Cikembar Sub-Watershed.

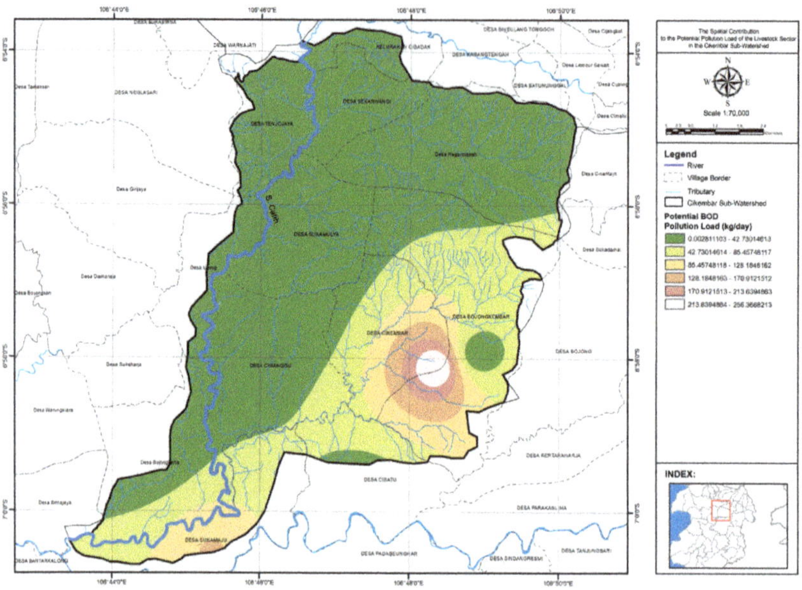

Figure 9. The Spatial Contribution to the Potential Pollution Load of the Livestock Sector in the Cikembar Sub-Watershed.

The discharge of animal waste, particularly from concentrated animal feeding operations (CAFOs) and large-scale livestock farms, is a significant source of river pollution in Indonesia, including in the Cikembar catchment. Animal manures containing minerals like nitrogen and phosphorus are mismanaged in the Cikembar catchment. Thus, they are washed into nearby rivers through rainfall runoff or leach into groundwater [24]. The other sources of contaminants from the livestock activities in Cikembar are pathogens and chemicals. Livestock produces pathogens, such as bacteria, viruses, and parasites, in their waste. In Cikembar, they are mismanaged and treated ineffectively, allowing these pathogens to contaminate nearby rivers, posing risks to human health and aquatic life. As for the chemicals, many farmers in Cikembar use chemicals and pesticides to control pests and parasites. These chemicals potentially enter rivers due to improper treatment of the chemical and pesticide waste [25].

3.2.4. Industrial and Commercial Sector

Based on available data, the potential pollution load for the industrial and commercial sectors is calculated using the number of employees and emission factor. The map of the industrial and commercial area in the Cikembar sub-watershed is shown in Figure 10.

Potential pollution load calculations for industries depend on the availability of respective emission factors. If some industries' emission factors are unavailable, the emission factor of a similar industry was used. For example, to calculate the potential pollution load for pharmaceutical industries, the emission factor for chemical industries was used due to the unavailability of the emission factor for pharmaceutical industries.

The results of the calculation of the potential pollution load of the industrial sector in the Cikembar sub-watershed are presented in Tables 14 and 15, and the contribution of each village to the potential pollution load of the industrial sector in the Cikembar sub-sub-watershed is shown in Figures 11–13.

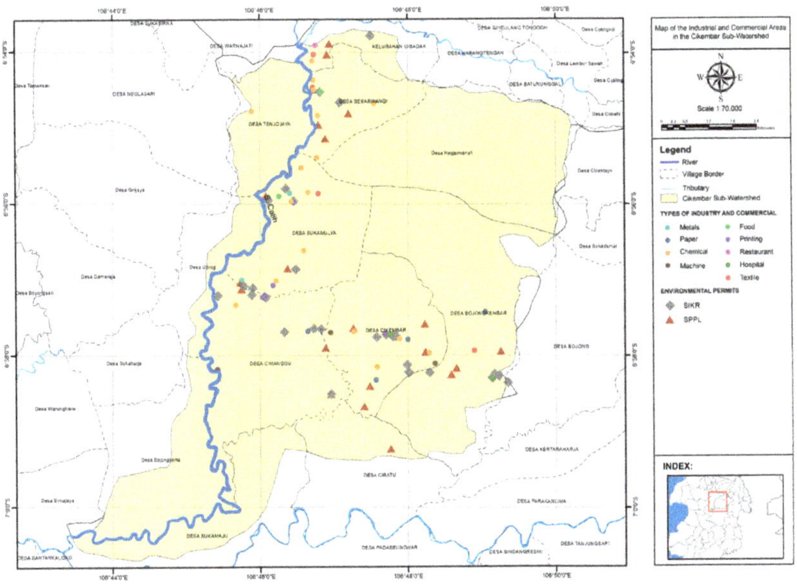

Figure 10. The Map of the Industrial and Commercial Areas in the Cikembar Sub-Watershed.

Table 14. Potential Pollution Load of the Industrial and Commercial Sector in the Cikembar Sub-watershed.

District	Village	Potential Pollution Load (kg/Day)	
		BOD	**COD**
Cibadak	Cibadak	366.31	549.47
	Sekarwangi	107.62	161.43
	Tenjojaya	28.47	42.71
Cikembar	Sukamulya	494.35	741.52
	Bojongkembar	38.15	57.22
	Cikembar	154.77	232.15
	Cimanggu	1.78	2.67
Total		1191.45	1787.18

Table 15. Potential Pollution Load of the Industrial and Commercial Sector in the Cikembar Sub-watershed Based on Industry Type.

Type of Industry	Potential Pollution Load (kg/Day)	
	BOD	**COD**
Metal goods	0.17	0.26
Paper	2.69	4.03
Chemical	540.99	811.48
Machine	0.13	0.20
Other foods	0.53	0.80
Printing	0.14	0.20
Restaurant	0.41	0.61
Hospital	8.36	12.55
Car & motorcycle parts	0.16	0.24
Textile	637.87	956.81
Total	1191.45	1787.18

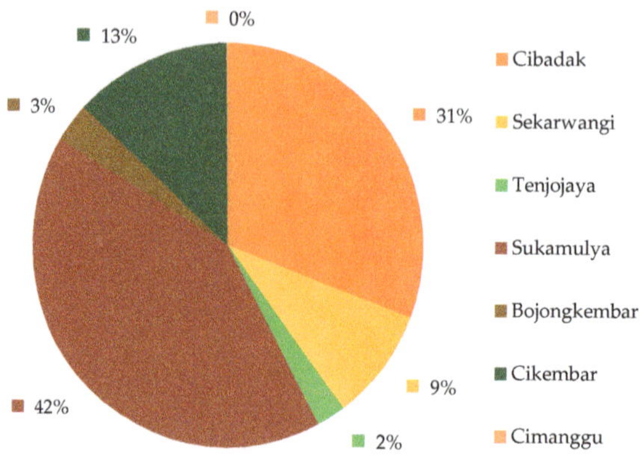

Figure 11. The Contribution of Each Village to the Potential Pollution Load of the Industrial and Commercial Sector in the Cikembar Sub-Watershed Based on Area.

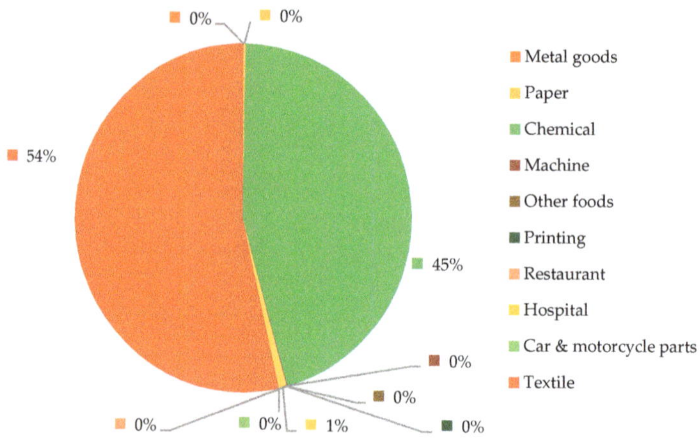

Figure 12. The Contribution of Each Village to the Potential Pollution Load of the Industrial and Commercial Sector in the Cikembar Sub-Watershed Based on Industry Type.

The potential pollution load generated from the industrial and commercial sectors is influenced by the number of employees and industry type; increase in employees results in increased potential pollution load. The results of the calculation show that Sukamulya Village and the textile industry are the biggest contributors to the potential pollution load of the industrial and commercial sectors in the Cikembar sub-watershed. However, due to limited data, only parameters BOD and COD are calculated and discussed in this manuscript.

Textile manufacturing in the Cikembar watershed, and generally in Indonesia, involves using dyes, chemicals, and finishing agents to color and treat fabrics. Wastewater generated from these processes contains various organic compounds and toxic substances. Various reasons may lead to improperly treated wastewater being discharged into rivers, significantly elevating BOD levels [19]. Also, these textile industries use various chemical additives, such as detergents, softeners, and bleach, which harm the environment when discharged to nearby rivers. Some of these chemicals are biodegradable, but others are not, and they can contribute to BOD pollution in the Cicatih River [26].

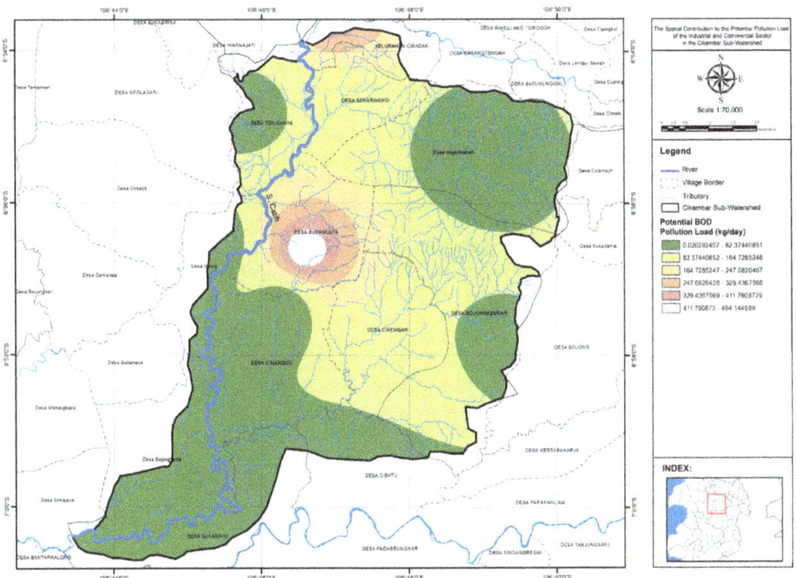

Figure 13. The Spatial Contribution to the Potential Pollution Load of the Industrial and Commercial Sector in the Cikembar Sub-Watershed.

The other major contributor to river pollution from the industrial sector comes from the chemical industries. In the Cikembar watershed, chemical industries can contribute to river pollution, including increasing BOD, in Indonesia through various processes and activities. Chemical manufacturing plants generate wastewater as a by-product of their processes, which contains various organic and inorganic compounds. When these effluents are discharged into nearby rivers without proper treatment, they significantly increase the BOD levels in the receiving water bodies. Further, accidental releases of chemicals or chemical spills from industrial facilities can introduce large quantities of toxic and organic compounds into rivers. These substances harm aquatic life and contribute to elevated BOD levels in the Cicatih River [25].

3.2.5. All Sector

The potential pollution load in the Cikembar sub-watershed is calculated based on four main sectors: domestic, agricultural, livestock, and industrial and commercial. Based on calculation results of the potential pollution load for the BOD parameter, the domestic sector is largest contributor to the pollution of the Cicatih River in the Cikembar sub-watershed. A total of 3604.46 kg of organic material presented in BOD value, enter the Cicatih river in the Cikembar sub-watershed daily from all sectors. The results of the calculation of the potential pollution load of all sectors in the Cikembar sub-watershed are shown in Table 16 and Figure 14 below.

The high levels of the TSS parameter are mainly caused by soil erosion eroding into water bodies. If not addressed, it will increase the turbidity of water and reduce the light that can enter the water, resulting in the death of aquatic organisms [25]. Elevated concentrations of Total-N pose a threat to water quality and certain aquatic organisms due to their potential toxicity. Conversely, heightened levels of Total-P precipitate algal blooms, causing oxygen depletion and light attenuation, thereby disrupting the ecological balance within the food chain and the overall aquatic ecosystem [27,28].

Table 16. The Potential Pollution Load of All Sectors in Cikembar Sub-watershed.

Type of Agriculture	Potential Pollution Load (kg/Day)				
	TSS	BOD	COD	Total-N	Total-P
Domestic	1802.93	1897.82	2609.50	92.52	9.96
Agricultural	0.20	66.94	100.41	5.97	2.98
Livestock	-	448.25	1071.09	1.26	0.60
Industrial and Commercial	-	1191.45	1787.18	-	-
Total	1803.13	3604.46	5568.17	99.74	13.54

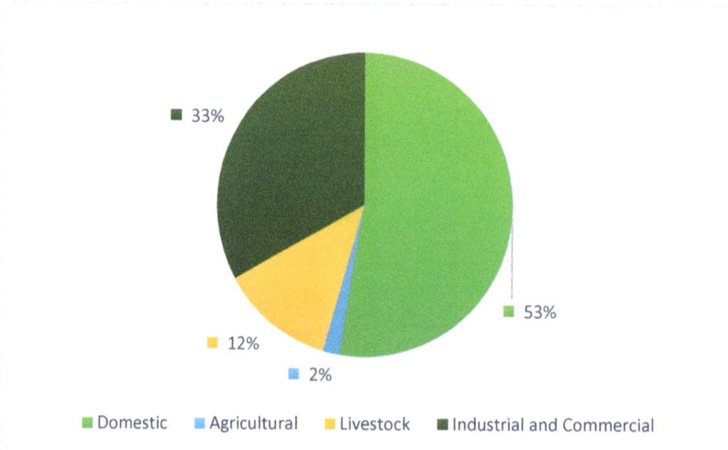

Figure 14. The Contribution of Each Sector to the Potential Pollution Load in the Cikembar Sub-watershed.

Other parameters, such as BOD and COD, are interrelated because they can be used to determine the oxygen content status in the water. The size of the oxygen content in the water can indicate whether the water is polluted [29]. A high BOD parameter value indicates polluted river water [18]. This also happened to the COD parameter. High values of BOD and COD affect the amount of oxygen needed to decompose organic matter that pollutes river water [30]. Increased oxygen consumption depletes the oxygen available in the river water, resulting in insufficient oxygen resources for aquatic organisms. If unresolved, the aquatic organisms will die, thus disrupting the food chain/ecosystem [29].

All explanations show the importance of domestic wastewater management. If the river water remains unattended, water quality deterioration is inevitable. Water quality is one of the most important aspects for the sustainability of water resources. Poor water quality will affect the environmental, social, health, and economic aspects [30], which is also exacerbated by climate change [31].

Bojongkembar Village needs special attention in managing the Cicatih River in the Cikembar sub-watershed because it is the largest contributor to the potential pollution load, especially in the domestic sector. As previously explained, river pollution from the domestic sector is strongly influenced by population. Therefore, discharging domestic wastewater directly or without treatment into the river must be prevented. Domestic waste management efforts can be conducted with a communal WWTP construction program. In addition, settlements near riverbanks must be prevented, and existing ones must be relocated. All these efforts must be made to reduce the potential pollution from the domestic sector and resolve other related problems, such as water-borne diseases, water availability shortage, and flood risk [16].

4. Conclusions

This research shows that the water quality status of the river in the Cikembar sub-watershed is slightly polluted, with BOD parameters exceeding the maximum threshold in all monitoring locations. Based on the results of the potential pollution load calculation for the BOD parameter, the domestic sector (53%) is the largest contributor to the pollution of the Cicatih River in the Cikembar sub-watershed, while the agricultural is the least (2%).

For the domestic sector, Bojongkembar village has the highest BOD contribution, with 16% of the total pollution load in the Cikembar sub-watershed. As for the agriculture sector, the highest pollution contributor is Bojongkerta village, which accounts for 30% of the total pollution load. For the livestock and industrial sectors, the highest contributors of BOD are Cikembar and Sukamulya villages, with BOD contributions of 57% and 42%, respectively. Based on the types of industries in the Cikembar sub-watershed, the main contributors of BOD are textile and chemical industries, with 54% and 45% contributions, respectively.

The potential pollution contribution of each village in the Cikembar sub-watershed to the Cicatih River will be used by the local government of Sukabumi Regency to develop programs to reduce the pollution load for each village.

To optimize the applicability of the findings outlined in this manuscript for diverse stakeholders, pollution load modeling in the Cicatih River will be performed. This will be accomplished through the utilization of Qual2K Version 5.1 and WASP Version 8.32 modeling software, considering variations in river flow during both rainy and dry seasons.

Author Contributions: Conceptualization, N.M.; Methodology, I.J. and A.D.S.; Software, A.L.I.; Validation, N.M.; Formal analysis, I.J. and A.D.S.; Investigation, I.J. and R.R.H.; Resources, I.J., A.S. and R.R.H.; Data curation, A.S. and A.L.I.; Writing—original draft, I.J. and R.R.H.; Writing—review & editing, I.J., N.M., A.L.I. and A.D.S.; Visualization, A.L.I.; Funding acquisition, A.S. All authors have read and agreed to the published version of the manuscript.

Funding: This work was supported by the Hibah Publikasi Terindeks Internasional (PUTI) Q1, Directorate of Research and Development, Universitas Indonesia NKB-521/UN2.RST/HKP.05.00/2022.

Data Availability Statement: The data presented in this study are available on request from the corresponding author.

Conflicts of Interest: The authors declare no conflict of interest.

References

1. Putri, F.K.; Hidayah, E.; Ma'ruf, M.F. Enhancing stormwater management with low impact development (LID): A review of the rain barrel, bioretention, and permeable pavement applicability in Indonesia. *Water Sci. Technol.* **2023**, *87*, 2345–2361. [CrossRef] [PubMed]
2. Capps, K.A.; Bentsen, C.N.; Ramírez, A. Poverty, urbanization, and environmental degradation: Urban streams in the developing world. *Freshw. Sci.* **2016**, *35*, 429–435. [CrossRef]
3. Singh, N.; Poonia, T.; Siwal, S.S.; Srivastav, A.L.; Sharma, H.K.; Mittal, S.K. Challenges of water contamination in urban areas. In *Current Directions in Water Scarcity Research*; Elsevier: Amsterdam, The Netherlands, 2022; Volume 6, pp. 173–202.
4. Sutadian, A.D.; Muttil, N.; Yilmaz, A.G.; Perera, B. Development of a water quality index for rivers in West Java Province, Indonesia. *Ecol. Indic.* **2018**, *85*, 966–982. [CrossRef]
5. Aroengbinang, B.W. Driving force analysis of landuse and cover changes in Cimandiri and Cibuni Watersheds. *Procedia Environ. Sci.* **2015**, *24*, 184–188. [CrossRef]
6. Government of Sukabumi Regency. *Kajian Lingkungan Hidup Strategis Kabupaten Sukabumi Tahun 2018*; Government of Sukabumi Regency: Sukabumi Regency, Indonesia, 2018.
7. Khan, M.H.R.B.; Ahsan, A.; Imteaz, M.; Shafiquzzaman, M.; Al-Ansari, N. Evaluation of the surface water quality using global water quality index (WQI) models: Perspective of river water pollution. *Sci. Rep.* **2023**, *13*, 20454. [CrossRef]
8. Varol, M.; Tokatlı, C. Evaluation of the water quality of a highly polluted stream with water quality indices and health risk assessment methods. *Chemosphere* **2023**, *311*, 137096. [CrossRef]
9. Juwana, I.; Rahardyan, N.A.; Permadi, D.A.; Sutadian, A.D. Uncertainty and Sensitivity Analysis of the Effective Implementation of Water Quality Improvement Programs for Citarum River, West Java, Indonesia. *Water* **2022**, *14*, 4077. [CrossRef]
10. Wang, B.; Wang, Y.; Wang, S. Improved water pollution index for determining spatiotemporal water quality dynamics: Case study in the Erdao Songhua River Basin, China. *Ecol. Indic.* **2021**, *129*, 107931. [CrossRef]

11. Lu, J.; Gong, D.; Shen, Y.; Liu, M.; Chen, D. An inversed Bayesian modeling approach for estimating nitrogen export coefficients and uncertainty assessment in an agricultural watershed in eastern China. *Agric. Water Manag.* **2013**, *116*, 79–88. [CrossRef]
12. Li, Y.; Xiao, H.; Zhao, Y.; Zhong, Y.; Fu, G.; Zhou, S.; Xu, Y.; Zhou, K. Study on total phosphorus pollution load estimation and prevention and control countermeasures in Dongting Lake. *Energy Rep.* **2023**, *9*, 294–305. [CrossRef]
13. Zhou, J.; Liu, X.; Liu, X.; Wang, W.; Wang, L. Assessing agricultural non-point source pollution loads in typical basins of upper Yellow River by incorporating critical impacting factors. *Process Saf. Environ. Prot.* **2023**, *177*, 17–28. [CrossRef]
14. Ministry of Environment and Forestry. *Guidelines for Determining Water Quality Status*; Ministry of Environment and Forestry: Nairobi, Kenya, 2002.
15. Herawati, E.Y.; Darmawan, A.; Valina, R.; Rachmanto, P.D.; Arinda, E.S.; Arsyaddirizali, A.; Khasanah, R.I. Analysis of water quality and primary productivity in Welang, Gembong, and Rejoso rivers, Pasuruan East Java, Indonesia. *Aquac. Aquar. Conserv. Legis.* **2023**, *16*, 508–522.
16. Yusuf, I. *Inventarisasi dan Identifikasi Sumber dan Beban Pencemaran Air*; Pusat Penelitian dan Pengembangan Sumber Daya Air Kementerian Pekerjaan Umum: Jakarta, Indonesia, 2010.
17. Kurniawan, B. *Inventarisasi dan Identifikasi Sumber Pencemar*; Kementerian Lingkungan Hidup dan Kehutanan: Jakarta, Indonesia, 2013.
18. Nugraha, W.D.; Hadi, S.P.; Sasongko, S.B.; Istirokhatun, T.; Faradiba, F.; Nopita, A.; Handayani, D.S.; Muhammad, F.I.; Susanto, H.; Budihardjo, M.A. Pollution load capacity assessment by utilizing QUAL2E modelling: A case study of Rambut River, Indonesia. *J. Ecol. Eng.* **2022**, *23*, 154–161. [CrossRef]
19. Djuwita, M.R.; Hartono, D.M.; Mursidik, S.S.; Soesilo, T.E.B. Pollution Load Allocation on Water Pollution Control in the Citarum River. *J. Eng. Technol. Sci.* **2021**, *53*, 210112. [CrossRef]
20. Pradana, H.A.; Novita, E.; Wahyuningsih, S.; Pamungkas, R. Analysis of deoxygenation and reoxygenation rate in the Indonesia River (a case study: Bedadung River East Java). *IOP Conf. Ser. Earth Environ. Sci.* **2019**, *243*, 012006. [CrossRef]
21. Fu, X.; Hou, R.; Yang, P.; Qian, S.; Feng, Z.; Chen, Z.; Wang, F.; Yuan, R.; Chen, H.; Zhou, B. Application of external carbon source in heterotrophic denitrification of domestic sewage: A review. *Sci. Total Environ.* **2022**, *817*, 153061. [CrossRef] [PubMed]
22. Shahid, A.; Khan, A.Z.; Malik, S.; Liu, C.-G.; Mehmood, M.A.; Syafiuddin, A.; Wang, N.; Zhu, H.; Boopathy, R. Advances in green technologies for the removal of effluent organic matter from the urban wastewater. *Curr. Pollut. Rep.* **2021**, *7*, 463–475. [CrossRef]
23. Wurtsbaugh, W.A.; Paerl, H.W.; Dodds, W.K. Nutrients, eutrophication and harmful algal blooms along the freshwater to marine continuum. *Wiley Interdiscip. Rev. Water* **2019**, *6*, e1373. [CrossRef]
24. Tian, M.; He, X.; Feng, Y.; Wang, W.; Chen, H.; Gong, M.; Liu, D.; Clarke, J.L.; van Eerde, A. Pollution by antibiotics and antimicrobial resistance in livestock and poultry manure in China, and countermeasures. *Antibiotics* **2021**, *10*, 539. [CrossRef]
25. Ahmad, K.; Iqhrammullah, M.; Rizki, D.R.; Aulia, A.; Mairizal, A.Q.; Purnama, A.; Qanita, I.; Abdulmadjid, S.; Puspita, K. Heavy metal contamination in aquatic and terrestrial animals resulted from anthropogenic activities in Indonesia: A review. *Asian J. Water Environ. Pollut.* **2022**, *19*, 1–8. [CrossRef]
26. Sarasi, V.; Primiana, I.; Harsanto, B.; Satyakti, Y. Sustainable supply chain of Indonesia's textile & apparel industry: Opportunities and challenges. *Res. J. Text. Appar.* **2023**. [CrossRef]
27. Asdak, C. *Hidrologi Dan Pengelolaan Daerah Aliran Sungai*; UGM Press: Depok, Indonesia, 2023.
28. Slamet, J.S. *Kesehatan Lingkungan*; Gajah Mada Universitiy Press: Depok, Indonesia, 1994.
29. Lestari, H.; Haribowo, R.; Yuliani, E. Determination of pollution load capacity using QUAL2Kw program on The Musi River Palembang. *Civ. Environ. Sci. J. (CIVENSE)* **2019**, *2*, 105–116. [CrossRef]
30. Sawyer, C.N.; McCarty, P.L.; Parkin, G.F. *Chemistry for Environmental Engineering and Science*; McGraw-Hill: New York, NY, USA, 2003.
31. Juwana, I.; Perera, B.; Muttil, N. Conceptual framework for the development of West Java water sustainability index. In Proceedings of the 18th World IMACS/MODSIM Congress, Cairns, QLD, Australia, 13–17 July 2009.

water

MDPI

Article

Characteristics and Changes in Water Quality Based on Climate and Hydrology Effects in the Cirata Reservoir

Panjie Wiranegara [1], Sunardi Sunardi [1,2,*], Dadan Sumiarsa [1,3] and Hafizan Juahir [4]

1 Department of Environmental Science, Postgraduate School, Universitas Padjadjaran, Jl. Dipati Ukur No. 35, Coblong, Bandung 40132, Indonesia; panjie17001@mail.unpad.ac.id (P.W.); dadan.sumiarsa@unpad.ac.id (D.S.)
2 Department of Biology, Faculty of Mathematics and Natural Science, Universitas Padjadjaran, Jl. Raya Bandung-Sumedang KM. 21 Jatinangor, Sumedang 45363, Indonesia
3 Department of Chemistry, Faculty of Mathematics and Natural Science, Universitas Padjadjaran, Jl. Raya Bandung-Sumedang KM. 21 Jatinangor, Sumedang 45363, Indonesia
4 East Coast Environmental Research Institute, Sultan Zainal Abidin University, Kuala Terengganu 21300, Malaysia; hafizanj@gmail.com
* Correspondence: sunardi@unpad.ac.id

Abstract: This research aimed to identify water quality changes in the Cirata Reservoir and the factors affecting them in terms of hydrology and climate. The sampling was carried out in both the rainy and dry seasons at 12 locations in the Cirata Reservoir. The Mann–Whitney *U*-test (different test) results showed that salinity, total suspended solids (TSS), the potential of hydrogen (pH), nitrate (NO_3-N), phosphate (PO_4), nitrate and phosphate content in the sediment were significantly different ($\alpha < 0.05$) between the rainy and dry seasons. The principal component analysis (PCA) results showed that the water quality characteristics in the Cirata Reservoir in the dry season were influenced by environmental conditions in the reservoir, especially by the floating cage aquaculture and climate conditions. The high solar radiation, low rainfall, and floating cage aquaculture increased the pH and amounts of dissolved oxygen (DO), ammonia (NH_3-N), PO_4, nitrate and phosphate in the sediment while decreasing transparency, salinity, TSS, and NH_3-N. During the rainy season, the high runoff from Citarum Watershed controlled the water quality characteristics of the Cirata Reservoir. In this season, transparency, salinity, pH, DO, NH_3-N, PO_4, nitrate and phosphate in the sediment increased, while TSS tended to be low. In general, the water volume addition decreased the nutrition and salinity concentration in the water body. However, a distinct phenomenon occurred in the Cirata Reservoir. The runoff from agriculture, settlement, livestock, and the Citatah Karst in the Upper Citarum Watershed increased nutrition and salinity in the reservoir. Land use in the Citarum Watershed and floating cage aquaculture had an important role in the reservoir water quality.

Keywords: Cirata Reservoir; Citarum Watershed; floating cage; dry season; rainy season

1. Introduction and Background

The maintenance and improvement of water quality in lentic systems is a challenge faced by the entire world. In particular, for artificial environments such as reservoirs, the water quality is affected by natural phenomena and human activities in the vicinity [1]. Changes in climate, season, and rainfall intensity affect reservoir water quality in tropical regions [2,3]. The prolonged dry season affects the water quality in terms of physical, chemical, and biological parameters [4,5]. In the rainy season, the reservoir water quality may deteriorate due to a combination of natural processes (rainfall intensity) and human activity [6]. In many cases, the deterioration of reservoir water quality is caused by complex pollution from both the watershed and the reservoir itself [7].

The Cirata Reservoir was built in 1988 for a 1008-megawatt (MW) hydropower plant, which supplied electricity for Java and Bali [8]. The other functions of the Cirata Reservoir are serving as a source of clean water and for irrigation, flood prevention, tourism, a site for aquaculture, and a conservation site for animals and plants [7]. The Cirata Reservoir is located in the middle of the Citarum Watershed [9] and is vulnerable to contamination by organic and inorganic pollutants from the Upstream Citarum Watershed [10].

For the last 30 years, the rapid economic growth in the Citarum Watershed spurred the conversion of green areas to open or built-up areas [11]. Deforestation, agricultural land clearing, urbanization, and construction lead to decreased water quality in the river [12]. Generally, the deterioration of water quality in the Citarum River is affected by agriculture, settlements, industry, livestock, and fisheries [13].

Agricultural activity is a main source of NO_3-N and phosphate (PO_4) flow to the Citarum Watershed [14]. Generally, farmers use urea (N) intensively, resulting in a very high nitrogen load in the Citarum Watershed [15]. Agricultural activity is also a main factor in soil erosion in the Upstream Citarum Watershed [16]. Land use conversions, such as forests into agricultural land and paddy fields, can increase the total suspended solids (TSS) in the Cirata Reservoir [17].

The high population in Bandung City contributes to a high nitrogen loading in the Citarum Watershed [15]. Domestic waste, such as feces and detergents, are the biggest sources of N and P [18]. The lack of an adequate sewage treatment system for domestic wastewater causes heavy flows of nitrogen, PO_4, and chemical oxygen demand (COD) into the Citarum River [19]. In addition, the increased settlement in the upper Citarum Watershed has had an effect on the dissolved oxygen (DO) decline [20].

It was estimated that 2700 industries were in the Citarum Watershed in 2018, which left 2800 tons of waste per day [21]. The disposal of textile industry effluent without a treatment process results in a high COD and biochemical oxygen demand (BOD) in the Citarum River [22]. This creates a dilemma for nearby inhabitants. For example, people in an industrial area in Majalaya compromise with poor water quality out of necessity because their income source relies on textile factories in their living area [23].

The cattle farm released 400 tons of cattle manure/day into the Citarum River [22]. Limited time and land are the farmers' reasons for not processing the cattle manure into biogas, fertilizer, or media for growing worms because several farmers do not have large areas of land [24]. Because of that, the cattle manure is transported by runoff to the river, and the nutrient concentration in the Citarum River increases.

Previously, the number of floating cage aquacultures in 1988 in the Cirata Reservoir was only 74 cages [25], but recently, it has increased to 85,393 cages [8]. The promising economic profits encourage fishers to feed their fish excessively to accelerate growth [26]. However, the uneaten pellets precipitate to the bottom of the reservoir [7], increasing PO_4 and eutrophication in the Cirata Reservoir [27].

The purpose of this study was to identify the characteristics and changes in water quality during the rainy and dry seasons in the Cirata Reservoir. The differences in water quality, climate, and hydrology parameters were tested using the Mann–Whitney *U*-test. A principal component analysis (PCA) was used to determine the water quality characteristics.

2. Materials and Methods

This research started with sampling and analysis of the water quality, climate, and hydrology in the Cirata Reservoir during the rainy and dry seasons. Afterward, different tests were used to identify significantly different parameters between the rainy and dry seasons. Then, each parameter was processed using the PCA method to obtain the water quality characteristics in each season. A flowchart of the research methodology is presented in Figure 1.

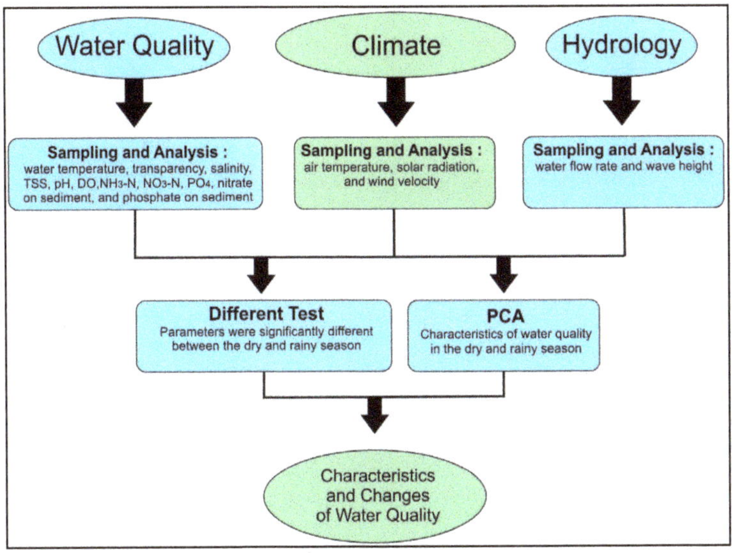

Figure 1. Flowchart of research methodology.

2.1. Research Area

The research area was located in the Cirata Reservoir, which is a part of the Middle Citarum Watershed. Administratively, the research area location is located in the Cianjur Regency (districts of Cikalong Kulon, Mande, Sukaluyu, Ciranjang, and Haurwangi), Purwakarta Regency (districts of Citamiang and Maniis), and Bandung Barat Regency (Cipeundeuy District). Geographically, the research location is located at 06°41′02.70″ S–06°47′51″ S and 107°14′42.13″ E–107°22′04.77″ E. A map of the research location is presented in Figure 2.

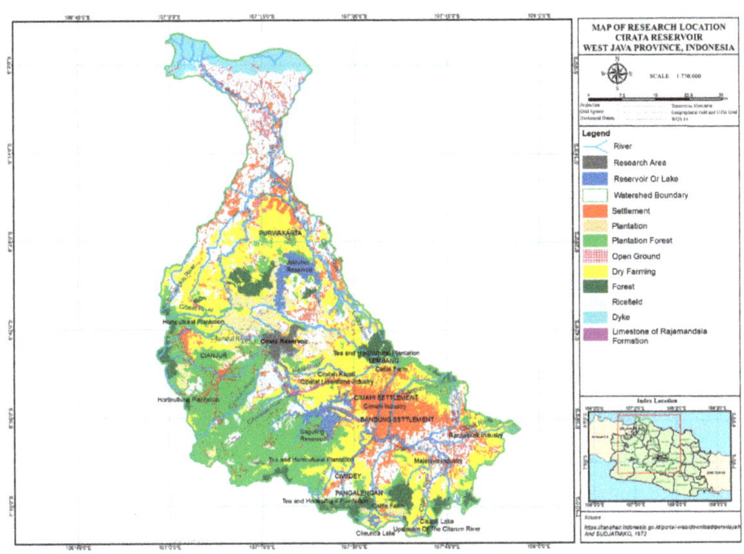

Figure 2. Map of the research area. Source: [28,29], modified by the authors.

2.2. Data Source

This research used primary data collected through sampling activity in the Cirata Reservoir. The 12 sampling sites were located in the Cirata Reservoir, where the samplings were carried out during the dry season (on 21 September 2019) and the rainy season (on 18 March 2020). The collected data consisted of water quality, climate, and hydrology parameters. The sampling sites are presented in Figure 3 and Table 1. Some images of the sampling locations during the survey activities are presented in Figure 4.

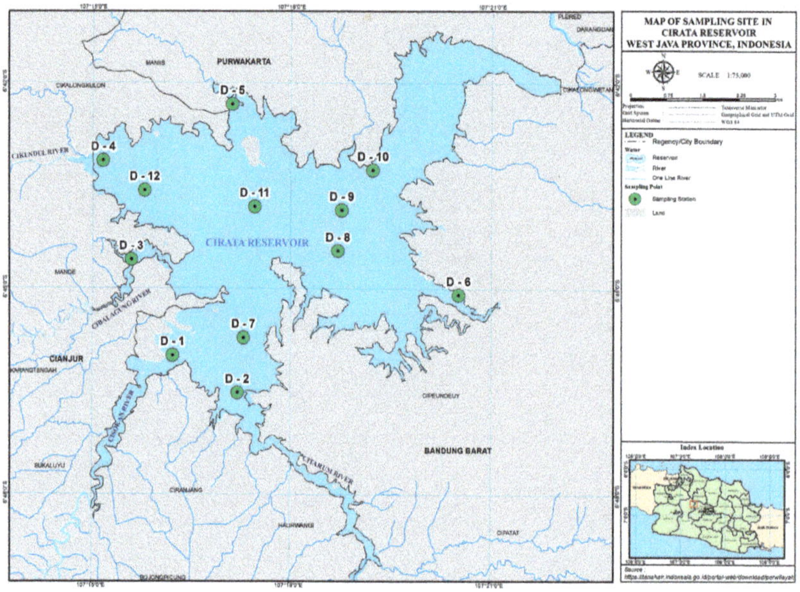

Figure 3. Map of sampling locations. Source: [28], modified by the authors.

Table 1. Information for sampling locations.

Site	Coordinate		Location	Sampling Location Characteristic
	Longitude	Latitude		
D-1	107°16′13.1″ E	06°45′59″ S	Cisokan River Mouth	River Mouth
D-2	107°17′11.9″ E	06°46′31.7″ S	Citarum River Mouth	River Mouth
D-3	107°15′35.0″ E	06°44′35.5″ S	Cibalagung River Mouth	River Mouth
D-4	107°15′06.1″ E	06°43′09″ S	Cikundul River Mouth	River Mouth
D-5	107°17′06.2″ E	06°42′19.1″ S	Cimanggu River Mouth	River Mouth
D-6	107°20′30.3″ E	06°45′05.5″ S	Nyenang	River Mouth
D-7	107°17′17.2″ E	06°45′43.5″ S	Pasir Pogor	Middle Reservoir
D-8	107°18′41.8″ E	06°44′27.1″ S	Sangkali	Middle Reservoir
D-9	107°18′45.2″ E	06°43′51.6″ S	Rawa Buaya	Middle Reservoir
D-10	107°19′13″ E	06°43′16.5″ S	Plumbon	Middle Reservoir
D-11	107°17′27.1″ E	06°43′48.7″ S	Plumbon Mosque	Middle Reservoir
D-12	107°15′47″ E	06°43′34.9″ S	Ciputri	Middle Reservoir

Figure 4. Sampling locations in the Cirata Reservoir. Note: (**A**) D-2 in the Citarum River mouth during the dry season; (**B**) D-7 in Pasir Pogor (middle reservoir) during the dry season; (**C**) D-11 in the Plumbon Mosque (middle reservoir) during the dry season; (**D**) D-2 in the Citarum River mouth during the rainy season; (**E**) D-7 in Pasir Pogor (middle reservoir) during the rainy season; and (**F**) D-11 in the Plumbon Mosque (middle reservoir) during the rainy season. The coordinates of each sampling location are presented in Table 1.

2.3. Analysis

This research analyzed the water quality, climate, and hydrology of the Cirata Reservoir. The water quality parameters consisted of transparency, salinity, TSS, pH, water temperature, DO, ammonia (NH_3-N), NO_3-N, PO_4, and nitrate and phosphate in the sediment. The climate parameters consisted of air temperature, solar radiation, and wind velocity. The hydrological parameters consisted of water flow rate and wave height.

Each water quality sample was tested in the field and then analyzed in the laboratory. The climate parameter measurements used a heat index WGT meter for air temperature, a Lutron LX—101A light meter for solar radiation, and an anemometer for air velocity. The measurement of the water flow rate parameter was performed using a current meter. The wave height parameter was measured using a HOBO water-level logger.

2.4. Statistical Analysis

The differences between water quality, hydrology, and climate parameters in the Cirata Reservoir were analyzed using the Mann–Whitney U-test method [30]. The Mann–Whitney U-test refers to equality below [31]:

$$U = \min\{U_1, U_2\}, \tag{1}$$

$$U_1 = n_1 n_2 + \frac{n_1(n_1+1)}{2} - R_1, \tag{2}$$

$$U_2 = n_1 n_2 + \frac{n_2(n_2+1)}{2} - R_2, \tag{3}$$

where:

U_1 = The total number of sample I observations preceding sample II observations;
U_2 = The total count of sample II observations preceding sample I observations;
n_1 or n_2 = The size of sample I (X_1) or sample II (X_2);
R_1 or R_2 = The rank sums of samples X_1 or X_2.

The water quality, climate, and hydrology data from each season were processed by PCA. The PCA results show the most useful parameters for explaining all data interpretations, reducing the data, and summarizing the statistical correlations between the parameters without reducing the original information [32]. The PCA equality is as follows [33]:

$$Z_{ij} = a_{i1}X_{1j} + a_{i2}X_{2j} + \cdots + a_{im}X_{mj} \tag{4}$$

where:

Z = The component score;
a = The component loading;
X = The measured value of the variable;
i = The component number;
j = The sample number;
m = The total number of variables.

3. Results

3.1. Water Quality

The variations in water temperature, transparency, salinity, TSS, pH, DO, NH_3-N, NO_3-N, nitrate and phosphate in the sediment during the dry season and rainy season in the Cirata Reservoir are presented in Tables 2 and 3, respectively. The water quality analysis showed that the average values of water temperature, salinity, DO, NH_3-N, and nitrate in the sediment were higher in the dry season. The average values of air temperature, salinity, DO, NH_3-N, and nitrate in the sediment in the dry season were 30.55 °C, 0.0225%, 4.21 mg/L, 0.043 mg/L, and 5.47 mg/kg, respectively, and in the rainy season, they were 29.91 °C, 0.0125%, 4.025 mg/L, 0.036 mg/L, and 2.46 mg/kg, respectively.

In contrast, transparency, pH, NO_3-N, PO_4, and phosphate in the sediment were higher during the rainy season. The average values of transparency, pH, NO_3-N, PO_4, and phosphate in the sediment in the rainy season were 74.16 cm, 8.27, 1.19 mg/L, 0.18 mg/L, and 9.23 mg/kg, respectively, and in the dry season, they were 66.33 cm, 7.57, 0.31 mg/L, 0.056 mg/L, and 6.72 mg/L, respectively.

Table 2. Analysis results of water quality, climate, and hydrology parameters in the Cirata Reservoir during the dry season.

Parameter	Unit	Site												Average
		D-1	D-2	D-3	D-4	D-5	D-6	D-7	D-8	D-9	D-10	D-11	D-12	
A. Water Quality														
Physical Parameters														
Water Temperature	°C	31	30.5	30.4	32.5	32.5	30	29.9	29.7	30.6	29.4	30.6	29.6	30.55833
Transparency	cm	62	70	50	15	45	92	75	85	85	80	80	57	66.33333
Salinity	%	0.03	0.02	0.02	0.03	0.02	0.02	0.03	0.02	0.02	0.02	0.02	0.02	0.0225
TSS	mg/L	2.9	3.3	6.8	5.4	3.6	6.5	7.1	10	4.3	3.3	3.6	5.9	5.225
Chemical Parameters														
pH	pH	7.28	7.48	7.15	7.41	8.22	7.24	7.65	7.24	7.96	7.45	8.21	7.59	7.573333
DO	mg/L	3.8	3.7	3.9	3.9	5.4	2	3.8	4.4	5.6	4	4.7	5.4	4.216667
NH_3-N	mg/L	0.007	0.027	0.013	0.024	0.132	0.013	0.04	0.013	0.078	0.025	0.119	0.03	0.043417
NO_3-N	mg/L	0.014	0.314	0.284	0.54	0.111	0.75	0.149	0.089	0.374	0.66	0.337	0.104	0.3105
Phosphate (PO_4)	mg/L	0.081	0.055	0.052	0.052	0.119	0.047	0.047	0.047	0.036	0.039	0.06	0.039	0.056167
Sediment Content														
Nitrate	mg/kg	4.503	5.311	5.282	5.664	4.752	5.117	6.092	5.443	5.726	6.21	5.668	5.904	5.472667
Phosphate	mg/kg	7.75	6.39	4.95	7.22	6.24	5.35	6.44	7.18	7.45	7.53	7.66	6.48	6.72
B. Climate														
Air Temperature	°C	32.7	33.6	31.9	39.4	37.6	29	35	31	34.7	33	36	30.5	33.7
Solar Radiation	lux	91,200	9700	88,000	42,000	42,600	1130	92,800	31,100	38,000	63,000	69,300	42,000	50,902.5
Wind Velocity	m/s	1.105	1.26	0.49	2.75	2.52	0.98	1.61	6.7775	4.37	4.05	3.61	4.985	2.875625
C. Hydrology														
Water Flow Rate	m/s	0.033	0.014286	0.05	0.5	0.02	0.009756	0.046875	0.05	0.1	0.06	0.046	0.037	0.080576
Wave Height	cm	0.5	1.4	0.3	0.3	0.5	0.5	0.6	4.9	3.8	1.5	4.6	1	1.658333

Table 3. Analysis results of water quality parameters in the Cirata Reservoir in the rainy season.

Parameter	Unit	Site												Average
		D-1	D-2	D-3	D-4	D-5	D-6	D-7	D-8	D-9	D-10	D-11	D-12	
A. Water Quality														
Physical Parameters														
Water Temperature	°C	31	31	29	28	33	30	30	30	28	29	28	32	29.91667
Transparency	cm	85	85	65	30	55	95	80	75	70	90	80	80	74.16667
Salinity	%	0.01	0.01	0	0.02	0.01	0.01	0.01	0.02	0.02	0.02	0.01	0.01	0.0125
TSS	mg/L	6.8	7.3	55	80	29	35	12	8.4	4.6	4.6	18	10	22.55833
Chemical Parameters														
pH	pH	7.92	8.81	7.92	8.03	8.56	8.1	8.2	7.84	8.44	7.55	8.83	9.05	8.270833
DO	mg/L	4.1	3.9	4.1	3.8	4.8	3.8	3.9	5.3	4.1	3	4.1	3.4	4.025
NH_3-N	mg/L	0.0054	0.0234	0.0054	0.0048	0.0246	0.016	0.0176	0.0105	0.0415	0.0091	0.092	0.1848	0.036258
NO_3-N	mg/L	1.051	0.577	1.051	1.276	0.938	0.818	1.156	1.456	1.449	1.464	1.637	1.449	1.1935
Phosphate (PO_4)	mg/L	0.0571	0.0727	0.0571	0.1143	0.1844	0.1844	0.1974	0.1948	0.2364	0.3195	0.2857	0.2468	0.179217
Sediment Content														
Nitrate	mg/kg	2.52	1.08	2.44	3.5	1.52	1.64	1.28	4.04	2.96	2.8	2.3	3.54	2.468333
Phosphate	mg/kg	10.74	7.16	10.4	9.62	8.12	6.3	5.78	17.16	11.6	10.04	7.46	6.38	9.23
B. Climate														
Air Temperature	°C	33.4	32	40.1	31	33.1	30.6	31.9	31	28.3	28.4	28.3	37	32.09167
Solar Radiation	lux	1140	1289	934	1088	672	265	499	437	441	470	326	1024	715.4167
Wind Velocity	m/s	1.36	3.04	0.685	1.35	1.25	1.275	1.295	3.94	4.065	5.05	0.7	1.25	2.105
C. Hydrology														
Water Flow Rate	m/s	0.071	0.1	0.036	0.2	0.027	0.055	0.0625	0.086	0.09	0.13	0.031	0.031	0.076625
Wave Height	cm	1.3	3.3	0.4	5.9	0.7	4.6	2.6	10.1	4.6	5.2	7.5	5.9	4.341667

The different test results showed that the parameters of salinity, TSS, pH, NO_3-N, PO_4, nitrate and phosphate in the sediment were significantly different ($\alpha < 0.05$) between the dry and rainy seasons. The different test results for water quality parameters are summarized in Table 4.

Table 4. Summary of the Mann–Whitney *U*-test results for water quality, climate, and hydrology parameters in the Cirata Reservoir.

Parameter	Unit	Average in Dry Season	Average in Rainy Season	α Significance ($\alpha < 0.05$)
Water Quality				
Water Temperature	°C	30.55	29.91	0.319
Transparency	cm	66.33	74.16	0.378
Salinity	%	0.0225	0.0125	0.001
TSS	mg/L	5.225	22.55	0.001
pH	pH	7.57	8.27	0.002
DO	mg/L	4.21	4.025	0.671
NH_3-N	mg/L	0.043	0.036	0.219
NO_3-N	mg/L	0.31	1.19	0.0001
PO_4	mg/L	0.056	0.179	0.0001
Nitrate in Sediment	mg/kg	5.47	2.46	0.0001
Phosphate in Sediment	mg/kg	6.27	9.23	0.045
Climate				
Air Temperature	°C	33.7	32.09	0.178
Solar Radiation	lux	50,902.5	715.41	0.0001
Wind Velocity	m/s	2.87	2.1	0.478
Hydrology				
Water Flow Rate	m/s	0.08	0.076	0.198
Wave Height	cm	1.65	4.34	0.014

Note: The gray color indicates significant differences in parameters between the two seasons.

3.2. Climate Conditions

The climate parameters in this research consisted of air temperature, solar radiation, and wind velocity in the Cirata Reservoir. The climate parameter conditions at each sampling site during the dry season and rainy season are presented in Tables 2 and 3, respectively. Generally, the average values of air temperature, solar radiation, and wind velocity in the dry season were higher than in the rainy season. The average values of air temperature, solar radiation, and wind velocity in the dry season were 33.70 °C, 50,902.5 lux, and 2.87 m/s, respectively, and in the rainy season, they were 32.09 °C, 715.41 lux, and 2.10 m/s, respectively.

The Mann–Whitney *U*-test showed that solar radiation was significantly different ($\alpha < 0.05$) between the dry and rainy seasons. The different test results for the climate parameters are summarized in Table 4.

3.3. Hydrological Conditions

The hydrological parameters in this research consisted of water flow rate and wave height. The hydrological parameter conditions in each location in the dry season and rainy season are presented in Tables 2 and 3, respectively. Generally, the average water flow rate was higher in the dry season: 0.08 m/s in the dry season and 0.07 m/s in the rainy season. In contrast, the average wave height was higher in the rainy season: 4.34 cm in the rainy season and 1.65 cm in the dry season.

Based on the different test results, the wave height was significantly different ($\alpha < 0.05$) between the dry and rainy seasons. The different test results for the hydrological parameters are summarized in Table 4.

3.4. Water Quality Characteristics in the Dry Season and Other Affected Parameters

PCA was used to determine the water quality characteristics in the Cirata Reservoir. Six principal components (PCs) were obtained that contained eigenvalue > 1 and explained 91.668% of the total variance, as shown in Table 5. In addition, the PCA results for the dry season are summarized in Table 6 and Figure 5.

Table 5. Eigenvalues of PCA in the dry season at the Cirata Reservoir.

Component	Eigen Value		
	Total	% Variance	Cumulative (%)
1	4.480	27.998	27.998
2	3.667	22.920	50.918
3	2.425	15.156	66.074
4	1.761	11.004	77.077
5	1.324	8.273	85.351
6	1.011	6.318	91.668
7	0.631	3.945	95.613
8	0.324	2.025	97.637
9	0.218	1.364	99.001
10	0.109	0.683	99.684
11	0.051	0.316	100.000

Note: The gray color shows components with eigenvalue > 1.

Table 6. Principal component (PC) loadings for the water quality, climate, and hydrology parameters of the Cirata Reservoir during the dry season.

Parameter	Component					
	1	2	3	4	5	6
Water Temperature	0.877					
Transparency	−0.637					
Salinity		−0.536				
TSS	−0.540				−0.605	
pH	0.641	0.627				
DO		0.687				
NH_3-N	0.664	0.566				
NO_3-N				−0.818		
PO_4	0.727		−0.566			
Nitrate in Sediment			0.648			
Phosphate in Sediment			0.460			0.553
Air Temperature	0.873					
Solar Radiation				0.704		
Wind Velocity		0.778				
Water Flow Rate			0.804			
Wave Height		0.842				

PC1 accounted for 27.998% of the total variance. PC1 showed significant positive loadings for water temperature, pH, NH_3-N, PO_4, and air temperature but significant negative loadings for transparency and TTS. The PC1 characteristics were affected by the dry season climate and floating cage aquaculture. The dry season climate was characterized by a high air temperature, high water temperature, and low rainfall. The low rainfall influenced the reduction of sedimentation and solid material transportation to the Cirata Reservoir, thus the TSS tended to be low. The uneaten pellets from floating cages increased the PO_4 in the reservoir water. The high PO_4 concentration stimulated algae growth that caused a greenish opaque color in the water, and thus, the water transparency tended to be low. Algal photosynthesis increases the pH in the reservoir water body. The incomplete degradation of uneaten pellets and dead fish promoted high NH_3-N levels in the reservoir water. Accordingly, the high NH_3-N concentration increased the pH in the reservoir water.

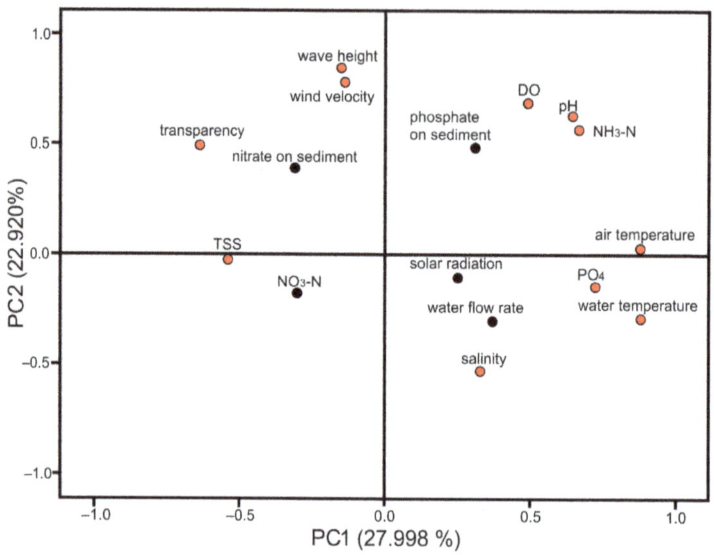

Figure 5. PCA diagram for the dry season. Note: Significant loadings are represented by red points, and black points are not significant loadings.

PC2 explained 22.920% of the total variance. PC2 showed significant positive loadings for pH, DO, NH_3-N, wind velocity, and wave height. In contrast, salinity had a significant negative loading. The dry season climate and floating cage aquaculture controlled the PC2 characteristics. The high wind velocity in the dry season increased the wave height in the Cirata Reservoir. The low rainfall caused low salinity due to the reduction in the inflow of potassium chloride (KCl) fertilizer residues from agricultural land and limestone dissolution from the Citatah Karst. The floating cage aquaculture caused uneaten pellet deposition and increased algae growth in the Cirata Reservoir. Algal photosynthesis increased the pH and DO in the reservoir water. The accumulation of uneaten pellets resulted in high NH_3-N levels in the reservoir water.

Approximately 15.156% of the total variance was described in PC3. A significant positive loading was shown in PC3 for the parameters of nitrate and phosphate in the sediment and water flow rate. In contrast, PO_4 had a significant negative loading. The PC3 characteristics were affected by the floating cage aquaculture and water movement. The uneaten pellets settled to the bottom of the floating cage, where they increased nitrate and phosphate in the sediment. The rapid water flow distributed the uneaten pellets around the reservoir bottom, which increased the nitrate and phosphate concentrations in the sediment. PO_4 tends to settle into reservoir sediments, and thus, its concentration decreases in the reservoir water.

PC4 described 11.004% of the total variance. PC4 showed a significant positive loading for solar radiation but a significant negative loading for NO_3-N. PC4 was controlled by the dry season climate. Sunny and cloudless weather resulted in high solar radiation around the Cirata Reservoir, and low rainfall in the dry season reduced NO_3-N input from the Citarum Watershed.

PC5 represented 8.273% of the total variance. In PC5, the TSS had a significant negative loading. The low TSS was affected by low rainfall during the dry season because rainfall is a carrier of sediment material to the Cirata Reservoir.

PC6 explained 6.318% of the total variance. PC6 showed a significant positive loading for phosphate in the sediment. The high phosphate levels in the sediment were due to the deposition of uneaten pellets from floating cage aquiculture during the dry season.

3.5. Water Quality Characteristics in the Rainy Season and Other Affected Parameters

In the PCA results, there were six PCs that had an eigenvalue > 1, which explained 91.879% of the total variance (Table 7). The PCA results of the rainy season are summarized in Table 8 and Figure 6.

Table 7. Eigenvalues of PCs for the rainy season at the Cirata Reservoir.

Component	Eigen Value		
	Total	% Variance	Cumulative (%)
1	5.036	31.477	31.477
2	3.279	20.493	51.970
3	2.163	13.517	65.486
4	1.789	11.181	76.668
5	1.386	8.665	85.333
6	1.047	6.546	91.879
7	0.550	3.436	95.315
8	0.406	2.535	97.850
9	0.220	1.374	99.224
10	0.084	0.526	99.750
11	0.040	0.250	100.000

Note: The gray color shows components with an eigenvalue > 1.

Table 8. Principal component (PC) loadings for the water quality, climate, and hydrology parameters of the Cirata Reservoir during the rainy season.

Parameter	Component					
	1	2	3	4	5	6
Water Temperature	−0.567					
Transparency		0.645	−0.561			
Salinity	0.884					
TSS		−0.670				
pH		0.593				0.497
DO				0.813		
NH_3-N		0.731	0.633			
NO_3-N	0.717					
PO_4	0.610	0.678				
Nitrate in Sediment	0.638		0.593			
Phosphate in Sediment	0.590			0.645		
Air Temperature	−0.704					
Solar Radiation	−0.509				0.675	
Wind Velocity	0.709					
Water Flow Rate	0.507	−0.626				
Wave Height	0.774					

PC1 accounted for 31.477% of the total variance. PC1 showed significant positive loadings for salinity, NO_3-N, PO_4, nitrate and phosphate in the sediment, wind velocity, water flow rate, and wave height, but significant negative loadings for water temperature, air temperature, and solar radiation. These patterns showed that the rainy season climate and the land use of the Citarum Watershed controlled PC1. In the rainy season, the high rainfall and wind velocity promoted increased water flow velocities and wave heights in the Cirata Reservoir. Agriculture, settlements, livestock, and the Citatah Karst in the Upstream Citarum Watershed affected the water quality of the Cirata Reservoir. The high rainfall flowed fertilizer residue, domestic waste, and livestock manure into the Cirata Reservoir, increasing NO_3-N, PO_4, nitrate and phosphate in the sediment, as well as its salinity. The increased salinity in the Cirata Reservoir is related to the contact of runoff with the Citatah Karst. Cloudy weather conditions in the rainy season decreased solar radiation, air temperature, and water temperature.

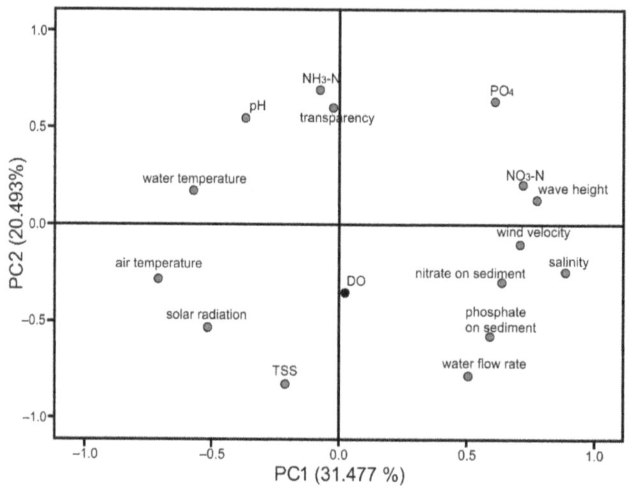

Figure 6. PCA diagram in the rainy season. Note: Significant loadings are represented by red points, and black points are not significant loadings.

PC2 described 20.493% of the total variance. PC2 had a significant positive loading for transparency, pH, NH_3-N, and PO_4. In contrast, TSS and water flow rate had significant negative loadings. This indicated that PC2 was related to runoff from the Citarum Watershed and increased water volumes during the rainy season. The runoff brought fertilizer residue, domestic waste, and livestock manure from the Citarum Watershed, which increased NH_3-N and PO_4 concentrations in the Cirata Reservoir. The runoff drove limestone dissolution in the Citatah Karst and transportation of calcium carbonate ($CaCO_3$), increasing the pH in the Cirata Reservoir. The increase in water volume from rainfall, runoff, and streamflow during the rainy season promoted dilution of the water body, which resulted in low TSS levels and high transparency.

PC3 represented 13.517% of the total variance and showed significant positive loadings for NH_3-N and nitrate in the sediment. PC3 was related to high rainfall and anthropogenic activities in the Citarum Watershed. Fertilizer residue, domestic waste, and livestock manure from the Citarum Watershed were pushed downstream by rainfall, increasing NH_3-N and nitrate in the sediment in the Cirata Reservoir.

PC4 explained 11.181% of the total variance. PC4 described significant positive loadings for DO and phosphate in the sediment. The high levels of DO and phosphate in the sediment were related to the land use in the Citarum Watershed and the rainy season climate. The high DO in the Cirata Reservoir was affected by the cold reservoir water and water inflow during the rainy season because oxygen in the atmosphere was more easily absorbed by cold water. Moreover, the high PO_4 concentration in the sediment was affected by fertilizer residue and domestic waste that were carried away by rainfall to the Cirata Reservoir.

PC5 explained 8.665% of the total variance and showed a significant positive loading for solar radiation. The high solar radiation showed that the samples collected at several locations were collected in sunny conditions.

PC6 described 6.546% of the total variance and had a significant positive loading for pH. The high pH in the rainy season was related to the Citatah Karst southeast of the Cirata Reservoir. The high runoff stimulated the limestone dissolution process in the Citatah Karst. Consequently, the $CaCO_3$ flowed to the Cirata Reservoir, which increased the pH.

4. Discussion

4.1. Water Quality Dynamics in the Dry Season

The water quality of the Cirata Reservoir in the dry season was controlled by the climate conditions and floating cage aquaculture. Solar radiation, air temperature, water temperature, wind velocity, water flow rate, wave height, pH, DO, NH_3-N, PO_4, and nitrate and phosphate in the sediment were high in the dry season. On the contrary, transparency, salinity, TSS, and NO_3-N were low.

The high solar radiation in the dry season increased the water temperature and air temperature in the Cirata Reservoir because solar radiation is the greatest source of heat for the reservoir [34]. The high wind velocity affected the increased water flow rate and wave height in the Cirata Reservoir. Furthermore, the low rainfall in the dry season influenced water quality in the Cirata Reservoir as a result of reduced runoff and material erosion from the Citarum Watershed.

The low TSS in the Cirata Reservoir during the dry season was related to the climate conditions, especially rainfall intensity. High rainfall distributes various solid materials into rivers and reservoirs, increasing the TSS [1,35]. The low rainfall in the dry season resulted in a low TSS level in the Cirata Reservoir because the rainfall acts as a medium for solid material transportation [36].

The low rainfall influenced the decrease in salinity in the Cirata Reservoir. The salinity concentration was related to the agricultural activity in the Upstream Citarum Watershed and the Citatah Karst located southeast of the Cirata Reservoir. Agricultural activities using KCI fertilizer could increase chloride (Cl^-) and salinity in the soil [37]. The Citatah Karst is composed of limestone [29], and its composition is dominated by $CaCO_3$ [36]. The inflow of Cl^- from agricultural activities and $CaCO_3$ from the Citatah Karst was decreased during runoff reduction in the dry season. Therefore, the salinity concentration in the dry season tended to be low.

During the dry season, the low rainfall was one of the factors that reduced the NO_3-N concentration in the Cirata Reservoir. The land use of the Upstream Citarum Watershed is dominated by agriculture and settlements [16]. Agricultural activities affect the NO_3-N concentrations in water [12]. The high rainfall can transport any material from agricultural land in the upstream watershed, which increased the nutrient levels in the reservoir water [1,2,6]. Conversely, the low rainfall decreased the NO_3-N distribution to the Cirata Reservoir. This pattern similarly happened in Thailand, where the low rainfall in the dry season resulted in low NO_3-N levels in the Kwan Phayao Reservoir [2].

The high PO_4 concentration during the dry season in the Cirata Reservoir was influenced by floating cage aquaculture. Most of the fisher continuously fed the fish until they were full [26]. The uneaten pellets accumulate as organic materials, which increase the PO_4 concentrations in the Cirata Reservoir [27].

The high PO_4 changed the Cirata Reservoir's trophic status to eutrophic [10], and it even reached a hypertrophic state in 2015 [27]. The eutrophic conditions drove algal blooms [4] because PO_4 is a key element for algal growth [1]. The rapid growth of algae increased organic turbidity [4], physically changing the water color to blue or greenish [38]. It caused low transparency in the Cirata Reservoir.

The high NH_3-N concentration in the dry season in the Cirata Reservoir was related to dense floating cages. The deposition of uneaten pellets disturbed aerobic and anaerobic bacteria activity, thus the incompletely degraded pellets produced NH_3-N, which is dangerous for aquatic organisms [39]. A large number of floating cages in the Cirata Reservoir also caused oxygen depletion in the water [8], and it potentially caused fish deaths. Fish death in the Cirata Reservoir occurs at a DO level under 2 mg/L, and it also increases the NH_3-N levels around the floating cages up to between 0.798 mg/L and 1.6 mg/L [40].

The increased pH and DO during the dry season in the Cirata Reservoir were influenced by the floating cage aquaculture. The deposition of uneaten pellets increased nutrient levels in the water, which drove algal growth [41]. During the photosynthesis process, algae consume carbon dioxide (CO_2), increasing the pH and DO in the water [42]. The photosynthesis processes of aquatic autotrophs are the main source of increasing oxygen levels in the water [43]. In addition, the accumulation of incompletely degraded pellets increases NH_3-N in the water [39]. Generally, NH_3-N is alkaline [44] and, therefore, increases the pH in the Cirata Reservoir.

The floating cage aquaculture and water current increased phosphate and nitrate levels in the sediment. The uneaten pellet deposits increase the sedimentation of organic material containing nitrogen and PO_4 onto the reservoir floor [7,39,41]. The water current can push sediment resuspension to quickly redeposit it around the reservoir floor [33], and the uneaten pellet deposits also spread to almost the entire reservoir bottom. Consequently, NO_3-N and PO_4 concentrations are increased in the reservoir sediments.

During the dry season, the conditions of the Cirata Reservoir area played an important role in the reservoir water quality. High solar radiation and low rainfall affected the water quality of the Cirata Reservoir. In addition, the floating cage aquaculture increased nutrient levels, pH, and DO and also decreased transparency.

4.2. *Water Quality Dynamics in the Rainy Season*

The climate conditions and water flow of the Citarum Watershed affected the water quality of the Cirata Reservoir in the rainy season. During this season, the transparency, salinity, pH, DO, NH_3-N, NO_3-N, PO_4, nitrate and phosphate in the sediment, wind velocity, water flow rate, and wave height were high. On the contrary, TSS, solar radiation, air temperature, and water temperature were low in the rainy season.

The cloudy weather during the rainy season decreased solar radiation, air temperature, and water temperature. The high rainfall and wind velocity increased the water flow rate and wave height in the Cirata Reservoir. The high DO was also influenced by the rainy season. The low air and water temperatures during the rainy season increased DO in the Cirata Reservoir. Generally, oxygen from the atmosphere is more easily absorbed by low-temperature water [34,44]. In addition, the high DO in the reservoir was also caused by the inflow of cold oxygenated waters from the watershed in the rainy season [45].

The high rainfall became a medium for waste transportation from the Citarum Watershed to the Cirata Reservoir [38]. Agriculture and settlements dominated the land use of the Citarum Watershed [16]. Generally, farmers in the Upstream Citarum Watershed use inorganic fertilizers that contain nitrogen and PO_4 [10]. Moreover, settlements along the Citarum River did not have adequate sanitation facilities [22], resulting in feces, urine, and detergents, which contain nitrogen (N) and phosphorus (P), to flow into the river [18]. The nutrient input was aggravated by livestock activities because about 57% of farmers in the Citarum Watershed dump their livestock manure directly without processing it [24]. The high rainfall transported agriculture, domestic, and livestock waste to the Cirata Reservoir, increasing NH_3-N, NO_3-N, PO_4, and nitrate and phosphate in the sediment in the Cirata Reservoir. This is similar to the patterns in the Kwan Phayao Reservoir, Thailand; the Adolfo López Mateos Reservoir, Mexico; and the Irapé Hydroelectric Power Plant Reservoir, Brazil. These areas also showed high nitrogen and PO_4 concentrations, which were related to agricultural runoff during the rainy season [1,2,6].

The high salinity during the rainy season was related to the agricultural activity in the Citarum Watershed and the Citatah Karst southeast of the Cirata Reservoir. The application of KCl fertilizer increased Cl^- and salinity in the soil [37]. The Citatah Karst is composed of limestone [29], and its composition is dominated by $CaCO_3$ [36]. The reaction of water with $CaCO_3$ resulted in calcium bicarbonate ($Ca(HCO3)_2$) [46]. Because of this, the confluence of runoff and limestone increased the salinity of the Citatah Karst. During the rainy season, the high runoff transported Cl^- from KCl fertilizer residues and $Ca(HCO3)_2$ from the Citatah Karst to the Cirata Reservoir, increasing the salinity. This is similar to the patterns in the Tombolo Dam, Tanzania, where the high salinity was affected by the limestone dissolution process and runoff from the upstream watershed in the rainy season [47].

The existence of the Citatah Karst was one of the factors for the high pH during the rainy season. The weathering of rocks that contain $CaCO_3$ drives carbonate transportation and pH increases in water bodies [48]. The high rainfall drove the Citatah Karst limestone weathering process and $CaCO_3$ transportation to the Cirata Reservoir, and increased its pH during the rainy season. This is similar to the situation in other tropical areas. For example, the pH was generally higher during the rainy season in a karst area in Yucatán, Mexico [49].

The high transparency and low TSS levels were affected by the water volume increase in the Cirata Reservoir during the rainy season. The high rainfall, runoff water, and streamflow diluted the water bodies, resulting in low TSS levels and high transparency. This is similar to the case in Bakun Reservoir, Malaysia [3]. The decrease in TSS concentrations and turbidity caused high transparency in the reservoir water. However, this is different from what occurred in the Atlanga Reservoir and Adolfo López Mateos Reservoir in Mexico. The high rainfall brought materials to the water body, increasing the TSS levels [1,35].

Land use in the Citarum Watershed and high rainfall were the main factors for water quality changes in the Cirata Reservoir during the rainy season. The inflow of water, fertilizer waste, livestock manure, domestic waste, and dissolved limestone from the Citarum Watershed to the Cirata Reservoir increased DO, nutrients, pH, and salinity and also decreased TSS levels in the reservoir.

The differences in water quality, climate, and hydrology between the rainy and dry seasons are summarized in Table 9. The table also illustrates some of the significantly different parameters and the loading contribution of each parameter.

Table 9. Summary of analysis results for water quality, climate, and hydrology parameters in the rainy and dry seasons.

Parameter	Season	Unit	Average	Significant Difference ($\alpha < 0.05$)	Loading Contribution for PC
A. Water Quality					
Physical Parameters					
Water Temperature	D R	°C	30.55833 29.91667	No	Positive Negative
Transparency	D R	cm	66.33333 74.16667	No	Negative Positive
Salinity	D R	%	0.0225 0.0125	Yes	Negative Positive
TSS	D R	mg/L	5.225 22.55833	Yes	Negative Negative
Chemical Parameters					
pH	D R	pH	7.573333 8.270833	Yes	Positive Positive
DO	D R	mg/L	4.216667 4.025	No	Positive Positive
NH_3-N	D R	mg/L	0.043417 0.036258	No	Positive Positive
NO_3-N	D R	mg/L	0.3105 1.1935	Yes	Negative Positive
PO_4	D R	mg/L	0.056167 0.179217	Yes	Positive Positive
Sediment Content					
Nitrate	D R	mg/kg	5.472667 2.468333	Yes	Positive Positive
Phosphate	D R	mg/kg	6.72 9.23	Yes	Positive Positive
B. Climate					
Air Temperature	D R	°C	33.7 32.09167	No	Positive Negative
Solar Radiation	D R	lux	50,902.5 715.4167	Yes	Positive Negative
Wind Velocity	D R	m/s	2.875625 2.105	No	Positive Positive
C. Hydrology					
Water Flow Rate	D R	m/s	0.080576 0.076625	No	Positive Positive
Wave Height	D R	cm	1.658333 4.341667	Yes	Positive Positive

Note: D is the rainy season and R is the dry season.

5. Conclusions

The water quality characteristics in the Cirata Reservoir during the dry season were affected by the environmental conditions in the reservoir. The high solar radiation stimulated an increase in water temperature. Furthermore, the low rainfall resulted in low salinity and concentrations of TSS and NO_3-N in the Cirata Reservoir. The deposition of pellets from floating cages directly or indirectly increased pH, DO, NH_3-N, PO_4, nitrate and phosphate in the sediments and decreased transparency.

The water quality characteristics in the Cirata Reservoir during the rainy season were highly related to the environmental dynamics outside the reservoir, especially the Citarum Watershed conditions. The high rainfall transported agriculture, livestock, and domestic waste from the Citarum Watershed to the Cirata Reservoir, increasing salinity, pH, NH_3-N, NO_3-N, PO_4, and nitrate and phosphate levels in the sediment. Furthermore, the increased water inflow from the Citarum Watershed resulted in a high DO, high transparency, and low TSS levels in the Cirata Reservoir.

The water quality conditions in the dry season and rainy season showed that the floating cage aquaculture and land use of the Citarum Watershed had negative effects on the water quality of the Cirata Reservoir. The fishers using floating cages should be involved in education and training on efficient feeding methods and floating cage limitations because dense floating cages and overfeeding decrease fishery productivity [8].

The complex activity in the Citarum Watershed area should be managed using an environmentally and economically advantageous approach. Farmers should be involved in education on manure processing. Furthermore, farmers should be encouraged to use livestock manure for fertilizer and should be educated about efficient fertilizer application. In addition, any stakeholder should plant commercial trees along the Citarum levee that can serve as sediment traps and increase economic value.

Therefore, integrated management between the Citarum Watershed and the Cirata Reservoir is needed to maintain environmental sustainability in the Cirata Reservoir to continuously provide benefits for hydropower plants, fishers' livelihoods, tourism, and a clean water supply.

Author Contributions: Research concept, P.W., S.S. and H.J.; methodology, P.W., S.S. and D.S.; field investigation, P.W.; laboratory analysis, P.W. and D.S.; article draft preparation, P.W. and S.S.; supervision, S.S. and H.J. All authors have read and agreed to the published version of the manuscript.

Funding: This research was partly funded by Universitas Padjadjaran.

Data Availability Statement: The authors have made the data accessible. Furthermore, the laboratory analysis results are available at the Center for Environment and Sustainability Science (CESS), Universitas Padjadjaran, Indonesia.

Acknowledgments: The authors wish to thank CESS for their support in the sampling and data analysis. The authors also wish to thank the East Coast Environmental Research Institute (ESERI), UniSZA, Malaysia, for research collaboration in the Cirata Reservoir.

Conflicts of Interest: The authors declare no conflict of interest.

References

1. Loaiza, J.G.; Rangel-Peraza, J.G.; Sanhouse-García, A.J.; Monjardín-Armenta, S.A.; Mora-Félix, Z.D.; Bustos-Terrones, Y.A. Assessment of Water Quality in a Tropical Reservoir in Mexico: Seasonal, Spatial and Multivariable Analysis. *Int. J. Env. Res. Public Health* **2021**, *18*, 7456. [CrossRef] [PubMed]
2. Ngearnpat, N.; Klayluk, B.; Kumla, A.; Ngamta, S.; Issakul, K. Phytoplankton Composition and Water Quality of Kwan Phayao Reservoir, Thailand, during Rainy and Cold Dry Seasons. *J. Food Health Bioenviron. Sci.* **2018**, *11*, 46–55.
3. Ling, T.; Gerunsin, N.; Soo, C.; Nyanti, L.; Sim, S.; Grinang, J. Seasonal Changes and Spatial Variation in Water Quality of a Large Young Tropical Reservoir and Its Downstream River. *J. Chem.* **2017**, *2017*, 8153246. [CrossRef]
4. Braga, G.G.; Becker, V.; Oliveira, J.N.P.D.; Junior, J.R.D.M.; Bezerra, A.F.D.M.; Torres, L.M.; Galvão, Â.M.F.; Mattos, A. Influence of Extended Drought on Water Quality in Tropical Reservoirs in a Semiarid Region. *Acta Limnol. Bras.* **2015**, *27*, 15–23. [CrossRef]

5. Arias-Rodriguez, L.F.; Duan, Z.; Sepúlveda, R.; Martinez-Martinez, S.I.; Disse, M. Monitoring Water Quality of Valle de Bravo Reservoir, Mexico, Using Entire Lifespan of MERIS Data and Machine Learning Approaches. *Remote Sens.* **2020**, *12*, 1586. [CrossRef]
6. Oliveira, K.L.D.; Ramos, R.L.; Oliveira, S.C.; Christofaro, C. Water Quality Index and Spatio-Temporal Perspective of a Large Brazilian Water Reservoir. *Water Supply* **2021**, *21*, 971–982. [CrossRef]
7. Sunardi, S.; Ariyani, M.; Agustian, M.; Withaningsih, S.; Parikesit, P.; Juahir, H.; Ismail, A.; Abdoellah, O.S. Water Corrosivity of Polluted Reservoir and Hydropower Sustainability. *Sci. Rep.* **2020**, *10*, 11110. [CrossRef]
8. Awaliyah, F.; Gunawan, W.; Tasrif, M. Analysis of Changes in Water Quality Used for Supporting Fish Productivity of Floating Cage/KJA (Keramba Jaring Apung) in Cirata Reservoir. *IOP Conf. Ser. Earth Environ. Sci.* **2019**, *306*, 012014. [CrossRef]
9. Masrofah, I.; Putro, B.A. Clustering of the Water Characteristics of the Cirata Reservoir Using the k-means clustering Method. *AIP Conf. Proc.* **2020**, *2217*, 030010. [CrossRef]
10. Ariyani, M.; Agustian, M.; Maharani, G.S.; Sunardi. Assessment of Seasonal Trophic State of Tropical Man-Made Lake, the Cirata Reservoir. *IOP Conf. Ser. Earth Environ. Sci.* **2019**, *277*, 012030. [CrossRef]
11. Kuntoro, A.A.; Cahyono, M.; Soentoro, E.A. Land Cover and Climate Change Impact on River Discharge: Case Study of Upper Citarum River Basin. *J. Eng. Technol. Sci.* **2018**, *50*, 364–381. [CrossRef]
12. Tromboni, F.; Dilts, T.E.; Null, S.E.; Lohani, S.; Ngor, P.B.; Soum, S.; Hogan, Z.; Chandra, S. Changing Land Use and Population Density Are Degrading Water Quality in the Lower Mekong Basin. *Water* **2021**, *13*, 1948. [CrossRef]
13. Mudjiardjo, A.S.U.; Moersidik, S.S.; Darmajanti, L. Analysis of Water Pollution Using the STORET Method in the Upper Citarum Watershed. *IOP Conf. Ser. Earth Environ. Sci.* **2021**, *716*, 012012. [CrossRef]
14. Marselina, M.; Wibowo, F.; Mushfiroh, A. Water Quality Index Assessment Methods for Surface Water: A Case Study of the Citarum River in Indonesia. *Heliyon* **2022**, *8*, e09848. [CrossRef] [PubMed]
15. Yoshida, K.; Tanaka, K.; Noda, K.; Homma, K.; Maki, M.; Hongo, C.; Shirakawa, H.; Oki, K. Quantitative Evaluation of Spatial Distribution of Nitrogen Loading in the Citarum River Basin, Indonesia. *J. Agric. Meteorol.* **2017**, *73*, 31–44. [CrossRef]
16. Chaidar, A.N.; Soekarno, I.; Wiyono, A.; Nugroho, J. Spatial Analysis of Erosion and Land Criticality of the Upstream Citarum Watershed. *Int. J. Geomate* **2017**, *13*, 133–140. [CrossRef]
17. Sunardi, S.; Nursamsi, I.; Dede, M.; Paramitha, A.; Arief, M.C.W.; Ariyani, M.; Santoso, P. Assessing the Influence of Land-Use Changes on Water Quality Using Remote Sensing and GIS: A Study in Cirata Reservoir, Indonesia. *Sci. Technol. Indones.* **2022**, *7*, 106–114. [CrossRef]
18. Suwarno, D.; Löhr, A.; Kroeze, C.; Widianarko, B. Fast Increases in Urban Sewage Inputs to Rivers of Indonesia. *Env. Dev. Sustain.* **2014**, *16*, 1077–1096. [CrossRef]
19. Yokosawa, R.; Mizunoya, T. Improving Water Quality in the Citarum River through Economic Policy Approaches. *Sustainability* **2022**, *14*, 5038. [CrossRef]
20. Nurhayati, S.A.; Sabar, A.; Marselina, M. The Effect of Land Use Changes to Discharge Extremities in Cimahi Watershed—Upper Citarum Watershed, West Java. *E3S Web. Conf.* **2020**, *148*, 07002. [CrossRef]
21. Sholeh, M.; Pranoto, P.; Budiastuti, S.; Sutarno, S. Analysis of Citarum River Pollution Indicator using Chemical, Physical, and Bacteriological Methods. *AIP Conf. Proc.* **2018**, *2049*, 020068. [CrossRef]
22. Musnansyah, A.; Kamil, A.A.; Marliana, L.; Widayati, E.; Zulfakriza. Assessment of Spatial Water Quality Observation of Citarum River Bandung Regency Using Multivariate Statistical Methods. *Int. J. Adv. Sci. Eng. Inf. Technol.* **2021**, *11*, 356–362. [CrossRef]
23. Bainus, A.; Darmawan, W.B.; Yulianti, D.; Husin, L.H. Between Fear and Survival: Human Security Issues in Citarum River Basin, Indonesia. *J. Hum. Secur.* **2021**, *17*, 4–14. [CrossRef]
24. Ariningsih, E.; Erwidodo; Irawan, A.R.; Saliem, H.P. Dairy Cattle Manure Utilization by Smallholder Dairy Farmers in West Java, Indonesia. *E3S Web. Conf.* **2022**, *361*, 03013. [CrossRef]
25. Garno, Y.S. Beban Pencemaran Limbah Perikanan Budidaya Dan Yutrofikasi Di Perairan Waduk Pada DAS Citarum. *J. Teknol. Lingkung.* **2002**, *3*, 112–120. [CrossRef]
26. Marselina, M.; Burhanudin, M. Phosphorus Load Concentration in Tropical Climates Reservoir for Each Water Quantity Class. *J. Water Land Dev.* **2018**, *36*, 99–140. [CrossRef]
27. Pratiwi, N.T.M.; Imran, Z.; Ayu, I.P.; Iswantari, A.; Wulandari, D.Y. The Phosphorus Load and the Variation of the Trophic States of Cirata Reservoir (West Java, Indonesia) from 1988 to 2017. *Biodiversitas* **2020**, *21*, 4176–4183. [CrossRef]
28. Badan Infomasi Geospasial. Geospasial Untuk Negeri. Available online: https://tanahair.indonesia.go.id/portal-web/download/perwilayah (accessed on 4 July 2023).
29. Sudjatmiko. *Peta Geologi Lembar Cianjur, Jawa*; Pusat Penelitian dan Pengembangan Geologi: Bandung, West Java, Indonesia, 1977.
30. Boyacioglu, H. Spatial Differentiation of Water Quality Between Reservoirs Under Anthropogenic and Natural Factors Based on Statistical Approach. *Arch. Environ. Prot.* **2014**, *40*, 41–50. [CrossRef]
31. Yue, S.; Wang, C.Y. The Influence of Serial Correlation on the Mann–Whitney Test for Detecting a Shift in Median. *Adv. Water Resour.* **2002**, *25*, 325–333. [CrossRef]
32. Helena, B.; Pardo, R.; Vega, M.; Barrado, E.; Fernandez, J.M.; Fernandez, L. Temporal Evolution of Groundwater Composition in an Alluvial Aquifer (Pisuerga River, Spain) by Principal Component Analysis. *Water Res.* **2000**, *34*, 807–816. [CrossRef]
33. Magyar, N.; Hatvani, I.G.; Székely, I.K.; Herzig, A.; Dinka, M.; Kovacs, J. Application of Multivariate Statistical Methods in Determining Spatial Changes in Water Quality in the Austrian Part of Neusiedler See. *Ecol. Eng.* **2013**, *55*, 82–92. [CrossRef]

34. Wetzel, R.G. *Limnology Lake and River Ecosystems*, 3rd ed.; Academic Press An Elsevier Science Imprint: San Diego, CA, USA, 2001; Volume 72, pp. 151–633.
35. Castilla-Hernández, P.; Torres-Alvarado, M.D.R.; Luis, J.A.H.; Cruz-López, N. Water Quality of a Reservoir and Its Major Tributary Located in East-Central Mexico. *Int. J. Environ. Res. Public Health* **2014**, *11*, 6119–6135. [CrossRef]
36. Nichols, G. *Sedimentology and Stratigraphy*, 2nd ed.; John Wiley & Sons Ltd.: Chichester, UK, 2009; Volume 5, p. 93.
37. Buvaneshwari, S.; Riotte, J.; Sekhar, M.; Sharma, A.K.; Helliwell, R.; Kumar, M.S.M.; Braun, J.J.; Ruiz, L. Potash Fertilizer Promotes Incipient Salinization in Groundwater Irrigated Semi-Arid Agriculture. *Sci. Rep.* **2020**, *10*, 3691. [CrossRef]
38. Luckyardi, S.; Gaol, T.V.L.; Oktafiani, D. Assessment of the Water Quality and Environmental Management in Jangari Reservoir Watershed Using Selected Physical, Chemical, and Biological Parameters. *J. Eng. Sci. Technol.* **2021**, *16*, 4518–4529.
39. Andriyani, N.; Mahdiana, A.; Dewi, R.; Subagyo, Y.; Siregar, A.S.; Junaidi, T.; Harisam, T. Biodiversity of Algae Potentially HABS (Harmfull Algae Blooms) in Reservoir Mrica, Banjarnegara. *E3S Web. Conf.* **2018**, *47*, 03001. [CrossRef]
40. Effendi, I.; Nirmala, K.; Saputra, U.H.; Sudrajat, A.O.; Zairin, M.; Kurokura, H. Water Quality Fluctuations Under Floating Net Cages for Fish Culture in Lake Cirata and Its Impact on Fish Survival. *Fish. Sci.* **2005**, *71*, 972–977. [CrossRef]
41. Marselina, M.; Burhanudin, M. Trophic Status Assessment of Saguling Reservoir, Upper Citarum Basin, Indonesia. *Air Soil Water Res.* **2017**, *10*, 1178622117746660. [CrossRef]
42. Melo, R.R.C.D.; Barbosa, I.M.B.R.; Ferreira, A.A.; Firmo, A.L.B.; Silva, S.R.D.; Cirilo, J.A.; Aquino, R.R.B.D. Influence of Extreme Strength in Water Quality of the Jucazinho Reservoir, Northeastern Brazil, PE. *Water* **2017**, *9*, 955. [CrossRef]
43. Ajayan, A.; Kumar, K.G.A. On the Seasonal Changes in the Surface Water Chemistry of Museum Lake, Thiruvananthapuram, Kerala, India. *Pollution* **2016**, *2*, 103–114. [CrossRef]
44. Cole, G.A. *Textbook of Limnology*, 4th ed.; Waveland Press Inc.: Long Grove, IL, USA, 1994; p. 243.
45. Klake, R.K.; Nartey, V.K.; Doamekpor, L.K.; Dinku, D.E. Seasonal Variation in Water Quality of the Weija Dam, Ghana. *J. Nat. Sci. Res.* **2015**, *5*, 20.
46. Jiang, J.; Huo, Z.; Feng, S.; Zhang, C. Effect of irrigation amount and water salinity on water consumption and water productivity of spring wheat in Northwest China. *Field Crops Res.* **2012**, *137*, 78–88. [CrossRef]
47. Shemsanga, C.; Muzuka, A.N.N.; Martz, L.; Komakech, H.C.; Elisante, E.; Kisaka, M.; Ntuza, C. Origin and Mechanisms of High Salinity in Hombolo Dam and Groundwater in Dodoma Municipality Tanzania, Revealed. *Appl. Water Sci.* **2017**, *7*, 2883–2905. [CrossRef]
48. Shabala, A.N.; Combrinck, L.; McCrindle, R. Effect of Farming Activities on Seasonal Variation of Water Quality of Bonsma Dam, KwaZulu-Natal. *S. Afr. J. Sci.* **2013**, *109*, 1–7. [CrossRef]
49. Long, D.T.; Pearson, A.L.; Voice, T.C.; Polanco-Rodríguez, A.G.; Sanchez-Rodríguez, E.C.; Xagoraraki, I.; Concha-Valdez, F.G.; Puc-Franco, M.; Lopez-Cetz, R.; Rzotkiewicz, A.T. Influence of Rainy Season and Land Use on Drinking Water Quality in a Karst Landscape, State of Yucatán, Mexico. *Appl. Geochem.* **2018**, *98*, 265–277. [CrossRef]

water

MDPI

Article

Combined Toxicity of Glyphosate (Faena®) and Copper to the American Cladoceran *Daphnia exilis*—A Two-Generation Analysis

Miriam Hernández-Zamora, Alma Rodríguez-Miguel, Laura Martínez-Jerónimo and Fernando Martínez-Jerónimo *

Instituto Politécnico Nacional, Escuela Nacional de Ciencias Biológicas, Lab. de Hidrobiología Experimental, Prol. de Carpio y Plan de Ayala s/n, Col. Santo Tomás, Mexico City 11340, Mexico

* Correspondence: fjeroni@ipn.mx

Abstract: Glyphosate and copper are common pollutants in water and soil. Glyphosate is the most used herbicide worldwide; despite being claimed to be a low-toxicity pesticide and easily degraded, several studies have demonstrated glyphosate's occurrence and toxicity in freshwater ecosystems. Copper is an essential micronutrient; however, at high concentrations, it becomes toxic, and it is a persistent contaminant discharged from agricultural and industrial activities. Both glyphosate and Cu are agrochemicals that can occur in aquatic environments and produce stress in aquatic biota. Cladocerans are important zooplankters, and their sensitivity to chemical stressors has been widely documented. In this study, the North American cladoceran *Daphnia exilis* was exposed to glyphosate (Faena®) and Cu mixtures. The effects were assessed in two generations to determine toxicity impairment in the parental (P_1) and filial (F_1) generations. The 48 h median lethal concentration (LC_{50}) of both chemicals was determined. After this, the generations P_1 and F_1 were exposed for 21 days to three concentrations of glyphosate and Cu mixtures (1.04 mg L^{-1} + 2.45 μg L^{-1}, 1.24 mg L^{-1} + 3.09 μg L^{-1}, and 1.57 mg L^{-1} + 4.31 μg L^{-1}), starting from neonates. Survival and reproduction were recorded, and macromolecule content and the size of neonates were measured in the progeny. The LC_{50} was 4.22 mg L^{-1} for Faena® and 13.45 μg L^{-1} for copper. Exposure to glyphosate and copper reduced fecundity and the number of clutches per female, delayed age at first reproduction, and increased the number of aborted eggs; these effects were more evident in the F_1 than in the P_1. In both generations, the concentration of carbohydrates and lipids was significantly reduced. The treatment and the brood number influenced the total length of neonates, but the total length did not differ significantly, whereas body length and body width decreased in F_1. Glyphosate and copper mixtures significantly increased toxicity in *D. exilis* in the F_1 generation, probably because the parents produced impaired progenies. Results warn about transgenerational effects in planktonic species chronically exposed to pollutants.

Keywords: zooplankton; herbicides; ecotoxicology; transgenerational effects; water pollution; toxic metals

1. Introduction

Freshwater ecosystems are susceptible to chemical pollution, which negatively affects the aquatic biota. In natural environments, organisms are rarely exposed to single pollutants, and the complex chemical interactions among pollutants, producing changes in their toxicity [1], usually are overridden. In an ecological risk assessment, studying the toxic effects of individual pollutants by evaluating reproductive and population effects using reference species is common. However, individual studies are insufficient to determine the toxicity of mixtures and their impact on biota because the action of substances may change due to chemical interactions that modify the endpoints [1,2]. Therefore, in recent years, the assessment of mixtures has become a new way of studying pollutants in the environment and in biological matrices. Nevertheless, it is true that to determine the combined effect of a mixture, it is necessary to previously know the toxicity of each component; therefore, the individual analysis of components as a traditional approach continues to be essential for the cumulative risk assessment [3].

Glyphosate is a herbicide widely used in agricultural and non-agricultural activities. It is marketed in different chemical formulations worldwide, including in Mexico, where it is sold under different trade names, with Faena® the most popular. Toxicological studies have demonstrated severe damage to aquatic organisms caused by glyphosate-based herbicide exposure, for instance, alteration in reproduction associated with endocrine disruption and oxidative stress [4], inhibition of enzymatic activity [5], and, more recently, cytotoxicity in rat and human cells [6,7]. Furthermore, glyphosate toxic effects have been associated with pesticide mixtures [8,9], but information about their interactions with metals is still scarce [10].

Copper is an oligo-nutrient for all living beings, but, as happens for all the essential metals, when quantities exceed the nutritional requirements, they exert toxic effects. Cu has been used to control fungi and bacteria in agriculture since the 19th century, and nowadays it is routinely used with other pesticides for integrated pest management [11,12]. For example, glyphosate is often applied with the Bordeaux mixture ($CuSO_4$) and may enter waterbodies through runoff and rainwater [13].

Copper is also used in industry and nanotechnology to produce metallic nanomaterials. Due to its characteristics, this metal is incorporated in a wide range of high-consumption products used daily, such as antibacterial agent paints [14], electrochemical sensors [15], lubricants, polymers, plastics, and metallic coatings [16]. This condition favors the release of copper as nanoparticles (NPs) or dissolved metal, increasing their occurrence in soil and water. Exposure to Cu ions is associated with physiologic alterations in aquatic species, such as tissue injury in fish [17], and acute toxicity and CuNP accumulation in microcrustaceans [18].

Because of its molecular characteristics, glyphosate was first patented as a metal chelator, which can form stable complexes with some divalent metal ions, including copper [10]. This chemical property can increase the potential damage that the combination of glyphosate and copper, as commonly used agrochemicals, can produce in aquatic biota.

In the United States, herbicides are nearly ubiquitous in lotic systems, with 90% of the streams located in agricultural, urban, and mixed lands [19]; for this reason, the coexistence of glyphosate and copper is widely found in waterbodies and in soil due to the widespread application of herbicides [13] and the use of Cu in agriculture, including organic horticulture [12].

Constant or periodic exposure to one or more pollutants can modify the population dynamics of aquatic biota in the exposed individuals, across generations, and in multiple generations [20,21]. Hence, the toxicity potential of pollutants could be more evident in the progeny and in future generations of exposed organisms [22]. In many cases, alterations could remain for subsequent generations as a result of the chronic, sub-lethal intoxication of parents. For this reason, it is necessary to determine whether the intoxication of a parental generation can produce measurable consequences in a second generation that was not directly exposed to toxic chemical pollutants. In ecotoxicology, a procedure using *Daphnia magna* as a test organism is suggested to measure the toxic effects in successive generations [23,24]. For other species, Rodriguez-Miguel et al. [21] studied the impact of glyphosate on two generations of the American cladoceran *Daphnia exilis*.

D. exilis is a North American freshwater cladoceran distributed from the north to the central part of Mexico. This species thrives from temperate to subtropical latitudes, and because of the similarities with the reference model *D. magna*, it is considered an alternative test organism for ecotoxicological studies in latitudes where *D. magna* is not distributed [25,26].

In daphnids, differences in the content of macromolecules and body size in the progeny can be modified by chemical stress and other environmental factors, such as temperature, photoperiod, type and quantity of food, and the chemistry of the water [21,27,28]; thus, macromolecule concentration and the size of exposed organisms can be used as biomarkers to assess the effects of toxic pollutants.

In this study, *D. exilis* was chronically exposed to a binary mixture of glyphosate (Faena®) and copper. We aimed to assess toxic effects in the parental generation and to determine if the filial generation can be impaired when the progeny is reared in media free of both pollutants. As endpoints, survival, reproduction, body size in neonates, and the content of the main macromolecules were assessed in the parental and filial generations.

2. Materials and Methods

2.1. Tested Chemicals

We used Faena® as the commercial presentation of the glyphosate-based herbicide (CAS: 38641-94-0; Monsanto Comercial, S. de R. L. de C. V., Zapopan, Jalisco, Mexico). Faena® contains 363 g L^{-1} of acid equivalent of the potassium salt of N-(phosphonomethyl) glycine (35.6% *w*/*w* of the active ingredient in the formulation). The other components in Faena® are adjuvants of unknown composition for the consumer (64.4% *w*/*w*). A stock solution of 500 mg L^{-1} of glyphosate using moderately hard reconstituted water (MHRW) as a medium free of any pollutant was prepared and freshly used in all the experiments and to renew the test solution. The actual concentration of glyphosate in solutions was determined using the Eurofins Abraxis glyphosate ELISA Plate kit, which is an immunoassay for the quantitative and sensitive screening of glyphosate in water samples.

Copper solutions were prepared using the salt $CuSO_4{\cdot}5H_2O$ (J. T. Baker® 98.5% purity, CAS no. 7758-99-8). A stock solution of 20 mg L^{-1} of copper was prepared using MHRW as a dilution medium and refrigerated at 4 °C. The actual concentrations of Cu^{+2} in the experiments were determined following the HACH 10238 method.

2.2. Test Organisms

A stock culture of *D. exilis* was started with females of known age according to the conditions stated by Martinez-Jerónimo et al. [26]. Briefly, they were reared in MHRW medium and fed with the green microalgae *Pseudokirchneriella subcapitata* at a concentration of 1×10^6 cells mL^{-1}. Incubation conditions were 25 °C and a photoperiod of 16:8 h (light:dark). Neonates used for the acute and chronic determinations were obtained, starting with the third clutch in the stock cultures.

2.3. Acute Toxicity Tests

The median lethal concentration (LC_{50}) for glyphosate and copper was determined at 48 h [29]. Glyphosate tested concentrations were 2, 3, 4, 5, 6, and 7 mg L^{-1} based on the glyphosate content in Faena®. The concentrations for copper were 13, 15, 17, 19, and 21 µg L^{-1}. MHRW was used as dilution water. Tested concentrations for both toxicants were determined after previously performing range-finding tests; this range enabled an accurate determination of the LC_{50} values.

Tests were prepared with 10 neonates (age < 24 h) distributed in 50 mL beakers containing 35 mL of test volume and three replicates per concentration; MHRW was used for the negative control and dilution water. The temperature for assays was 25 °C and the photoperiod was 16:8 h (light:dark). Immobilization and death of test organisms were recorded at 24 and 48 h to calculate the LC_{50} of both toxicants.

2.4. Sub-Chronic Test

Sub-chronic experiments were performed to determine the combined effect of glyphosate (Faena®) and copper in the progenitors (P_1 generation) and in the progeny (filial generation F_1) of *D. exilis* for 21 days. The combined assays of both toxicants were prepared based on equivalent concentrations of the LC values previously obtained. Three mixtures of both toxicants were assayed by combining the same toxic units of each one in every mixture, i.e., 50% of the $LC_{0.1}$, LC_1, or LC_{10} corresponding to each toxicant, estimated through the acute assays.

Neonates of *D. exilis* obtained from stock cultures were used to prepare the experiments for the P_1, whereas for the F_1 assays, the neonates obtained in each treatment were placed in the same concentrations as for the parents used to obtain the F_1 generation. The filial generation test (F_1) was prepared using the third clutch obtained from the parental generation (P_1). In all cases, one neonate was distributed, for each of 10 replicates per treatment, in 150 mL beakers containing 100 mL of test volume. MHRW was used as negative control and dilution medium in all assays. For both bioassays, females were fed with *Pseudokirchneriella subcapitata* (8×10^5 cells mL^{-1}). Incubation conditions were 25 °C and a 16:8 h (light:dark) photoperiod. Test solutions were fully renewed every 48 h.

2.4.1. Population Parameters

The population responses regarding survival, fecundity (measured as the accumulated progeny), the start of reproduction, and the number of clutches released during the 21-day period were determined through daily observations; these parameters were recorded in both generations (P_1 and F_1). Additionally, abortions (the non-viable organisms expulsed out of the brood chamber before they completed development) were recorded in both bioassays.

2.4.2. Macromolecule Biomarkers

From every clutch obtained in P_1 and F_1 tests, a sample of 10 neonates was selected, all the water was removed, and 1 mL of potassium-phosphate buffer (100 mM, pH 7.2) was added to preserve the samples at −20 °C. For the macromolecule analysis, samples were thawed and processed with a tissue grinder.

Total protein quantification was determined according to Bradford's method [30], adding 1 mL of Bradford´s reagent to 100 µL of the homogenized sample. Processed samples were read at 595 nm, using bovine serum albumin as the standard for the calibration curve.

The content of total carbohydrates was determined according to the method described by Dubois et al. [31]. Briefly, 5% phenol and concentrated sulfuric acid were added to 200 µL of the homogenized sample. Samples were read at 490 nm. Anhydrous serum glucose was used for the calibration curve.

Lipid concentration was determined with the sulfo-phospho-vanillin method [32]. A sample of chloroform:methanol solution (2:1) was added to 500 µL of the homogenized sample; 200 µL of the obtained organic phase was supplemented with concentrated sulfuric acid and evaporated at 90 °C; finally sulfo-phospho-vanillin solution was added to this sample and read at 525 nm, using cholesterol as standard.

2.4.3. Neonates' Size

The body size of the progeny was recorded during both bioassays (P_1 and F_1). From each concentration and for the negative control, 20 neonates were selected randomly and conserved in potassium-phosphate buffer (100 mM, pH 7.2). The total length, body length, and body width were measured using a stereo microscope according to the criteria established in [21] using the software CellsSens® v.1.0.

2.5. Data Analysis

The Probit LC_{50} was determined using the Risk Hazard Assessment Tools, v. 1.0 software. Kaplan–Meier's test was used for the analysis of survival. One way-analysis of variance (ANOVA) was used to compare reproduction parameters, morphological measures, and macromolecule content; normality of data (Shapiro–Wilk test) and homoscedasticity (Bartlett's Test for Homogeneity of Variances) were confirmed before performing the ANOVAs. Multiple pairwise comparisons were conducted according to Fisher's least significant difference (LSD) to determine significant differences ($p < 0.05$) among treatments and control in the P_1 and F_1 generations. To establish significant differences of treatments with respect to the control, Dunnett's test was performed ($p < 0.05$), specifically for the effects of the treatments on the size of the progeny. The software packages Statistica v. 10.0, GraphPad Prism v. 6.0, and Sigma Plot v. 11.0 were used for the statistical analysis.

3. Results

3.1. Acute Toxicity of Glyphosate and Copper to Daphnia Exilis

Seven acute toxicity tests for glyphosate were performed, with the coefficients of determination (R^2) for the Probit equation values ranging from 0.85 to 0.99. Six acute toxicity bioassays were performed for copper, obtaining R^2 values from 0.84 to 0.96. For both toxicants, the fit to the Probit equation was highly significant. The average 48 h LC_{50} was 4.22 mg L^{-1} (3.47–5.43 mg L^{-1} 95% confidence limits), expressed as glyphosate content in the Faena® formulation. The average median lethal concentration (LC_{50}) for copper was 13.45 μg L^{-1} (95% limits: 11.48–14.82 μg L^{-1}). Additional values of lethal concentration and 95% limits are shown in Table 1.

Table 1. Lethal concentration values for *D. exilis* exposed to mixtures of glyphosate and copper. Values in brackets indicate the 95% confidence limits.

Lethal Concentration (LC)	Glyphosate (mg L^{-1})	Copper (g L^{-1})
0.1	2.08 (0.76–2.72)	4.90 (3.03–6.22)
1	2.48 (1.02–3.06)	6.18 (4.10–7.47)
10	3.14 (1.63–3.64)	8.62 (6.28–9.79)
50	4.22 (3.47–5.43)	13.45 (11.48–14.82)

3.2. Survival in Chronic Exposure to Glyphosate and Copper

Figure 1A,B show survival curves of *D. exilis* after 21 days of exposure to glyphosate and copper mixtures in generations P_1 and F_1. Survival in the control groups of both generations was 100%, and was reduced but not significantly in P_1 at all concentrations of glyphosate and copper mixtures. Nevertheless, survival decreased significantly in F_1 at all the concentrations of glyphosate and copper (Figure 1B); in the treatment of 1.57 mg L^{-1} glyphosate + 4.31 μg L^{-1} copper, 90% mortality was documented on day 21.

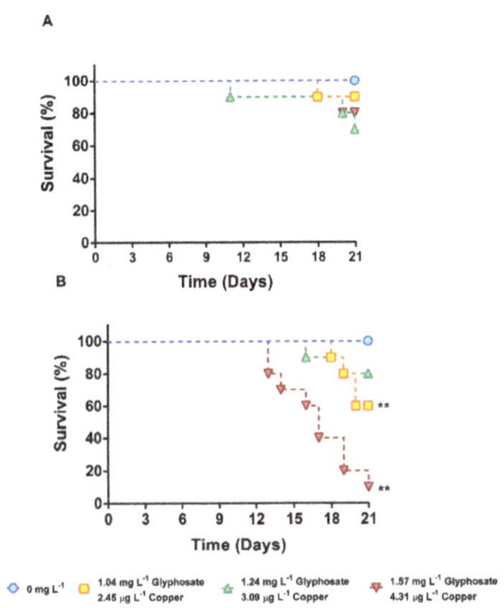

Figure 1. Survival of *D. exilis* exposed to sublethal concentrations of glyphosate and copper mixtures (1.04 mg L^{-1} and 2.45 μg L^{-1}; 1.24 mg L^{-1} and 3.09 μg L^{-1}; 1.57 mg L^{-1}, and 4.31 μg L^{-1}) for 21 days: (**A**) parental generation (P_1), (**B**) filial generation (F_1). Differences with respect to the control are marked with asterisks (Kaplan-Meier test, ** $p < 0.001$).

3.3. *Reproductive Effects in D. exilis Sub-Chronically Exposed to Glyphosate and Copper*

Figure 2A shows the total fecundity (measured as the accumulated progeny) determined in P_1 and F_1 generations of *D. exilis* stressed with glyphosate + copper mixtures for 21 days. The fecundity in the parental generation was not significantly affected in all the treatments; nonetheless, in the filial generation compared with the control, the fecundity was reduced by 44.1, 26.9, and 75.7%, respectively, at the concentrations of glyphosate + copper of 1.04 mg L^{-1} + 2.45 μg L^{-1}, 1.24 mg L^{-1} + 3.09 μg L^{-1}, and 1.57 mg L^{-1} + 4.31 μg L^{-1}.

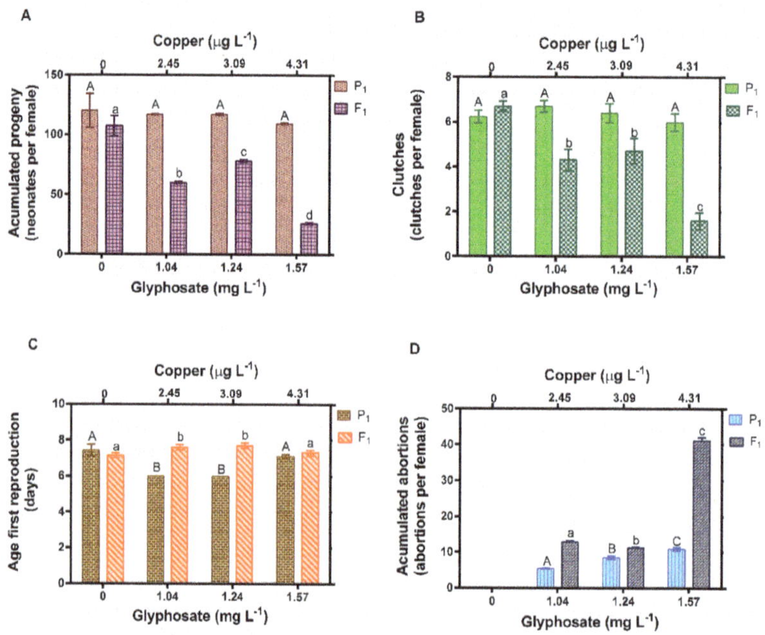

Figure 2. Effects in reproductive responses of *D. exilis* exposed to glyphosate and copper mixtures (1.04 mg L^{-1} and 2.45 μg L^{-1}; 1.24 mg L^{-1}, and 3.09 μg L^{-1}; 1.57 mg L^{-1}, and 4.31 μg L^{-1}) in the parental (P_1) and filial (F_1) generations: (**A**) accumulated progeny, (**B**) number of clutches, (**C**) age at first reproduction, and (**D**) abortions. Average values ± standard error limits. Different uppercase and lowercase letters are used for significative differences in the P_1 and F_1 generations, respectively, according to the post hoc LSD test ($p < 0.05$).

The number of clutches per female in P_1 and F_1 generations of *D. exilis* is shown in Figure 2B. In the parental generation, the number of clutches was not reduced compared to the control ($p < 0.05$) and was higher than that observed in the filial generation. Compared with the control, the number of clutches was significantly reduced in the F_1, observing a remarkable reduction in fecundity (76%) in the 1.57 mg L^{-1} glyphosate + 4.31 μg L^{-1} copper mixture.

The age at first reproduction in P_1 and F_1 is shown in Figure 2C. In the progenitors' generation, females started reproduction at a younger age at the two lowest concentrations of glyphosate + copper treatment than in the F_1. In contrast, in the filial generation, the start of reproduction was delayed by 6.6 and 8% with treatments of 1.04 mg L^{-1} glyphosate + 2.45 μg L^{-1} copper and 1.24 mg L^{-1} glyphosate + 3.09 μg L^{-1} copper, respectively, in comparison to the control.

Abortions were recorded in both P_1 and F_1 generations at all glyphosate + copper treatments but not in the controls (Figure 2D). The number of abortions increased as the concentration of toxicants increased ($p < 0.05$); the highest number of abortions was recorded at 1.57 mg L^{-1} glyphosate + 4.31 μg L^{-1} copper in the F_1 generation.

Overall, adverse effects on reproduction increased in the filial generation compared with the outcomes observed in the progenitors.

3.4. Macromolecules Content in *D. exilis* Exposed to Glyphosate and Copper Mixtures

The contents of proteins, lipids, and carbohydrates measured in the progeny in the P_1 and F_1 generations of *D. exilis* after exposure to glyphosate and copper mixtures are shown in Figure 3. Protein content was affected only in the filial generation (F_1) in the mixture of 1.24 mg L^{-1} of glyphosate + 3.09 μg L^{-1} copper (Figure 3A). In the parental and filial generations, lipids were significantly reduced in the two lowest combinations of glyphosate and copper, but the carbohydrates decreased at all concentrations with respect to the control. In P_1, the reduction in carbohydrates was 68, 95, and 94% for the glyphosate and copper mixtures (1.04 mg L^{-1} + 2.45 μg L^{-1}, 1.24 mg L^{-1} + 3.09 μg L^{-1}, and 1.57 mg L^{-1} + 4.31 μg L^{-1}, respectively). The content of carbohydrates was higher in F_1 than in P_1 at the two highest concentrations of toxicants.

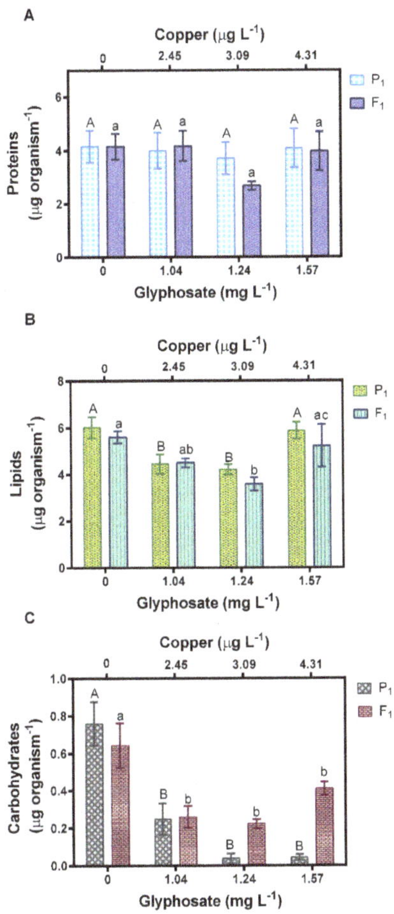

Figure 3. Macromolecule concentration determined in neonates of *D. exilis* exposed for 21 days to glyphosate and copper mixtures (1.04 mg L^{-1} and 2.45 μg L^{-1}; 1.24 mg L^{-1}, and 3.09 μg L^{-1}; 1.57 mg L^{-1}, and 4.31 μg L^{-1}) in the parental (P_1) and filial (F_1) generations: (**A**) proteins, (**B**) lipids, and (**C**) carbohydrates. Average values ± standard error limits. Different uppercase and lowercase letters are used for significative differences in the P_1 and F_1 generations, respectively, according to the post hoc LSD test ($p < 0.05$).

3.5. Effects of Glyphosate and Copper Mixtures on the D. exilis Neonate Size

The body measures of neonates produced in consecutive clutches during the P1 and F1 generations are presented in Figure 4A–C; missing information is for clutches without enough neonates to be measured. TL and BL of neonates in the parental generation displayed reduced variation in the controls. In the controls of F_1, a similar situation was observed, except for clutch number four, in which neonates were significantly larger. According to Dunnett's test, TL in both generations (P_1 and F_1) did not reveal significant differences with respect to their control groups, except for some clutches (Figure 4A).

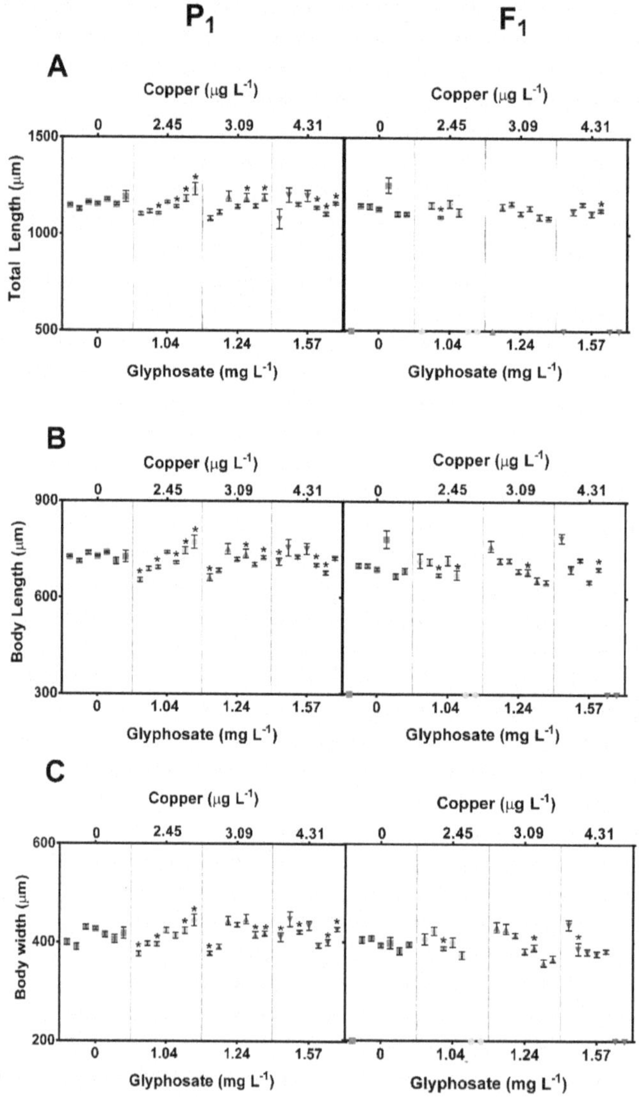

Figure 4. Body size of the neonates produced in progenitors exposed to glyphosate and copper mixtures in generations P_1 and F_1 (left and right, respectively): (**A**) total length (TL), (**B**) body length (BL), and (**C**) body width (BW). Significant differences comparing with the controls are represented by asterisks (Dunnett's test; $p < 0.05$). Mean values ± standard error limits.

The body length of neonates in the parental generation (P_1) was larger in some clutches in the three tested treatments; however, this effect was not observed in the filial generation. In F_1, the BL decreased as the number of the clutch event increased. The BW (Figure 4C) varied similarly to BL (Figure 4B).

4. Discussion

Glyphosate and copper are common agricultural products that can be found as toxic pollutants in aquatic environments. More than 130 commercial formulations of the herbicide glyphosate are currently commercialized for application in different crops in agriculture [33]. The present study documented significant adverse effects of glyphosate and copper on the reproduction and survival of *Daphnia exilis*. For glyphosate, with the formulation known as Faena®, similar results to those observed here were reported by Rodríguez-Miguel et al. [21] in *D. exilis* (LC_{50}: 4.22 mg L^{-1}), and Domínguez-Cortinas et al. [5] in *D. magna* (LC_{50}: 4.1 mg L^{-1}). Lares et al. [34] observed that Faena® was 12 times more toxic to *D. exilis* than to *Alona guttata*. Reno et al. [35,36] determined the 48 h LC_{50} for *D. magna* exposed to the four glyphosate-based formulations, Escoba®, Roundup®, Ultramax Panzer Gold®, and Sulfosato Touchdown®; the obtained values were 29.48, 11.68, 2.12, and 1.62 mg L^{-1}, respectively. The differences in the acute toxicity of glyphosate (as Faena®) in the present work and the other commercial formulations could be attributed to the active ingredients and the type and concentration of surfactants and other adjuvants used in the different formulations. In this respect, Lares et al. [34] reported that the toxicity of commercial formulations in *D. magna* and *C. dubia* exceeded the toxicity of active ingredients by 77-fold. Tsui and Chu [37] reported that, in Roundup®, the surfactant POEA (polyoxyethylene amine) was more toxic to *C. dubia* than the formulation and the active ingredient. Nevertheless, most of the toxicity information on glyphosate and its formulations refers to short-term, acute toxicity measurements.

Cladocerans have been proved to be the most sensitive of zooplankters to the toxicity of copper [38]. For copper, the LC_{50} determined in *D. exilis* was 13.45 μg L^{-1}. Fan et al. [39] reported that, in *D. magna*, the 24 h LC_{50} for this metal was 45 μg L^{-1}. The 48 h LC_{50} value for *D. carinata* exposed to Cu was 14.79 μg L^{-1} [40] and 37.3 μg L^{-1} [41]. Banks et al. [42] determined that the 48 h LC_{50} value for copper in *Ceriodaphnia dubia* was 12.0 μg L^{-1} (8–18.8 μg L^{-1}). Differences in toxicity among cladocerans could be due to species-specific differences related to different capabilities to accumulate, metabolize, and eliminate the toxicants [34].

The mixture of the herbicide Faena® and copper produced effects in the content of macromolecules in both generations of *D. exilis* that are related to physiological alterations; this impairment finally affected survival and reproduction. We observed that the effect of this mixture at all the tested concentrations increased in the filial generation with respect to the parental generation. Other studies have reported the single effect of glyphosate and copper on the survival of cladocerans. Rodriguez-Miguel et al. [21] described that the survival in the parental generation decreased by 50% when exposed to 2.49 mg L^{-1} glyphosate (Faena®). Survival in *D. magna* decreased significantly at 4.05 mg L^{-1} of Roundup® [43]. Furthermore, for *D. magna*, Fan et al. [39] documented that the survival rate decreased from 100% to 0% as the Cu concentration increased from 10 to 100 μg L^{-1}. Koivisto et al. [44] reported that cladoceran sensitivity to copper is species-specific and that mortality increased at low food concentrations (2000 to 20,000 *Scenedesmus* cells mL^{-1}, depending on the species), indicating that the toxicity of this metal also depends on feeding conditions.

However, the effect of binary mixtures as tested here has not been assayed often in cladocerans; for example, Dong et al. [40] reported that in *D. carinata* the mortality increased 6–7% on the first 5 days of exposure with 3 μg L^{-1} of copper combined with microplastics. Mahar and Watzin [45] found that the binary mixture of Cu and Zn affected the survival and reproduction of *C. dubia*. Banks et al. [42] found that the survival of *C. dubia* was affected by the concentrations of copper and diazinon in the mixture. In the present study, survival in the filial generation (F_1) decreased 20, 40, and 90% with glyphosate and copper mixtures (1.04 mg L^{-1} + 2.45 μg L^{-1}, 1.24 mg L^{-1} + 3.09 μg L^{-1}, and 1.57 mg L^{-1} + 4.31 μg L^{-1}, respectively). The increase in sensitivity observed in the F_1 demonstrated transgenerational effects; this increase in the toxic response of the progeny could also mean risks for subsequent generations that could affect the persistence of the population in natural conditions [21].

The toxic effects of Faena® and copper mixtures found in the present study demonstrated reproductive impairment in *D. exilis*, evidenced by reduced fecundity, decreased number of clutches per female, delayed age at first reproduction, and increased number of aborted eggs; all these effects were mainly observed in the F_1. Similar results have been described for copper and glyphosate tested separately. For example, refs. [46–48] reported that, in *D. magna, D. pulex-pulicaria,* and *D. longispina,* reproduction was delayed when exposed to Cu. In *Bosmina longirostris* and *Chidorus sphaericus*, the total number of offspring was significantly reduced when cladocerans were exposed to Cu [44].

We observed adverse effects on the start of reproduction, and on the number and size of released clutches, mainly in F_1 of *D. exilis*. Concerning the toxicity of glyphosate on cladocerans, when *D. magna* was exposed to 2.92, 4.38, 6.57, 9.86, and 14.86 mg glyphosate L^{-1}, age at first reproduction was delayed with all concentrations, and at 14.86 mg L^{-1} this delay was up to two days [49]. In *D. magna* exposed to Roundup®, the fecundity and the number of neonates per brood in the maternal generation (I) and in generations II–IV were reduced significantly compared with the control [50]. Villarroel et al. [51] observed that offspring in *D. magna* exposed to tetradifon decreased in F_1. The number of broods of *Daphnia carinata* exposed to copper and microplastics (3 μg L^{-1} and 0.25 mg L^{-1}, respectively) was reduced by 9% [40]. Delay in the sexual maturity of organisms has been observed in daphnids under chemical stress [34,52].

Reduction in clutch size does not seem to be a typical response to sub-lethal toxic metal stress, and the lifetime reduction in offspring production is generally a consequence of increased mortality [44,53,54]. Toxic effects of Faena® and copper in *D. exilis* might lead to impaired vital functions, leading to increased chronic mortality in adults and reduced fecundity. Under stress conditions, organisms face a tradeoff to endure the harmful effects of toxic chemical stressors, maintaining some functions at a basic level and devoting energy to ameliorate the effects of stressors, resulting in a reduction in the energy available for reproduction and for the reserves and growth in the progeny [55].

In our study, the number of abortions increased as the concentration of glyphosate and copper in the mixture increased, both in the parental and filial generations. Similarly, Rodriguez-Miguel et al. [21] documented aborted eggs in *D. exilis* at all the tested glyphosate (Faena®) concentrations, with the highest number of abortions for 3.15 mg L^{-1} glyphosate. Cuhra et al. [43] and Gill et al. [56] reported that 1.35 mg L^{-1} of Roundup® delayed the start of reproduction and produced 100% abortions in *D. magna*.

Glyphosate could act as a chelating agent to form stable complexes with copper [57]. These complexes are stable in water and can modify the toxicity of glyphosate and copper as single toxicants. However, scarce information is available about the aquatic toxicity of these chemical aggregations [10] and the alteration they produce in biochemical processes, such as macromolecules' synthesis. In this study, we demonstrate that the most affected macromolecule was the concentration of carbohydrates, followed by the contents of lipids and proteins, in both generations. Rodriguez-Miguel et al. [21] mentioned that proteins and lipids decreased at all sub-lethal Faena® concentrations assessed. Similarly, Villarroel [58] observed a decrement in the three main macromolecules in neonates of *D. magna* produced by exposure to the herbicide propanil. The reduction in lipids and proteins under conditions of toxic stress could also be due to different physiological mechanisms that affect the formation of lipoproteins, which are used to repair the damage in cells and tissues [59]. In our study, a change in macromolecule content could have resulted from toxic effects produced by glyphosate and copper mixtures.

Chemical toxic stress triggers compensatory processes that require energy; this response forces a redistribution of the available energy resources, reducing the energy destined for reproduction and growth [60]. Thus, the exposure of cladocerans to chemical stressors will produce negative effects at metabolic levels due to decrements in energy reserves and damage to growth and reproduction [61]. In *D. exilis*, reduction in the levels of carbohydrates and lipids as the primary energy sources drive organisms to use their available energy to stay alive, leading to less reproduction and growth; in this way, this cladoceran manages to counteract the toxic effects that the glyphosate and copper mixture is causing.

In polyembryonic organisms, such as the cladocerans *Simocephalus* and *Daphnia*, during the life cycle the size of neonates is typically variable, with the first reproductions being of smaller neonates; the number of neonates in each clutch is also variable [62]. Reno et al. [63] mentioned that when organisms are exposed to pollutants, cladocerans invest energy to survive at the cost of other biological functions such as reproduction or growth. During Cu exposure, reduced weight and body length were observed in *Daphnia*, resulting in an impaired growth rate [64]. In chronic exposure of *D. magna* to glyphosate and Roundup®, Cuhra et al. [43] documented a significant reduction in juvenile size even at the lowest concentrations they evaluated. Our results demonstrated that, in the F_1 generation, the size of *D. exilis* neonates was significantly smaller than that of the neonates produced in P_1. Papchenkova et al. [50] observed similar results for morphological growth, reporting that the growth in juveniles of *D. magna* exposed to glyphosate was significantly lower compared with the controls and in successive generations. Rodriguez-Miguel et al. [21] found that exposure of *D. exilis* to Faena® provoked smaller adults in F_1 than in P_1. The reduced size of cladocerans can result from the chemical stress produced by active ingredients and, in commercial formulations (v. gr. Faena®), other ingredients such as surfactants, adjuvants, and solubilizers also contribute to this effect.

Cladocerans, one of the three main zooplankton groups in freshwater bodies, are exposed to chemical stressors in polluted environments. In particular, the extended use of glyphosate in all the commercial formulations currently available adds to the use of other agrochemicals, such as copper salts, which can be found together in agricultural effluents and runoffs that can reach freshwater environments. The multi-generational and transgenerational effects of chemical pollutants are of great concern now because wild populations could be affected, threatening their persistence and modifying the structure of aquatic communities. Our results demonstrate that the combination of one glyphosate formulation with copper produced significant adverse effects in the filial generation of *D. exilis*. This information confirms impairments of different magnitudes and levels, ranging from biochemical to reproductive affectations caused by the tested pollutants.

5. Conclusions

The sensitivity of *D. exilis* to glyphosate and copper mixtures was evaluated at very low concentrations (1.04–1.57 mg L^{-1} for glyphosate and 2.45–4.31 μg L^{-1} for copper), which can be found in aquatic environments. These chemicals affected survival and reproduction in both generations (P_1 and F_1) and induced abortions. The biochemical changes in *D. exilis* during exposure to both toxicants can be associated with the effects observed in reproduction and growth. The glyphosate and copper mixtures caused increased toxic effects in the F_1 generation of D. *exilis*, probably because, in the parents, the damages were counter-balanced by using energy stores to endure the toxic effects. However, the sensitivity in the progeny was affected, probably due to a reduction in lipid reserves or a modification in the balance of the main macromolecules, caused by the intoxication of the progenitors. Although the effect of each toxicant was not assessed separately in the sub-chronic tests, and it was not possible to determine if chemical interactions between glyphosate and copper happened, toxic transgenerational effects were documented in *D. exilis*. However, it also should be considered that because the ingredients and concentrations in the commercial formula of glyphosate are unknown, the effect of the varying concentrations of adjuvants and other constituents in the dilutions of Faena® could also modify the toxicity of the active ingredient. Nevertheless, it is relevant for the environment that the biological effects of the mixture of toxicants could be determined despite the possible chemical interactions, as the test organisms were able to integrate the effects of all the toxicants in the biological responses they exhibited.

Author Contributions: Conceptualization, M.H.-Z. and F.M.-J.; Formal analysis, M.H.-Z. and F.M.-J.; Funding acquisition, F.M.-J.; Investigation, M.H.-Z., A.R.-M., L.M.-J. and F.M.-J.; Methodology, M.H.-Z., A.R.-M., L.M.-J. and F.M.-J.; Writing—original draft, M.H.-Z., A.R.-M. and F.M.-J.; Writing—review and editing, F.M.-J. All authors have read and agreed to the published version of the manuscript.

Funding: This research received no external funding.

Data Availability Statement: All the data supporting the results presented here are available under request.

Acknowledgments: This study was supported by the Instituto Politécnico Nacional (SIP), Secretaría de Investigación y Posgrado. F. Martínez-Jerónimo thanks the COFAA-IPN and the EDI-IPN for grant support. Thanks to three anonymous reviewers because their comments and suggestions improved this paper.

Conflicts of Interest: The authors declare that they have no known competing financial interests or personal relationships that could have appeared to influence the work reported in this paper.

References

1. Uwizeyimana, H.; Wang, M.; Chen, W.; Khan, K. The eco-toxic effect of pesticide and heavy metal mixtures towards earthworms in soil. *Environ. Toxicol. Pharm.* **2017**, *55*, 20–29. [CrossRef] [PubMed]
2. Altenburger, R.; Walter, H.; Grote, M. What contributes to combined effect of a complex mixture? *Environ. Sci. Technol.* **2004**, *38*, 6353–6362. [CrossRef] [PubMed]
3. Hernández, A.F.; Gil, F.; Lacasaña, M. Toxicological interactions of pesticides mixtures: An update. *Arch. Toxicol.* **2017**, *91*, 3211–3223. [CrossRef] [PubMed]
4. Tamsyn, M.; Webster, U.; Laing, L.V.; Florance, H.; Santos, E.M. Effects of glyphosate and its formulation, roundup, on reproduction in zebrafish (*Danio rerio*). *Environ. Sci. Technol.* **2014**, *48*, 1271–1279. [CrossRef]
5. Domínguez-Cortinas, G.; Mejía-Saavedra, J.; Santos-Medrano, G.E.; Rico-Martínez, R. Analysis of the toxicity of glyphosate and Faena® using the freshwater invertebrates *Daphnia magna* and *Lecane quadritentata*. *Toxicol. Environ. Chem.* **2008**, *90*, 377–384. [CrossRef]
6. Defarge, N.; Takács, E.; Lozano, V.L.; Mesnage, R.; Spiroux de Vendomois, J.; Serálini, G.E.; Székács, A. Co-formulants in glyphosate-based herbicides disrupt aromatase activity in human cells below toxic levels. *Int. J. Environ. Res. Public Health* **2016**, *13*, 264. [CrossRef] [PubMed]
7. Vanlaeys, A.; Dubuisson, F.; Seralini, G.E.; Travert, C. Formulants of glyphosate-based herbicides have more deleterious impact than glyphosate on TM4 Sertoli cells. *Toxicol. In Vitro* **2018**, *52*, 14–22. [CrossRef]
8. Lajmanovich, R.C.; Junges, C.M.; Attademo, A.M.; Peltzer, P.M.; Cabagna-Zenklusen, M.C.; Basso, A. Individual and mixture toxicity of commercial formulations containing glyphosate, metsulfuron-methyl, bispyribac-sodium, and picloram on *Rhinella arenarum* tadpoles. *Water Air Soil Pollut.* **2013**, *224*, 1404. [CrossRef]
9. Roustan, A.; Aye, M.; De Meo, M.; Di Giorgio, C. Genotoxicity of mixtures of glyphosate and atrazine and their environmental transformation products before and after photoactivation. *Chemosphere* **2014**, *108*, 93–100. [CrossRef]
10. Hansen, L.R.; Roslev, P. Behavioral responses of juvenile *Daphnia magna* after exposure to glyphosate and glyphosate-copper complexes. *Aquat. Toxicol.* **2016**, *179*, 36–43. [CrossRef]
11. La Torre, A.; Iovino, V.; Caradonia, F. Copper in plant protection: Current situation and prospects. *Phytopathol. Mediterr.* **2018**, *57*, 201–236.
12. Tamm, L.; Thuerig, B.; Apostolov, S.; Blogg, H.; Borgo, E.; Corneo, P.E.; Fittje, S.; de Palma, M.; Donko, A.; Experton, C.; et al. Use of Copper-Based Fungicides in Organic Agriculture in Twelve European Countries. *Agronomy* **2022**, *12*, 673. [CrossRef]
13. Liu, N.; Zhong, G.; Zhou, J.; Liu, Y.; Pang, Y.; Cai, H.; Wu, Z. Separate and combined effects of glyphosate and coper on growth and antioxidative enzymes in *Salvina natans* (L.) All. *Sci. Total Environ.* **2019**, *655*, 1448–1456. [CrossRef] [PubMed]
14. Palza, H.; Nuñez, M.; Bastías, R.; Delgado, K. In situ antimicrobial behavior of materials with copper-based additives in a hospital environment. *Int. J. Antimicrob. Agents* **2018**, *51*, 912–917. [CrossRef] [PubMed]
15. Xu, G.; Huo, D.; Hou, C.; Zhao, Y.; Bao, J.; Yang, M.; Fa, H. A regenerative and selective electrochemical aptasensor based on copper oxide nano flowers-single walled carbon nanotubes nanocomposite for chlorpyrifos detection. *Talanta* **2018**, *178*, 1046–1052. [CrossRef] [PubMed]

16. Arratia, F.; Olivares-Ferretti, P.; García-Rodríguez, A.; Marcos, E.; Carmona, E.R. Comparative toxic effects of copper-based nanoparticles and their microparticles in Daphnia magna by using natural freshwater media. *N. Z. J. Mar. Freshw. Res.* **2019**, *53*, 460–469. [CrossRef]
17. Griffitt, R.J.; Weil, R.; Hyndman, K.A.; Denslow, N.D.; Powers, K.; Taylor, D.; Barber, D.S. Exposure to copper nanoparticles causes gill injury and acute lethality in zebrafish (*Danio rerio*). *Environ. Sci. Technol.* **2007**, *41*, 8178–8186. [CrossRef] [PubMed]
18. Xiao, Y.; Vijver, M.G.; Chen, G.; Peijnenburg, W.J.G.M. Toxicity and accumulation of Cu and ZnO nanoparticles in *Daphnia magna*. *Environ. Sci. Technol.* **2015**, *49*, 4657–4664. [CrossRef]
19. Gilliom, R.J.; Barbash, J.E.; Crawford, C.G.; Hamilton, P.A.; Martin, J.D.; Nakagaki, N.; Nowell, L.H.; Scott, J.C.; Stackelberg, P.E.; Thelin, G.P.; et al. Pesticides in the nation's streams and ground water, 1992–2001. In *The Quality of Our Nation's Waters*; U.S. Geological Survey: Reston, VA, USA, 2006; p. 184.
20. Bal, N.; Kumar, A.; Du, J.; Nugegoda, D. Multi-generational effects of two glucocorticoids (prednisolone and dexamethasone) on life-history parameters of crustacean *Ceriodaphnia dubia* (Cladocera). *Environ. Pollut.* **2017**, *225*, 569–578. [CrossRef]
21. Rodríguez-Miguel, A.; Hernández-Zamora, M.; Martínez-Jerónimo, L.; Martínez-Jerónimo, F. Exposure to sublethal concentrations of the glyphosate- based herbicide Faena increases sensitivity in the progeny of the American cladoceran *Daphnia exilis* (Herrick, 1895). *Enviromental Sci. Pollut. Res.* **2021**, *28*, 38094–38105. [CrossRef]
22. Li, Z.; Al, F.; Zhang, J.; Yu, Z.; Yin, D. Using *Caenorhabditis elegans* for studying trans- and multi-generational effects of toxicants. *J. Vis. Exp.* **2019**, *29*, e59367. [CrossRef]
23. Barata, C.; Campos, B.; Rivetti, C.; LeBlanc, G.A.; Eytcheson, S.; McKnight, S.; Tobor-Kaplon, M.; de Vries, B.S.; Choi, S.; Choi, J.; et al. Validation of a two-generational reproduction test in *Daphnia magna*: An interlaboratory exercise. *Sci. Total Environ.* **2017**, *579*, 1073–1083. [CrossRef] [PubMed]
24. OECD. Test, No. 202: Daphnia sp., acute immobilisation test. In *OECD Guidelines for the Testing of Chemicals, Section 2*; OECD Publishing: Paris, France, 2004; p. 12. [CrossRef]
25. Hairston, N.G.; Perry, L.J., Jr.; Bohonak, A.J.; Fellows, M.Q.; Kearns, C.M. Population biology of a failed invasion: Paleolimnology of *Daphnia exilis* in upstate New York. *Limnol. Oceanogr.* **1999**, *44*, 477–486. [CrossRef]
26. Martínez-Jerónimo, F.; Rodríguez-Estrada, J.; Martínez-Jerónimo, L. *Daphnia exilis* Herrick, 1985 (Crustacea: Cladocera). Una especie zooplanctónica potencialmente utilizable como organismo de prueba en bioensayos de toxicidad aguda en ambientes tropicales y subtropicales. *Rev. Int. Contam. Ambient.* **2008**, *24*, 153–159.
27. Fontoura, N.F.; Agostinho, A.A. Growth with seasonally varying temperatures: An expansion of the von Bertalanffy growth model. *J. Fish Biol.* **1996**, *48*, 569–584. [CrossRef]
28. Martínez-Jerónimo, F. Description of the individual growth of *Daphnia magna* (Crustacea: Cladocera) through the von Bertalanffy growth equation. Effect of photoperiod and temperature. *Lymnology* **2012**, *13*, 65–71. [CrossRef]
29. US Environmental Protection Agency. *Methods for Measuring the Acute Toxicity of Effluents and Receiving Waters to Freshwater and Marine Organisms*, 5th ed.; Office of Research and Development: Cincinnati, OH, USA, 2002.
30. Bradford, M.M. A rapid and sensitive method for the quantitation of microgram quantities of protein utilizing the principle of protein-dye binding. *Anal. Biochem.* **1976**, *72*, 248–254. [CrossRef]
31. Dubois, M.; Gille, K.A.; Hamilton, J.K.; Rebers, P.A.; Smith, F. Colorimetric method for determination of sugar and related substances. *Anal. Chem.* **1956**, *28*, 350–356. [CrossRef]
32. Zöllner, N.; Kirsch, K. Microdetermination of lipids by sulphophosphovanillin reaction. *Z. Für Die Gesamte Exp. Med.* **1962**, *135*, 545–561. [CrossRef]
33. Guyton, K.; Loomis, D.; Grosse, Y.; El Ghissassi, F.; Benbrahim-Tallaa, L.; Guha, N.; Scoccianti, C.; Mattoc, H.; Straif, K. Carcinogenicity of tetrachlorvinphos, parathion, malathion, diazinon, and glyphosate. *Lancet Oncol.* **2015**, *16*, 490–491. [CrossRef]
34. Lares, B.A.; Vignatti, A.M.; Echaniz, S.A.; Gutiérrez, M.F. Effects of glyphosate on Cladocera: A synthetic Review. *Aquat. Toxicol.* **2022**, *249*, 106232. [CrossRef] [PubMed]
35. Reno, U.; Doyle, S.R.; Momo, F.R.; Regaldo, L.; Gagneten, A.M. Effects of glyphosate formulations on the population dynamics of two freshwater cladoceran species. *Ecotoxicology* **2018**, *27*, 784–793. [CrossRef] [PubMed]
36. Reno, U.; Gutierrez, M.F.; Longo, M.; Vidal, E.; Regaldo, L.; Negro, A.; Mariani, M.; Zalazar, C.; Gagneten, A.M. Microcrustaceans: Biological models to evaluate a remediation process of glyphosate-based formulations. *Water Air Soil Pollut.* **2015**, *226*, 226–349. [CrossRef]
37. Tsui, M.T.K.; Chu, L.M. Aquatic toxicity of glyphosate-based formulations: Comparison between different organisms and the effects of environmental factors. *Chemosphere* **2003**, *52*, 1189–1197. [CrossRef] [PubMed]
38. Bondarenko, O.; Juganson, K.; Ivask, A.; Kasemets, K.; Mortimer, M.; Kahru, A. Toxicity of Ag, CuO and ZnO nanoparticles to select environmentally relevant test organism and mammalian cells in vitro: A critical review. *Arch. Toxicol.* **2013**, *87*, 1181–1200. [CrossRef]
39. Fan, W.; Zhang, Y.; Liu, S.; Li, X.; Li, J. Alleviation of copper toxicity in *Daphnia magna* by hydrogen nanobubble water. *J. Hazard. Mater.* **2020**, *389*, 122155. [CrossRef]
40. Dong, D.T.; Miranda, A.; Trestrail, C.; De Souza, H.; Dinh, K.V.; Nugegoda, D. Antagonistic effects of copper and microplastics in single and binary mixtures on development and reproduction in the freshwater cladoceran *Daphnia carinata*. *Environ. Technol. Innov.* **2021**, *24*, 102045. [CrossRef]

41. Cooper, N.L.; Bidwell, J.R.; Kumar, A. Toxicity of copper, lead, and zinc mixtures to *Ceriodaphnia dubia* and *Daphnia carinata*. *Ecotoxicol. Environ. Saf.* **2009**, *72*, 1523–1528. [CrossRef]
42. Banks, K.E.; Wood, S.H.; Matthews, C.; Thuesen, K. Joint acute toxicity of diazinon and copper to *Ceriodaphnia dubia*. *Environ. Toxicol. Chem.* **2003**, *22*, 1562–1567. [CrossRef]
43. Cuhra, M.; Traavik, T.; Bøhn, T. Clone- and age-dependent toxicity of a glyphosate commercial formulation and its active ingredient in *Daphnia magna*. *Ecotoxicology* **2013**, *22*, 251–262. [CrossRef]
44. Koivisto, S.; Ketola, M.; Wall, M. Comparison of five cladoceran species in short- and log-term cooper exposure. *Hydrobiologia* **1992**, *248*, 125–136. [CrossRef]
45. Mahar, A.M.; Watzin, M.C. Effects of metal and organophospate mixtures on *Ceriodaphnia dubia* survival and reproduction. *Environ. Toxicol. Chem.* **2005**, *24*, 1579–1586. [CrossRef] [PubMed]
46. Agra, A.R.; Soares, A.M.V.M.; Barata, C. Life-history consequences of adaptation to pollution. *Daphnia longispina clones historically exposed to copper. Ecotoxicology* **2011**, *20*, 552–562. [CrossRef] [PubMed]
47. Pereira, C.M.S.; Deruytter, D.; Blust, R.; De Schamphelaere, K.A.C. Effect of temperature on chronic toxicity of copper, zinc, and nickel to *Daphnia magna*. *Environ. Toxicol. Chem.* **2017**, *36*, 1909–1916. [CrossRef]
48. Taylor, N.S.; Kirwan, J.A.; Johnson, C.; Yan, N.D.; Viant, M.R.; Gunn, J.M.; McGeer, J.C. Predicting chronic copper and nickel reproductive toxicity to *Daphnia pulex-pulicaria* from whole-animal metabolic profiles. *Environ. Pollut.* **2016**, *212*, 325–329. [CrossRef]
49. Silva, M.L.N.; Nogueira, D.J.; Köerich, J.S.; Vaz, V.P.; Justino, N.M.; Schmidt, J.R.A.; Matias, W.G. Multi-generational toxic effects on *Daphnia magna* induced by silver nanoparticles and glyphosate mixture. *Environ. Toxicol. Chem.* **2020**, *39*, 28–29. [CrossRef]
50. Papchenkova, G.A.; Golovanova, I.L.; Ushakova, N.V. The parameters of reproduction, sizes, and activities of hydrolases in *Daphnia magna* Straus of successive generations affected by Roundup herbicide. *Inland Water Biol.* **2009**, *2*, 286–291. [CrossRef]
51. Villarroel, M.J.; Ferrando, M.D.; Sancho, E.; Andreau, E. Effects of tetradifon on *Daphnia magna* during chronic exposure and alterations in the toxicity to generations pre-exposed to the pesticide. *Aquat. Toxicol.* **2000**, *49*, 39–47. [CrossRef]
52. Campos, B.; Jordao, R.; Rivetti, C.; Lemos, M.F.L.; Soares, A.M.V.M.; Barata, C. Two-generational effects of contaminants in *Daphnia magna*: Effects of offspring quality. *Environ. Toxicol. Chem.* **2016**, *35*, 1470–1477. [CrossRef]
53. Ingersoll, C.G.; Winner, R.W. Effect on *Daphnia pulex* (de Geer) of daily pulse exposures to cooper or cadmium. *Environmental Toxicol. Chem.* **1982**, *1*, 321–327. [CrossRef]
54. Winner, R.W. Evaluation of the relative sensitivies of 7-D *Daphnia magna* and *Ceriodaphnia dubia* toxicity tests for cadmium and sodium pentachlorophenate. *Environ. Toxicol. Chem.* **1988**, *7*, 153–159. [CrossRef]
55. Kimberly, D.A.; Salice, C.J. Multi-generational contaminant exposures produce non-monotonic transgenerational responses in *Daphnia magna*. *Environ. Pollut.* **2015**, *207*, 176–182. [CrossRef] [PubMed]
56. Gill, J.P.K.; Sethi, N.; Mohan, A.; Datta, S.; Girdhar, M. Glyphosate toxicity for animals. *Environ. Chem. Lett.* **2018**, *16*, 401–426. [CrossRef]
57. Zhou, C.F.; Wang, Y.J.; Li, C.C.; Sun, R.J.; Yu, C.; Zhou, D.M. Subacute toxicity of copper and glyphosate and their interaction to earthworm (*Eisenia foetida*). *Environ. Pollut.* **2013**, *180*, 71–77. [CrossRef]
58. Villarroel, U.M.J. Alteraciones Fisiológicas en el Crustáceo *Daphnia magna* por Exposición a Plaguicidas. Ph.D. Thesis, Universidad de Valencia, Valencia, Spain, 2004. Available online: https://roderic.uv.es/handle/10550/15079 (accessed on 23 May 2023).
59. Rambabu, J.P.; Rao, M.B. Effects of organochlorine and three organophosphate pesticides on glucose, glycogen, lipids and proteins contents in tissues of the freshwater snail *Bellamya dissimilis* (Müller). *Bull. Environ. Contam. Toxicol.* **1994**, *53*, 142–148. [PubMed]
60. Beyers, D.W.; Rice, J.A.; Clements, W.H.; Henry, C.J. Estimating physiological cost of chemical exposure: Integrating energetics and stress to quantify toxic effects in fish. *Can. J. Fish. Aquat. Sci.* **1999**, *56*, 814–822. [CrossRef]
61. Calow, P.; Silby, M.M. A physiological basis of population processes: Ecotoxicological implications. *Funct. Ecol.* **1990**, *4*, 283–288. [CrossRef]
62. Green, J. Size reproduction in *Daphnia magna* (Crustacea: Cladocera). *Proc. Zool. Soc. Lond.* **1954**, *124*, 535–545. [CrossRef]
63. Reno, U.; Regaldo, L.; Vidal, E.; Mariani, M.; Zalazar, C.; Gagneten, A.M. Water polluted with glyphosate formulations: Effectiveness of a decontamination process using *Chlorella vulgaris* growing as bioindicator. *J. Appl. Phycol.* **2016**, *28*, 2279–2286. [CrossRef]
64. Knops, M.; Altenburger, R.; Segner, H. Alterations of physiological energetics, growth and reproduction of *Daphnia magna* under toxicant stress. *Aquat. Toxicol.* **2001**, *53*, 79–90. [CrossRef]

water

MDPI

Article

SARS-CoV-2 Surveillance in Hospital Wastewater: CLEIA vs. RT-qPCR

Supranee Thongpradit [1], Suwannee Chanprasertyothin [1], Ekawat Pasomsub [2], Boonsong Ongphiphadhanakul [1,3] and Somsak Prasongtanakij [1,*]

[1] Research Center, Faculty of Medicine Ramathibodi Hospital, Mahidol University, Bangkok 10400, Thailand
[2] Department of Pathology, Faculty of Medicine Ramathibodi Hospital, Mahidol University, Bangkok 10400, Thailand
[3] Department of Medicine, Faculty of Medicine Ramathibodi Hospital, Mahidol University, Bangkok 10400, Thailand
* Correspondence: somsak.pra@mahidol.edu; Tel.: +66-9557-09888

Abstract: The utilization of wastewater as a community surveillance method grew during the COVID-19 epidemic. COVID-19 hospitalizations are closely connected with wastewater viral signals, and increases in wastewater viral signals can serve as an early warning indication for rising hospital admissions. While reverse transcriptase quantitative polymerase chain reaction (RT-qPCR) is the most often used approach for detecting SARS-CoV-2 in wastewater, chemiluminescence enzyme immunoassay (CLEIA) is an alternative automated method. In two assays, 92 wastewater grab samples from a hospital were investigated for the presence of SARS-CoV-2, expected for continuous and monitoring SARS-CoV-2 surveillance. One was in the RT-qPCR nucleic acid test, and another was in the CLEIA assay quantitative antigen test. In 24/92 (26.09%) of the wastewater samples, RT-qPCR identified at least two SARS-CoV-2 genes (ORF1ab, N, or S genes). CLEIA, on the other hand, detected SARS-CoV-2 antigen in 39/92 (42.39%) of the samples. CLEIA demonstrated a low sensitivity and specificity of sensitivity of 54.2% (95% CI: 44.0–64.3%) and 61.8% (95% CI: 51.8–71.7%), respectively, as compared to RT-qPCR. The κ coefficient indicated slight agreement between assay. Then, the CLEIA assay cannot replace molecular-based testing like RT PCR for determining SARS-CoV-2 in hospital wastewater.

Keywords: COVID-19; SARS-CoV-2; hospital wastewater; CLEIA; RT-qPCR

1. Introduction

The severe acute respiratory syndrome coronavirus, also known as SARS-CoV, caused a sudden epidemic outbreak of coronavirus illness in 2019 (COVID-19), which has raised significant concerns among the general public, scientific community, and healthcare professionals worldwide, and has had a major detrimental effect on people's health, the economy, and society. COVID-19 hospitalizations are strongly linked to wastewater viral signals, and increases in wastewater viral signals can serve as an early warning indicator for growing hospital admissions [1,2]. Several papers have revealed the presence of SARS-CoV-2 RNA in stools from COVID-19 patients, as well as the presence of SARS-CoV-2 in wastewaters around the world [3–6]. Another avenue for COVID-19 transfer into water and wastewater is through the widespread usage of face masks by the general population, patients, and health personnel worldwide. Following their use, those face masks were discarded without treatment or disinfection, raising worries about potential health risks and affecting the environment [7]. A transmission channel through the sanitary (or wastewater) plumbing system may be responsible for the spread of COVID-19 within communities and for environmental contamination. Reports of SARS-CoV-2 in wastewater and water have been established [8,9]. In general, monitoring viral infections in wastewater presents various obstacles. The varied composition of wastewater matrices, the low concentration nature of biomarkers in wastewater, the difficulty of obtaining good sample locations, and the requirement for effective virus-concentrating technologies often restrict this method's ability to produce quantitative predictions using viral RNA [10,11]. Despite these obstacles, several studies have detected SARS-CoV-2 in the feces of COVID-19 patients and in wastewater [3,5,12]. Furthermore, if high temporal resolution is followed by timely analysis and reporting, SARS-CoV-2 wastewater surveillance has the potential to serve as an early warning system for disease outbreaks in certain localities [13,14]. It is a non-invasive, community-wide surveillance technology that decreases selection bias by detecting subclinical illnesses [15]. Thus, the United States Centers for Disease Control and Prevention has advised the use of SARS-CoV-2 wastewater surveillance for assessing the prevalence and transmission of SARS-CoV-2 in communities [16]. Besides toxic substances and pathogenic microorganisms, hospital wastewater contains 2–3 times the amount of chemical and biological pollutants as urban wastewater, especially chemical demand of oxygen (COD), biological demand of oxygen (BOD), and suspended solids (SS) [17]. In addition, hospital wastewater contains a diverse range of micropollutants and macropollutants [17]. Micropollutants included hormones, detergents, antiseptics, antibiotics (such as sulfamethoxazole, ciprofloxacin, and paracetamol), absorbable organic halogens (AOX), contrast substances, phenols, analgesics, cytostatic, as well as heavy metals (such as iron, zinc, cadmium, chrome, copper, nickel, and lead). Macropollutants have physical-chemical parameters such as pH, total demand of oxygen (TOC), biological demand of oxygen, chemical demand of oxygen, ammonium ions and chloride, suspended solids, microbiological contaminants as coliforms, bacteria (enterococcus, shigella, salmonella), and viruses, especially SARS-CoV-2. This wastewater is discharged from various hospital units such as operating rooms, laboratories, laundries, kitchens, patient rooms, and research labs [17,18]. Recent studies have stated that SARS-CoV-2 should be monitored in community wastewater and hospital wastewater to help prevent COVID-19 outbreaks [3,5,19,20]. Therefore, wastewater surveillance for SARS-CoV-2 was implemented.

The identification of SARS-CoV-2 genetic targets in numerous kinds of samples using the molecular technique of reverse transcriptase quantitative polymerase chain reaction (RT-qPCR) is the current benchmark for COVID-19 diagnosis [21]. We presented a method for detecting SARS-CoV-2 in wastewater during the initial stages of the pandemic, when the prevalence of reported COVID-19 cases in Thailand was negligible [22]. Although this method is sensitive, the use of RT-qPCR to detect SARS-CoV-2 in wastewater requires specialized laboratory equipment and qualified staff and can take time. Several sensitive and user-friendly techniques for detecting SARS-CoV-2 have been investigated. Among them, chemiluminescence enzyme immunoassay (CLEIA) was developed which is fully automated and sensitive for the identification of SARS-CoV-2 N-Protein using a unique two-reaction. The sample and sample treatment solution are incorporated with an anti-SARS-CoV-2 monoclonal antibody-coated magnetic particle solution and incubated for 10 min at 37 °C to enable the formation of particular antigen–antibody immunocomplexes. An alkaline phosphatase-labelled anti-SARS-CoV-2 monoclonal antibody solution is included to the second reaction (available after washing) and incubated at 37 °C for 10 min in order to permit selective binding to the antigen of the aforementioned immunocomplexes and the production of subsequent immunocomplexes. Lastly, a substrate solution is mixed in and incubated for 5 min at 37 °C prior to the chemiluminescence signals being automatically read by the instrument and utilized in determining the amount of SARS-CoV-2 antigen in the sample using interpolation with a SARS-CoV-2 Ag calibrator curve [23]. Based on a recent study, CLEIA has the ability to detect and measure SARS-CoV-2 nucleocapsid protein from both nasopharyngeal swabs and saliva samples [23–26]. CLEIA is widely used for COVID-19 screening in Japanese airports, and it has been used for community and population screening to identify SARS-CoV-2 nucleocapsid protein in specimens (nasopharyngeal swabs) [27]. Furthermore, we established the potential role of CLEIA for identifying SARS-CoV-2 antigen in fresh market wastewater [6].

The goal of this work was finding SARS-CoV-2 in hospital wastewater by implementing RT-qPCR and CLEIA assay. The antigen results of CLEIA were then compared to the results of an RT-qPCR test targeting SARS-CoV-2 genomic RNA to evaluate these assays.

2. Materials and Methods

The study protocols were implemented in accordance with World Health Organization biosafety guidelines [28]. Likewise, the research project received authorization by Mahidol University's Institution Biosafety Committee (MU 2021-002).

(1) Sample collection

SARS-CoV-2 wastewater surveillance was implemented for two buildings, the Administration Building and the Research and Welfare Building, in Ramathibodi Hospital from February to December 2022. Weekly grab samples of wastewater were obtained in sanitized bottles from each building, delivered to the laboratory on ice, and maintained at 4 °C until further investigation.

(2) Sample preparation and concentration

To remove particles, a subsample (100–400 mL) of each collected wastewater grab sample was centrifuged at 3000× *g* for 10 min at room temperature. The supernatant was then filtered through a mixed cellulose ester membrane filter (pore size, 0.45 µm; diameter, 47 mm; GE Healthcare, Chicago, IL, USA) attached to a disposable Millicup™-FLEX filtration unit (Merck Ltd., Darmstadt, Germany). A vacuum pump was connected to the assembly filtration to filter the sample. Then, membrane filter was removed and set in a sterile 5 mL tube. Each sample tube received 1 mL of DNA/RNA Shield™ and 0.1 g of ZR BashingBeads (Zymo Research, Sigma, Irvine, CA, USA) prior to being maintained at −80 °C until further investigations.

(3) SARS-CoV-2 identification and quantification via RT-qPCR

SARS-CoV-2 RNA extraction and RT-qPCR were performed to detect the ORF1ab, spike (S), and nucleocapsid (N) regions, as described in our previous research [22].

To elute the viral RNA, the prepared solution was first mixed 10 times (60 s each) with a vortex mixer at near-maximum speed. After that, the solution was then transferred in 400 µL to a fresh nuclease-free tube. The viral RNA was then extracted using the Viral RNA Mini Kit (Qiagen, Hilden, Germany) allowing compliance with the manufacturer's protocol. The sample was loaded onto the QIAamp Mini spin column which RNA bind to the membrane, whereas impurities were removed using two separate wash buffers (AW1 and AW2) in two brief centrifugations. The purified RNA free of protein, nucleases, other contaminants, and inhibitors was eluted in a special RNase-free buffer, ready for direct use or safe storage. A NanodropTM (Thermo Fisher Scientific, Waltham, MA, USA) was used to determine the purity and concentration of the extracted RNA. The nucleic acid purity was assessed using absorbance values at 260 and 280 nm (260/280 ratio). The general range of the 260/280 ratio was 1.9–2.1.

TaqManTM 2019 nCoV Assay Kit v1 (Thermo Fisher Scientific) was used for RT-qPCR. Each 25 µL RT-PCR reaction mixture included 6.25 µL of 4×TaqPath™ 1-Step RT-qPCR Master Mix, 1.25 µL of COVID-19 Real-Time PCR Assay Multiplex Solution, 12.5 µL of nuclease-free water, and 5 µL of extracted RNA. The RT-qPCR experiment was performed on a ViiA 7 Real-Time PCR apparatus (Applied Biosystems, Waltham, MA, USA) with the following thermocycling conditions: 2 min at 25 °C for UNG incubation to remove amplicon carryover, 15 min at 50 °C for reverse transcription, 2 min at 95 °C for predenaturation, and 40 cycles of 3 s at 95 °C and 30 s at 60 °C for denaturation, annealing, and extension. As a positive control and internal positive control, the TaqPath COVID-19 control (Thermo Fisher Scientific) and MS2 phage control were utilized, respectively. A negative control was DNase/RNase-free water. The results were considered positive for SARS-CoV-2 detection if the cycle threshold (Ct) value for two or more SARS-CoV-2 target genes was less than 37.

(4) Detection of SARS-CoV-2 antigen

In parallel with detection and quantification by RT-qPCR, 100 μL of concentrated samples was analyzed for the antigen quantification. The Lumipulse G1200 automated immunoassay analyzer (Fujirebio) was used to assess a specific chemiluminescence-based immunoassay technique [23]. The results were categorized as positive for SARS-CoV-2 detection by applying the manufacturer's advised antigen concentration cutoff for nasopharyngeal samples as 1.34 pg/mL.

(5) Statistical analysis

Sensitivity and specificity values were calculated using the RT-qPCR results as the reference and with the exclusion of inconclusive samples determined by CLEIA assay. Cohen's kappa (κ) coefficient of results between the two tests with 95% confidence intervals (CI) was calculated.

3. Results

The numbers of reported cases and deaths are shown in Supplementary Figure S1. The number of in-hospital cases and deaths with confirmed SARS-CoV-2 detection during this period, compared to Thailand and Bangkok province, increased in early February 2022 and peaked in early May 2022. The number of reported cases then decreased to less than 100 by mid-July. According to Thailand's relaxation of COVID-19 restrictions on 1 October 2022 and current regulation of hospitals, the numbers of confirmed cases are not being disclosed [29].

Wastewater samples collected from February to December 2022 from the Administration Building and the Research and Welfare Building at Ramathibodi Hospital were tested for SARS-CoV-2 by RT-qPCR and CLEIA (Table 1). A total of 92 wastewater grab samples were tested. In total, 24/92 (26.09%) of the samples tested positive for SARS-CoV-2 by RT-qPCR, defined as a Ct value of less than 37 for two or more SARS-CoV-2 target genes. By contrast, 39/92 (42.39%) of the samples tested positive in the SARS-CoV-2 CLEIA antigen test, defined according to the manufacturer's antigen cutoff concentration (1.34 pg/mL) for nasopharyngeal swab samples.

As shown in Table 2, the RT-PCR results for 92 wastewater grab samples are compared to those produced by the CLEIA assay. Taking RT-qPCR as the reference assay, CLEIA returned 13 true-positive, 26 false-positive, 11 false-negative, and 42 true-negative results. Thus, the overall agreement rate of CLEIA was 59.8% (95% CI: 49.8–69.8%) with a sensitivity of 54.2% (95% CI: 44.0–64.3%) and specificity of 61.8% (95% CI: 51.8–71.7%). The κ coefficient = 0.1325 indicated slight agreement between CLEIA and RT-qPCR.

Table 1. SARS-CoV-2 detections in wastewater grab samples from Ramathibodi Hospital from February to December 2022.

Date Collected	Administration Building						Research and Welfare Building					
	RT-qPCR (Ct Value)				CLEIA		RT-qPCR (Ct Value)				CLEIA	
	N	ORF1ab	S	Interpretation	Antigen Concentration (pg/mL)	Interpretation	N	ORF1ab	S	Interpretation	Antigen Concentration (pg/mL)	Interpretation
15-Feb-2022	UD	UD	UD	neg	11.73	pos	UD	UD	UD	neg	0.65	neg
22-Feb-2022	31.77	33.4	33.34	pos	0.42	neg	UD	UD	UD	neg	1.53	pos
1-Mar-2022	33.73	35.01	37.11	pos	0.42	neg	37.25	UD	UD	neg	0.56	neg
8-Mar-2022	UD	36.5	37.99	neg	0.39	neg	UD	UD	UD	neg	0.41	neg
15-Mar-2022	UD	36.54	34.86	pos	0.48	neg	UD	UD	UD	neg	0.06	neg
22-Mar-2022	UD	35.59	UD	neg	0.35	neg	UD	38	UD	neg	0.57	neg
29-Mar-2022	34.09	33.6	32.96	pos	3.2	pos	29.12	30.31	28.89	pos	2.14	pos
5-Apr-2022	38.36	38	33.96	neg	0.38	neg	31.08	32.49	31.49	pos	0.58	neg
11-Apr-2022	UD	UD	34.4	neg	2.38	pos	33.76	33.69	27.22	pos	5.79	pos
19-Apr-2022	UD	34.28	29.33	pos	2.34	pos	UD	35.32	UD	neg	0.22	neg
26-Apr-2022	30.34	32.99	32.05	pos	32.86	pos	UD	UD	UD	neg	15.09	pos
3-May-2022	UD	38	UD	neg	0.17	neg	UD	UD	37.69	neg	1.13	neg
10-May-2022	UD	UD	UD	neg	1.15	neg	UD	UD	UD	neg	0.86	neg
17-May-2022	UD	34.69	34.27	pos	0.69	neg	UD	38	UD	neg	0.53	neg
24-May-2022	31.42	33.28	33.24	pos	0.61	neg	UD	UD	UD	neg	1.28	neg
31-May-2022	34.22	38	35.89	pos	0.94	neg	UD	UD	UD	neg	1.1	neg
7-Jun-2022	37.63	33.452	UD	neg	1.05	neg	UD	UD	UD	neg	1.55	pos
14-Jun-2022	UD	37.96	UD	neg	0.7	neg	UD	UD	UD	neg	1.09	neg
21-Jun-2022	UD	35.93	UD	neg	1.93	pos	UD	UD	UD	neg	1.11	neg
28-Jun-2022	34.34	32.85	32.79	pos	0.46	neg	UD	UD	33.97	neg	2.21	pos

Table 1. *Cont.*

Date Collected	Administration Building						Research and Welfare Building					
	RT-qPCR (Ct Value)				CLEIA		RT-qPCR (Ct Value)				CLEIA	
	N	ORF1ab	S	Interpretation	Antigen Concentration (pg/mL)	Interpretation	N	ORF1ab	S	Interpretation	Antigen Concentration (pg/mL)	Interpretation
5-Jul-2022	34.11	34.38	28.73	pos	1.04	neg	UD	UD	UD	neg	2.99	pos
12-Jul-2022	UD	34.99	UD	neg	2.7	pos	UD	UD	UD	neg	2.73	pos
19-Jul-2022	32.19	35.16	UD	pos	2.66	pos	34.72	UD	UD	neg	1.13	neg
26-Jul-2022	UD	UD	UD	neg	1.21	neg	UD	36.28	UD	neg	0.65	neg
2-Aug-2022	35.23	35.48	UD	pos	0.42	neg	36.86	UD	UD	neg	0.82	neg
9-Aug-2022	UD	UD	UD	neg	0.88	neg	UD	UD	UD	neg	1	neg
16-Aug-2022	UD	UD	UD	neg	0.82	neg	UD	UD	UD	neg	1.45	pos
23-Aug-2022	UD	37.98	UD	neg	0.7	neg	UD	UD	UD	neg	0.71	neg
30-Aug-2022	UD	UD	UD	neg	0.3	neg	UD	UD	34.18	neg	0.39	neg
6-Sep-2022	UD	UD	UD	neg	0.07	neg	UD	UD	UD	neg	6.01	pos
13-Sep-2022	UD	35.73	38.11	neg	1.10	neg	UD	UD	UD	neg	4.82	pos
19-Sep-2022	39.94	39.85	UD	neg	1.15	neg	UD	UD	UD	neg	4.05	pos
26-Sep-2022	UD	37.97	UD	neg	2.8	pos	UD	UD	UD	neg	1.48	pos
3-Oct-2022	39.63	39.38	UD	neg	1.41	pos	UD	UD	UD	neg	2	pos
11-Oct-2022	33.31	UD	UD	neg	0.75	neg	34.29	UD	UD	neg	4.46	pos
18-Oct-2022	31.99	34.7	UD	pos	4.82	pos	UD	UD	UD	neg	3.9	pos
25-Oct-2022	31.98	34.96	32.89	pos	3.85	pos	33.99	UD	UD	neg	3.94	pos
1-Nov-2022	32.9	UD	33.82	pos	1.99	pos	UD	UD	UD	neg	1.22	neg
8-Nov-2022	36.52	39.35	UD	neg	1.58	pos	24.78	24.22	UD	pos	3.02	pos
15-Nov-2022	33.14	35.32	UD	pos	1.15	neg	UD	UD	UD	neg	2.52	pos

Table 1. *Cont.*

Date Collected	Administration Building						Research and Welfare Building					
	RT-qPCR (Ct Value)				CLEIA		RT-qPCR (Ct Value)				CLEIA	
	N	ORF1ab	S	Interpretation	Antigen Concentration (pg/mL)	Interpretation	N	ORF1ab	S	Interpretation	Antigen Concentration (pg/mL)	Interpretation
22-Nov-2022	UD	UD	UD	neg	3.29	pos	34.79	35.01	UD	pos	2.14	pos
29-Nov-2022	37.1	35.25	36.76	pos	2.65	pos	UD	UD	UD	neg	1.29	neg
6-Dec-2022	UD	35.19	UD	neg	2.21	pos	UD	UD	UD	neg	1.3	neg
13-Dec-2022	36.83	35.65	UD	pos	2.33	pos	UD	UD	UD	neg	0.38	neg
20-Dec-2022	UD	UD	UD	neg	1.15	neg	UD	UD	UD	neg	0.89	neg
27-Dec-2022	UD	UD	UD	neg	0.96	neg	UD	UD	UD	neg	1.67	pos

Notes: CLEIA, chemiluminescence enzyme immunoassay; Ct, cycle threshold; RT-qPCR, reverse transcriptase quantitative polymerase chain reaction; UD, undetermined. Positive results were defined as antigen cutoffs greater than 1.34 pg/mL by CLEIA and a Ct value < 37 for two or more SARS-CoV-2 target genes by RT-qPCR.

Table 2. Comparison of the CLEIA antigen test and RT-qPCR results for SARS-CoV-2 detection in wastewater grab samples.

	RT-qPCR+	RT-qPCR-	Total
CLEIA + (≥1.34 pg/mL)	13	26	39
CLEIA − (<1.34 pg/mL)	11	42	53
Total	24	68	92

4. Discussion

Wastewater surveillance is crucial for detecting and evaluating pathogens and viruses, especially SARS-CoV-2. The use of wastewater as a community surveillance tool has been widely recognized and accepted because it is non-intrusive and cost-effective. Infectious SARS-CoV-2 has not been isolated from either raw or processed wastewater effluents. High decay rates in raw sewage, resulting in low amounts of live virus particles, will make detection challenging using typical filtration-elution processes. SARS-CoV-2 survival has been studied in filtered and unfiltered raw wastewater, as well as secondary effluent at room temperature, in order to better understand the persistence of SARS-CoV-2 and developing variations in wastewater. The time required to inactivate 90% of SARS-CoV-2 was 10.4, 10.8, and 18.3 h for unfiltered raw, filtered raw, and secondary effluent, respectively. Following first order kinetics, there was a steady decrease in virus infectivity in different wastewater matrices [30].

We previously reported that RT-qPCR [22] and CLEIA [6] could be used to detect SARS-CoV-2 in wastewater. In the present study, we determined the presence of SARS-CoV-2 in hospital wastewater samples using RT-qPCR to identify SAR-CoV-2 viral RNA and CLEIA to identify SAR-CoV-2 viral antigen.

The association between increased vaccination coverage and alterations in dominant variations of concern was explored. There will be no vaccine-derived shedding from the current commercial supply of COVID-19 vaccinations, which may complicate the interpretation of SARS-CoV-2 wastewater-based surveillance data. Additionally, emerging infections from people who have been immunized contribute significantly to wastewater signals and ought to be assessed in accordance with the evolving patterns of shedding from novel varieties of concern [31].

The collection method for wastewater samples plays a crucial factor in the detection of the SARS-CoV-2 RNA virus. In our study, we employed a specific point-in-time approach, using single grab samples for COVID-19 wastewater surveillance. This method was chosen due to its rapidity, efficiency, cost-effectiveness, and safety, and it does not require automated equipment. In other words, the study attempts to conduct the simple and non-invasive surveillance for detection of SARS-CoV-2. This study focused on observing the discharge of COVID-19 virus antigen in wastewater representing at a specific time, rather than continuously monitoring throughout the day. If there is a need for consistent monitoring at different times throughout the day, a more reliable method, such as composite sampling, for the collection of hospital wastewater samples should be utilized. Previous studies have shown that the collection of grab samples, as opposed to composite samples, provides reliable identification of similarities and sufficient information regarding the presence of the SARS-CoV-2 virus in wastewater samples [32,33]. Furthermore, a recent study has suggested that performing grab sampling between 8 a.m. and 10 a.m. exhibits less variability in viral RNA concentrations [34].

PCR-based wastewater surveillance of SARS-CoV-2 RNA does not have the same biases as clinical surveillance and can be used as a tool to monitor changes in a region's overall SARS-CoV-2 prevalence [20,35]. However, it has unique limitations/biases of its own. Thus, working in tandem as complementary sources of data can help to provide a comprehensive view of disease transmission in communities. The wastewater-based surveillance's determining limitation was being unable to track virus lineage prevalence or establish epidemiological transmission links, which require understanding of the genome sequence [36]. The number of COVID-19 cases needed in a population to generate a detectable viral RNA signal in wastewater was indicated. Li et al. (2023) gathered and analyzed a unique and big dataset composed of a significant number of diagnostic tests for COVID-19 and wastewater analyses spanning the first three waves of COVID-19 in Alberta, Canada. Results showed that a minimum of 4–17 (median 8), 9–43 (median 18), and 17–97 (median 38) daily reported new COVID-19 cases per 100,000 population were requirements for SARS-CoV-2 RNA to be found in community wastewater with 50%, 80%, and 99% probability, respectively [37]. In a previous study, we demonstrated the performance of RT-qPCR to detect three different SARS-CoV-2 genes (ORF1ab, the N protein gene, and the S protein gene) in wastewater from a market and a hospital even in high ambient temperature and relatively low prevalence of COVID-19 [22]. In the current study, we also applied this method to detect SARS-CoV-2 and confirmed that the virus can be detected in hospital wastewater. However, the interpreted positive results were not significantly correlated with newly confirmed COVID-19 cases on each water sampling date. It is possible that chemicals that are used in the hospital impair the detection efficiency. Zhang et al. reported that negative SARS-CoV-2 RNA results could be attributed to increased sodium hypochlorite supplementation used for disinfection [38]. Other disinfection methods, such as applying chlorine, chlorine dioxide, ozone, or UV irradiation, might also impair the detection efficiency [17]. Furthermore, the total volume of daily wastewater generated in buildings is not entirely correlated with the amount of people employing the wastewater system. Similar to our results, Tiacharoen et al. quantified SARS-CoV-2 RNA concentrations in hospital wastewater using a commercial clinical kit and RT-qPCR detection and found no significant correlation between SARS-CoV-2 RNA concentrations and newly confirmed COVID-19 cases on each wastewater collection date, pointing to a small number of SARS-CoV-2-positive samples [39]. Peng et al. demonstrate the high correlation between the viral signal of SARS-CoV-2 in wastewater with a time-varying relationship. In addition, the 15-day time lag between the average of SARS-CoV N1 and N2 gene concentrations and COVID-19 hospitalizations is adjusted for vaccination efforts [40]. The systematic review identified correlation between SARS-CoV-2 RNA concentration in wastewater and clinically confirmed cases, as demonstrated by Li et al. Among the 133 correlation coefficients, the systematic study report ranges from −0.38 to 0.99. Indeed, SARS-CoV-2 RNA concentration and new cases (whether daily new, weekly new, or future cases) had a stronger correlation than active cases and cumulative cases. Furthermore, environmental and epidemiological circumstances, as well as the wastewater-based epidemiology sampling strategy, could have an impact on these association coefficients. Wider changes in air temperature and clinical testing coverage, as well as an increase in catchment size, all had a significant negative impact on the connection between SARS-CoV-2 RNA concentration and COVID-19 case numbers [41].

The analytical capabilities of the CLEIA assay were determined in a previous investigation [6]. SARS-CoV-2 was identified in wastewater by the CLEIA assay downward to the 10^4 spike cell, whereas RT-qPCR detected SARS-CoV-2 in wastewater downward to the 10^2 spike cell. In addition, we used CLEIA to examine SARS-CoV-2 in 14 fresh market wastewater grab samples. The assay detected SARS-CoV-2 in the samples during an outbreak with 66.7% specificity and 100% sensitivity [6]. Therefore, in the present study, we used CLEIA to test for SARS-CoV-2 antigen in hospital wastewater. Using the manufacturer's recommended antigen threshold concentration for nasopharyngeal samples (1.34 pg/mL), we classified our data as positive for SARS-CoV-2 detection. The CLEIA results were compared with the RT-qPCR results as a standard control. SARS-CoV-2 was detected by CLEIA in 39/92 (42.39%) of samples, achieving a sensitivity and specificity of 54.2% (95% CI: 44.0–64.3%) and 61.8% (95% CI: 51.8–71.7%), respectively. Whenever we changed the cut-off level, the sensitivity and specificity did not improve. The newly identified SARS-CoV-2 genome mutations and a number of important variations, including multiple variants of concern (VOC), may influence to sensitivity and specificity of CLEIA. However, many reports were confirmed that the protein region detected by the CLEIA assay is not altered by mutations. Roy established that the alpha and delta versions are distinguished by spike protein mutations; hence, CLEIA efficiency was unaffected [42]. Similar with Osterman et al., the CLEIA assay has applications for identifying alpha (B.1.1.7) or beta (B.1.351) [43]. Whereas CLEIA has been shown by Gandolfo et al. who succeeded to figure out the variant type and demonstrate amino acid substitutions within or near the functional N antigenic epitope, such as B.1.351, B.1.258, B.1.177.75, B.1.1.420, B.1.1.34, B.1.1.7, and P.1 [44]. The World Health Organization named omicron variants in November 2021, which contain mutations in spike proteins as well, and these variants will have no impact on the CLEIA results [45]. There was no statistically significant difference between the results of strains with mutations and SARS-CoV-2 antigen quantities [46,47].

The low specificity may have been owing to nonspecific obstructions from substances in the hospital wastewater, which might be alleviated by dilution [48], or it may have been affected by environmental factors such as temperature, pH, or the presence of chemical pollution [38]. CLEIA employs monoclonal antibodies directed against the SARS-CoV-2 N protein and is designed to perform double-antibody immunodiagnostic tests (sandwich assay) for the detection and quantification of SARS-CoV-2 antigen. However, the disinfectant substances used in the hospital may have significantly impacted the assay. The findings in this report are subject to at least four limitations. First, wastewater monitoring cannot offer a comprehensive view of disease transmission. Second, outpatient department interpretation of the results is constrained. Third, weekly grab wastewater cannot be representative of all due to hospitals generating large volumes of wastewater. Finally, pretreatment of wastewater surveillance can affect test results.

Recently, the growth and advancement of the digital revolution, including artificial intelligence, evolutionary computational, data science, big data, quantum science, bioinformatics, nanotechnology, internet of things (IoT), financial technology, and blockchain, create opportunities in the delivery of goods and services with higher revenue and a greater opportunity to combat the COVID-19 crisis. Predictive analytics based on data approaches (statistical analysis, machine learning, deep learning, and predictive models and algorithms) are a critical component of COVID-19 pandemic prediction and decision-making tools. Moreover, biosensors and smart sanitation technology are being developed to prevent, identify, and monitor new pollutants with global health potential at the communal (disease surveillance) and personal (for diagnostics) levels [49].

5. Conclusions

The presence of SARS-CoV-2 in hospital wastewater was confirmed by our findings. We found that 24/92 (26.09%) of hospital wastewater samples were positive for at least two SARS-CoV-2 genes (the ORF1ab, N, or S genes). However, unexpectedly, CLEIA had a low sensitivity and specificity of 54.2% (95% CI: 44.0–64.3%) and 61.8% (95% CI: 51.8–71.7%), respectively, as compared to RT-qPCR. The κ coefficient indicated slight agreement between assays. Based on the findings of our study, the CLEIA assay cannot replace molecular-based testing like RT PCR to determine SARS-CoV-2 in hospital wastewater, suspectedly due to hospital disinfectant substances. However, wastewater surveillance is and will continue to be a helpful tool for monitoring SARS-CoV-2 and providing an early warning indication. The application of computational modeling methods and big data analytics should be involved for the effective management of COVID-19.

Supplementary Materials: The following supporting information can be downloaded at: https://www.mdpi.com/article/10.3390/w15132495/s1, Figure S1: Number of reported cases and deaths. References [50–52] are cited in the supplementary materials.

Author Contributions: S.T., S.C. and S.P. contributed to samples collection and preparation, performed experiments, and wrote the manuscript. S.P., E.P. and B.O. contributed to designing the study concept, supervision, and critically revising the manuscript. All authors have read and agreed to the published version of the manuscript.

Funding: This research project is supported by Faculty of Medicine Ramathibodi Hospital Mahidol University (RF 65134).

Data Availability Statement: The data that support the findings of this study are available from the corresponding author upon reasonable request.

Conflicts of Interest: The authors declare that they have no conflicts of interest to disclose.

References

1. Galani, A.; Aalizadeh, R.; Kostakis, M.; Markou, A.; Alygizakis, N.; Lytras, T.; Adamopoulos, P.G.; Peccia, J.; Thompson, D.C.; Kontou, A.; et al. SARS-CoV-2 wastewater surveillance data can predict hospitalizations and ICU admissions. *Sci. Total Environ.* **2022**, *804*, 150151. [CrossRef] [PubMed]
2. D'Aoust, P.M.; Mercier, E.; Montpetit, D.; Jia, J.J.; Alexandrov, I.; Neault, N.; Baig, A.T.; Mayne, J.; Zhang, X.; Alain, T.; et al. Quantitative analysis of SARS-CoV-2 RNA from wastewater solids in communities with low COVID-19 incidence and prevalence. *Water Res.* **2021**, *188*, 116560. [CrossRef]
3. Ahmed, W.; Angel, N.; Edson, J.; Bibby, K.; Bivins, A.; O'Brien, J.W.; Choi, P.M.; Kitajima, M.; Simpson, S.L.; Li, J.; et al. First confirmed detection of SARS-CoV-2 in untreated wastewater in Australia: A proof of concept for the wastewater surveillance of COVID-19 in the community. *Sci. Total Environ.* **2020**, *728*, 138764. [CrossRef]
4. Mecenas, P.; Bastos, R.T.d.R.M.; Vallinoto, A.C.R.; Normando, D. Effects of temperature and humidity on the spread of COVID-19: A systematic review. *PLoS ONE* **2020**, *15*, e0238339. [CrossRef]
5. Nemudryi, A.; Nemudraia, A.; Wiegand, T.; Surya, K.; Buyukyoruk, M.; Cicha, C.; Vanderwood, K.K.; Wilkinson, R.; Wiedenheft, B. Temporal Detection and Phylogenetic Assessment of SARS-CoV-2 in Municipal Wastewater. *Cell Rep. Med.* **2020**, *1*, 100098. [CrossRef] [PubMed]
6. Thongpradit, S.; Prasongtanakij, S.; Srisala, S.; Chanprasertyothin, S.; Pasomsub, E.; Ongphiphadhanakul, B. The Detection of SARS-CoV2 Antigen in Wastewater Using an Automated Chemiluminescence Enzyme Immunoassay. *Int. J. Environ. Res. Public Health* **2022**, *19*, 7783. [CrossRef]
7. Tran, H.N.; Le, G.T.; Nguyen, D.T.; Juang, R.S.; Rinklebe, J.; Bhatnagar, A.; Lima, E.C.; Iqbal, H.M.N.; Sarmah, A.K.; Chao, H.P. SARS-CoV-2 coronavirus in water and wastewater: A critical review about presence and concern. *Env. Res* **2021**, *193*, 110265. [CrossRef]
8. Holst, M.M.; Person, J.; Jennings, W.; Welsh, R.M.; Focazio, M.J.; Bradley, P.M.; Schill, W.B.; Kirby, A.E.; Marsh, Z.A. Rapid Implementation of High-Frequency Wastewater Surveillance of SARS-CoV-2. *ACS EST Water* **2022**, *2*, 2201–2210. [CrossRef]
9. Shah, S.; Gwee, S.X.W.; Ng, J.Q.X.; Lau, N.; Koh, J.; Pang, J. Wastewater surveillance to infer COVID-19 transmission: A systematic review. *Sci. Total Environ.* **2022**, *804*, 150060. [CrossRef]
10. Wade, M.J.; Lo Jacomo, A.; Armenise, E.; Brown, M.R.; Bunce, J.T.; Cameron, G.J.; Fang, Z.; Farkas, K.; Gilpin, D.F.; Graham, D.W.; et al. Understanding and managing uncertainty and variability for wastewater monitoring beyond the pandemic: Lessons learned from the United Kingdom national COVID-19 surveillance programmes. *J. Hazard. Mater.* **2022**, *424*, 127456. [CrossRef]
11. Abdeldayem, O.M.; Dabbish, A.M.; Habashy, M.M.; Mostafa, M.K.; Elhefnawy, M.; Amin, L.; Al-Sakkari, E.G.; Ragab, A.; Rene, E.R. Viral outbreaks detection and surveillance using wastewater-based epidemiology, viral air sampling, and machine learning techniques: A comprehensive review and outlook. *Sci. Total Environ.* **2022**, *803*, 149834. [CrossRef] [PubMed]
12. Ahmed, W.; Bertsch, P.M.; Angel, N.; Bibby, K.; Bivins, A.; Dierens, L.; Edson, J.; Ehret, J.; Gyawali, P.; Hamilton, K.A.; et al. Detection of SARS-CoV-2 RNA in commercial passenger aircraft and cruise ship wastewater: A surveillance tool for assessing the presence of COVID-19 infected travellers. *J. Travel Med.* **2020**, *27*, taaa116. [CrossRef] [PubMed]
13. Assoum, M.; Lau, C.L.; Thai, P.K.; Ahmed, W.; Mueller, J.F.; Thomas, K.V.; Choi, P.M.; Jackson, G.; Selvey, L.A. Wastewater Surveillance Can Function as an Early Warning System for COVID-19 in Low-Incidence Settings. *Trop. Med. Infect. Dis.* **2023**, *8*, 211. [CrossRef] [PubMed]
14. Maryam, S.; Ul Haq, I.; Yahya, G.; Ul Haq, M.; Algammal, A.M.; Saber, S.; Cavalu, S. COVID-19 surveillance in wastewater: An epidemiological tool for the monitoring of SARS-CoV-2. *Front. Cell. Infect. Microbiol.* **2022**, *12*, 978643. [CrossRef]
15. Thompson, J.R.; Nancharaiah, Y.V.; Gu, X.; Lee, W.L.; Rajal, V.B.; Haines, M.B.; Girones, R.; Ng, L.C.; Alm, E.J.; Wuertz, S. Making waves: Wastewater surveillance of SARS-CoV-2 for population-based health management. *Water Res.* **2020**, *184*, 116181. [CrossRef]
16. Center for Disease Control and Prevention. National Wastewater Surveillance System (NWSS). Available online: https://www.cdc.gov/coronavirus/2019-ncov/cases-updates/wastewater-surveillance.html (accessed on 7 February 2022).
17. Achak, M.; Alaoui Bakri, S.; Chhiti, Y.; M'Hamdi Alaoui, F.E.; Barka, N.; Boumya, W. SARS-CoV-2 in hospital wastewater during outbreak of COVID-19: A review on detection, survival and disinfection technologies. *Sci. Total Environ.* **2021**, *761*, 143192. [CrossRef]
18. Debnath, B.; Birawat, R.K.; Birawat, K.K.; Modak, A. Occurrence of COVID-19 virus in hospital wastewater: Treatment pathways and sustainability aspects. In *Hospital Wastewater Treatment: Global Scenario and Case Studies*; Khan, N.A., Vambol, V., Vambol, S., Mozaffari, N., Mozaffari, N., Eds.; IWA Publishing: London, UK, 2022.

19. Medema, G.; Heijnen, L.; Elsinga, G.; Italiaander, R.; Brouwer, A. Presence of SARS-Coronavirus-2 RNA in Sewage and Correlation with Reported COVID-19 Prevalence in the Early Stage of the Epidemic in The Netherlands. *Environ. Sci. Technol. Lett.* **2020**, *7*, 511–516. [CrossRef]
20. Randazzo, W.; Truchado, P.; Cuevas-Ferrando, E.; Simón, P.; Allende, A.; Sánchez, G. SARS-CoV-2 RNA in wastewater anticipated COVID-19 occurrence in a low prevalence area. *Water Res.* **2020**, *181*, 115942. [CrossRef]
21. Wang, W.; Xu, Y.; Gao, R.; Lu, R.; Han, K.; Wu, G.; Tan, W. Detection of SARS-CoV-2 in Different Types of Clinical Specimens. *JAMA* **2020**, *323*, 1843–1844. [CrossRef]
22. Thongpradit, S.; Prasongtanakij, S.; Srisala, S.; Kumsang, Y.; Chanprasertyothin, S.; Boonkongchuen, P.; Pitidhammabhorn, D.; Manomaipiboon, P.; Somchaiyanon, P.; Chandanachulaka, S.; et al. A Simple Method to Detect SARS-CoV-2 in Wastewater at Low Virus Concentration. *J. Environ. Public Health* **2022**, *2022*, 4867626. [CrossRef]
23. Menchinelli, G.; Bordi, L.; Liotti, F.M.; Palucci, I.; Capobianchi, M.R.; Sberna, G.; Lalle, E.; Romano, L.; De Angelis, G.; Marchetti, S.; et al. Lumipulse G SARS-CoV-2 Ag assay evaluation using clinical samples from different testing groups. *Clin. Chem. Lab. Med.* **2021**, *59*, 1468–1476. [CrossRef]
24. Murakami, K.; Iwasaki, S.; Oguri, S.; Tanaka, K.; Suzuki, R.; Hayasaka, K.; Fujisawa, S.; Watanabe, C.; Konno, S.; Yokota, I.; et al. SARS-CoV-2 Omicron detection by antigen tests using saliva. *J. Clin. Virol. Plus* **2022**, *2*, 100109. [CrossRef] [PubMed]
25. Aoki, K.; Nagasawa, T.; Ishii, Y.; Yagi, S.; Okuma, S.; Kashiwagi, K.; Maeda, T.; Miyazaki, T.; Yoshizawa, S.; Tateda, K. Clinical validation of quantitative SARS-CoV-2 antigen assays to estimate SARS-CoV-2 viral loads in nasopharyngeal swabs. *J. Infect. Chemother. Off. J. Jpn. Soc. Chemother.* **2021**, *27*, 613–616. [CrossRef]
26. Kobayashi, R.; Murai, R.; Moriai, M.; Nirasawa, S.; Yonezawa, H.; Kondoh, T.; Saeki, M.; Yakuwa, Y.; Sato, Y.; Katayama, Y.; et al. Evaluation of false positives in the SARS-CoV-2 quantitative antigen test. *J. Infect. Chemother. Off. J. Jpn. Soc. Chemother.* **2021**, *27*, 1477–1481. [CrossRef]
27. Gili, A.; Paggi, R.; Russo, C.; Cenci, E.; Pietrella, D.; Graziani, A.; Stracci, F.; Mencacci, A. Evaluation of Lumipulse®G SARS-CoV-2 antigen assay automated test for detecting SARS-CoV-2 nucleocapsid protein (NP) in nasopharyngeal swabs for community and population screening. *Int. J. Infect. Dis. IJID Off. Publ. Int. Soc. Infect. Dis.* **2021**, *105*, 391–396. [CrossRef]
28. World Health Organization (WHO). Laboratory Biosafety Guidance Related to Coronavirus Disease (COVID-19): Interim Guidance. Available online: https://www.who.int/publications/i/item/WHO-WPE-GIH-2021.1 (accessed on 28 June 2021).
29. Available online: https://www.thaiembassy.com/travel-to-thailand/thailand-travel-restrictions (accessed on 21 February 2023).
30. Sherchan, S.; Thakali, O.; Ikner, L.A.; Gerba, C.P. Survival of SARS-CoV-2 in wastewater. *Sci. Total Environ.* **2023**, *882*, 163049. [CrossRef] [PubMed]
31. Armas, F.; Chandra, F.; Lee, W.L.; Gu, X.; Chen, H.; Xiao, A.; Leifels, M.; Wuertz, S.; Alm, E.J.; Thompson, J. Contextualizing Wastewater-Based surveillance in the COVID-19 vaccination era. *Env. Int* **2023**, *171*, 107718. [CrossRef] [PubMed]
32. Curtis, K.; Keeling, D.; Yetka, K.; Larson, A.; Gonzalez, R. Wastewater SARS-CoV-2 Concentration and Loading Variability from Grab and 24 Hour Composite Samples. *MedRxiv* **2021**. [CrossRef]
33. Kmush, B.L.; Monk, D.; Green, H.; Sachs, D.A.; Zeng, T.; Larsen, D.A. Comparability of 24-hour composite and grab samples for detection of SARS-2-CoV RNA in wastewater. *FEMS Microbes* **2022**, *3*, xtac017. [CrossRef]
34. Augusto, M.R.; Claro, I.C.M.; Siqueira, A.K.; Sousa, G.S.; Caldereiro, C.R.; Duran, A.F.A.; de Miranda, T.B.; Bomediano Camillo, L.M.; Cabral, A.D.; de Freitas Bueno, R. Sampling strategies for wastewater surveillance: Evaluating the variability of SARS-CoV-2 RNA concentration in composite and grab samples. *J. Environ. Chem. Eng.* **2022**, *10*, 107478. [CrossRef]
35. Karthikeyan, S.; Ronquillo, N.; Belda-Ferre, P.; Alvarado, D.; Javidi, T.; Longhurst, C.A.; Knight, R. High-Throughput Wastewater SARS-CoV-2 Detection Enables Forecasting of Community Infection Dynamics in San Diego County. *mSystems* **2021**, *6*, 10–1128. [CrossRef]
36. Karthikeyan, S.; Levy, J.I.; De Hoff, P.; Humphrey, G.; Birmingham, A.; Jepsen, K.; Farmer, S.; Tubb, H.M.; Valles, T.; Tribelhorn, C.E.; et al. Wastewater sequencing reveals early cryptic SARS-CoV-2 variant transmission. *Nature* **2022**, *609*, 101–108. [CrossRef] [PubMed]
37. Le, C. Sensitivity of wastewater surveillance: What is the minimum COVID-19 cases required in population for SARS-CoV-2 RNA to be detected in wastewater? *J. Environ. Sci.* **2023**, *125*, 851–853. [CrossRef] [PubMed]
38. Zhang, D.; Ling, H.; Huang, X.; Li, J.; Li, W.; Yi, C.; Zhang, T.; Jiang, Y.; He, Y.; Deng, S.; et al. Potential spreading risks and disinfection challenges of medical wastewater by the presence of Severe Acute Respiratory Syndrome Coronavirus 2 (SARS-CoV-2) viral RNA in septic tanks of Fangcang Hospital. *Sci. Total Environ.* **2020**, *741*, 140445. [CrossRef] [PubMed]
39. Tiacharoen, V.; Denpetkul, T.; Kosoltanapiwat, N.; Maneekan, P.; Thippornchai, N.; Saeoueng, A.; Jittmittraphap, A.; Sattabongkot, J.; Leaungwutiwong, P. Detection of SARS-CoV-2 and Variants in Hospital Wastewater in a Developing Country. *Water* **2022**, *14*, 3798. [CrossRef]
40. Peng, K.K.; Renouf, E.M.; Dean, C.B.; Hu, X.J.; Delatolla, R.; Manuel, D.G. An exploration of the relationship between wastewater viral signals and COVID-19 hospitalizations in Ottawa, Canada. *Infect. Dis. Model.* **2023**, *8*, 617–631. [CrossRef]
41. Li, X.; Zhang, S.; Sherchan, S.; Orive, G.; Lertxundi, U.; Haramoto, E.; Honda, R.; Kumar, M.; Arora, S.; Kitajima, M.; et al. Correlation between SARS-CoV-2 RNA concentration in wastewater and COVID-19 cases in community: A systematic review and meta-analysis. *J. Hazard. Mater.* **2023**, *441*, 129848. [CrossRef]
42. Roy, U. Comparative structural analyses of selected spike protein-RBD mutations in SARS-CoV-2 lineages. *Immunol. Res.* **2022**, *70*, 143–151. [CrossRef]

43. Osterman, A.; Iglhaut, M.; Lehner, A.; Späth, P.; Stern, M.; Autenrieth, H.; Muenchhoff, M.; Graf, A.; Krebs, S.; Blum, H.; et al. Comparison of four commercial, automated antigen tests to detect SARS-CoV-2 variants of concern. *Med. Microbiol. Immunol.* **2021**, *210*, 263–275. [CrossRef]
44. Gandolfo, C.; Morecchiato, F.; Pistello, M.; Rossolini, G.M.; Cusi, M.G. Detection of SARS-CoV-2 N protein allelic variants by rapid high-throughput CLEIA antigen assay. *J. Clin. Virol.* **2021**, *142*, 104942. [CrossRef]
45. King, K.L.; Wilson, S.; Napolitano, J.M.; Sell, K.J.; Rennert, L.; Parkinson, C.L.; Dean, D. SARS-CoV-2 variants of concern Alpha and Delta show increased viral load in saliva. *PLoS ONE* **2022**, *17*, e0267750. [CrossRef] [PubMed]
46. Sato, Y.; Murai, R.; Kobayashi, R.; Togashi, A.; Fujiya, Y.; Kuronuma, K.; Takahashi, S. Factors affecting the sensitivity of quantitative severe acute respiratory syndrome coronavirus 2 antigen test. *J. Infect. Chemother. Off. J. Jpn. Soc. Chemother.* **2023**, *29*, 754–758. [CrossRef] [PubMed]
47. Yin, N.; Debuysschere, C.; Daubie, V.; Hildebrand, M.; Martin, C.; Curac, S.; Ponthieux, F.; Payen, M.-C.; Vandenberg, O.; Hallin, M. Evaluation and Modelling of the Performance of an Automated SARS-CoV-2 Antigen Assay According to Sample Type, Target Population and Epidemic Trends. *Diagnostics* **2022**, *12*, 447. [CrossRef] [PubMed]
48. Andreasson, U.; Perret-Liaudet, A.; van Waalwijk van Doorn, L.J.; Blennow, K.; Chiasserini, D.; Engelborghs, S.; Fladby, T.; Genc, S.; Kruse, N.; Kuiperij, H.B.; et al. A Practical Guide to Immunoassay Method Validation. *Front. Neurol.* **2015**, *6*, 179. [CrossRef]
49. Matheri, A.N.; Belaid, M.; Njenga, C.K.; Ngila, J.C. Water and wastewater digital surveillance for monitoring and early detection of the COVID-19 hotspot: Industry 4.0. *Int. J. Environ. Sci. Technol.* **2023**, *20*, 1095–1112. [CrossRef]
50. Zhao, L.; Zou, Y.; David, R.E.; Withington, S.; McFarlane, S.; Faust, R.A.; Norton, J.; Xagoraraki, I. Simple methods for early warnings of COVID-19 surges: Lessons learned from 21 months of wastewater and clinical data collection in Detroit, Michigan, United States. *Sci. Total Environ.* **2023**, *864*, 161152. [CrossRef]
51. Amin, N.; Haque, R.; Rahman, M.Z.; Rahman, M.Z.; Mahmud, Z.H.; Hasan, R.; Islam, M.T.; Sarker, P.; Sarker, S.; Adnan, S.D.; et al. Dependency of sanitation infrastructure on the discharge of faecal coliform and SARS-CoV-2 viral RNA in wastewater from COVID and non-COVID hospitals in Dhaka, Bangladesh. *Sci. Total Environ.* **2023**, *867*, 161424. [CrossRef]
52. Núñez-Delgado, A.; Bontempi, E.; Coccia, M.; Kumar, M.; Farkas, K.; Domingo, J.L. SARS-CoV-2 and other pathogenic microorganisms in the environment. *Environ Res* **2021**, *201*, 111606. [CrossRef]

water

MDPI

Article

Multivariate Analysis of Heavy Metals and Human Health Risk Implications Associated with Fish Consumption from the Yangtze River in Zhenjiang City, China

Peter Kaba [1,2,*], Sato Shushi [3], Eric Gyimah [4], Mansuur Husein [2] and Abdelfatah Abomohra [5]

1 The United Graduate School of Agricultural Sciences, Ehime University, Matsuyama 791-0295, Ehime Prefecture, Japan
2 School of the Environment and Safety Engineering, Jiangsu University, Zhenjiang 212013, China; mhusein@tatu.edu.gh
3 Department of Agricultural Sciences, Kochi University, Monobe, Nankoku 783-8502, Kochi Prefecture, Japan; syu@kochi-u.ac.jp
4 Department of Environmental and Safety Engineering, University of Mines and Technology, Tarkwa P.O. Box 237, Ghana; egyimah@umat.edu.gh
5 Department of Environmental Engineering, School of Architecture and Civil Engineering, Chengdu University, Chengdu 610106, China; abomohra@cdu.edu.cn
* Correspondence: kaba.peter20@gmail.com or s20dre15u@kochi-u.ac.jp; Tel.: +81-080-6574-0164

Abstract: The purpose of this study was to analyze levels of heavy metals and human health risk implications associated with fish consumption from the Yangtze River. A total of 60 fish muscles were taken from six different fish species—*Hypophthalmichthys molitrix*, *Ctenopharyngodon idellus*, *Blicca bjoerkna*, *Mylopharyngodon piceus*, *Carassius carassius* and *Pelteobagrus fulvidraco*—and digested using standard protocols. Contents of lead (Pb), cadmium (Cd), zinc (Zn), aluminum (Al) cobalt (Co), manganese (Mn), chromium (Cr), and copper (Cu), were analyzed using an Atomic Absorption Spectroscophotometer ((ZEEnit 700 P Zeeman)). Based on consumer health risk indicators, the health implications to children and adults upon consuming the analyzed fish species were assessed. Findings revealed that Zn recorded the highest mean concentration of 9.87 μg/g in *Carassius carassius* followed by Mn (7.97 μg/g) in *Pelteobagrus fulvidraco*, Cu (2.07 μg/g) in *Mylopharyngodon piceus*, Pb (1.04 μg/g) in *Hypophthalmichthys molitrix*, Cr (0.63 μg/g) in *Hypophthalmichthys molitrix*, Cd (0.19 μg/g) in *Blicca bjoerkna* and Ni (0.16 μg/g) (w/w) in *Pelteobagrus fulvidraco*. In addition, the health risk assessments revealed that children are at heightened non-carcinogenic risk for Pb, Cd, and Co upon consuming the examined fish species.

Keywords: heavy metal; water pollution; bioaccumulation; risk assessment; fish

1. Introduction

Heavy metals in the aquatic environment are of concern due to their accumulation in aquatic habitats, and the possible toxicity to both humans and other wildlife along the food chain [1,2]. Heavy metals may enter the aquatic environment from diverse natural and anthropogenic sources, including discharge from industrial or domestic sewage, storm runoff, leaching from landfills/dumpsites, and atmospheric deposits [3]. The menace of metal pollution has become more alarming because some industries often discharge wastes that contain these metallic contaminants without proper treatment into the environment which exceeds the permissible limit [4]. Some metals such as copper (Cu), manganese (Mn), nickel (Ni), and zinc (Zn) are biologically essential and natural constituents of aquatic ecosystems but could exert toxic effects at concentrations that are relatively higher than their recommended levels [5]. For example, Zn and Cu are responsible for the skeletal formation, maintenance of colloidal systems, regulation of acid–base equilibrium, and biologically important compounds such as hormones and enzymes, but in excess amounts, Cu or Zn could elicit toxicities to some extent [6]. On the other hand, elements such as Cd and Pb have no known biological importance and pronounce toxicity even at relatively low levels [5,7,8].

The global fish consumption per capita has increased from an average of 9.9 kg in the 1960s to 20.5 kg in 2017, which shows a staggering increasing trend [9]. Despite the known nutritional values, fish accumulate both essential and toxic metals in the aquatic ecosystem. In order not to trade nutritional value for future adversities, it is therefore important for the routine monitoring of contaminants in fish, especially in the most edible parts of which is the muscle tissue [10]. This not only informs food safety but is also a signal of the water quality [11]. Owing to significance, the accumulation of heavy metals in fish due to anthropogenic activities has become a global problem, not only because of the threat to fish but also the health risks associated with fish consumption. Significant studies have raised public awareness of the potential accumulation of heavy metals in edible fish and possibly ending up in the human diet through the food chain, resulting in health risks from metal toxicity [12]. To these effects, health problems such as kidney and skeletal damage, neurological disorders, endocrine disruption, cardiovascular dysfunction, and carcinogenic effects have been reported in both epidemiological and animal studies [13,14]. Hence, the levels of trace elements as well as toxic metals in tissues of edible fish warrant significant attention [15]. Noteworthily, the most edible part of fish is the muscle, which is reported in literature to have low metal accumulation compared to other organs such as the liver, kidney, and gills [15,16]. In addition, fish are considered as good indicators of toxicity of heavy metals in freshwater environments as fish inhabit various trophic levels [17].

Anthropogenic activities such as agricultural, industrial, and economic development have threatened the quality of the Yangtze River which has been very topical and has attracted increasing attention [18]. It is estimated that more than 25 billion tons of wastewater from the sewage of nearby cities and their economic activities is discharged into the Yangtze River annually. Approximately sixty percent of the length of the main river channel of the Yangtze River is affected by pollution [19]. Intriguingly, a significant amount (~80%) of the effluents discharged into the Yangtze River basin is untreated [19]. A number of studies have been carried out to examine the risk assessment of metal levels in water and sediment, seasonal variations, and spatial distribution along the Yangtze River [20–22]. Despite the significant dataset indicating the pollution of the Yangtze River, the fate of consumable fishes from the Yangtze River remains vague. Evidence from previous studies supports the hypothesis that fish highly concentrate pollutants in surface water; hence, it is important to investigate the human health risk implications associated with fish consumption from the Yangtze River in Zhenjiang city, China.

To assess the quality of edible fish the from Yangtze River, the levels of copper (Cu), zinc (Zn), manganese (Mn), iron (Fe), cadmium (Cd), nickel (Ni), chromium (Cr) and lead (Pb) were measured in the muscle of fish species (*Hypophthalmichthys molitrix*, *Ctenopharyngodon idellus*, *Blicca bjoerkna*, *Mylopharyngodon piceus*, *Carassius carassius*, and *Pelteobagrus fulvidraco*) based on their dominance in the river and their consumption by humans. Furthermore, the observed levels of heavy metals were compared with available certified safety guidelines proposed by FAO, WHO, and China criteria. In addition, the potential health risk to consumers was estimated using the USEPA proposed models. Furthermore, a multivariate technique was used to predict the possible sources of metal pollution in the Yangtze River.

2. Materials and Methods

2.1. The Study Area

As shown in Figure 1, the Yangtze River is situated in Jiangsu Province of China, and it serves as a source of drinking water to proximate inhabitants and China at large. Zhenjiang city is located in the southwestern part of Jiangsu Province, on the south bank of the lower reaches of the Yangtze River, 31°37′–32°19′ North Latitude, and 118°58′–119°58′ Longitude East [23]. The maximum straight-line distance from east to west is 95.5 km, and the maximum straight-line distance from north to south is 76.9 km. The city's total land area is 3847 square kilometers, accounting for 3.7% of the province [23]. Furthermore, water from the river is immensely used for purposes such as irrigation.

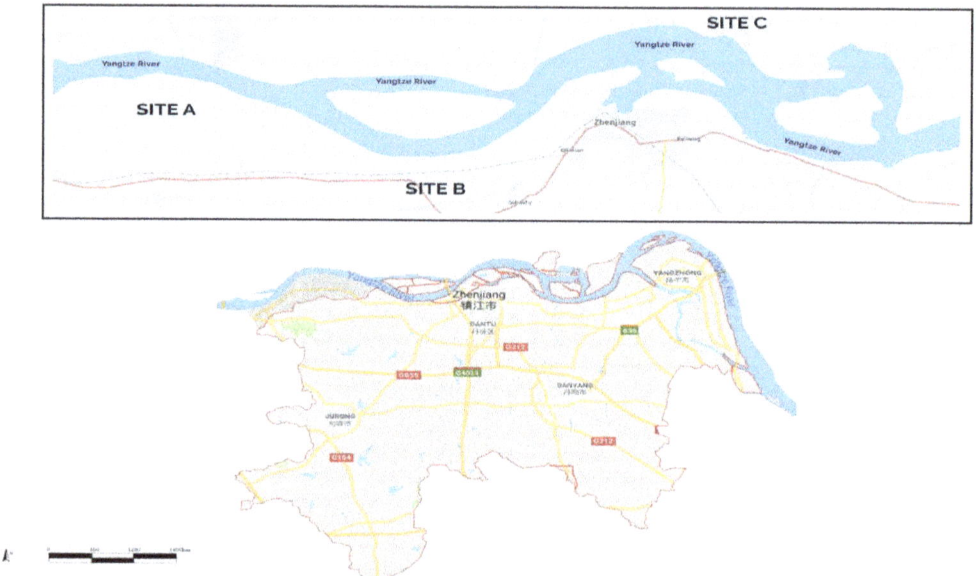

Figure 1. Location of the Yangtze River within Zhenjiang city, China; Source: Google map: https://www.google.com.sg/maps/place/Zhenjiang,+Jiangsu,+China/@32.175422,119.2942119,11z/data=!3m1!4b1!4m6!3m5!1s0x35b42ad8a70e54a3:0x3d27805b2c802e4c!8m2!3d32.1895899!4d119.425!16zL20vMDI3dl82?entry=ttu (accessed on 28 May 2019).

Along the Yangtze River of Jiangsu Province, there are highly polluting industries such as electroplating factories, coal-fired plants, cement plants, and steel plants [24], and most of them discharge their wastewater into the river without proper treatment. This means that through anthropogenic activities, Yangtze River has undergone serious contamination by heavy metals. Accordingly, a total of 60 fish samples spread over the study sites at predominant known offshore selling spots near the river were selected.

2.2. Sampling and Sample Preparation

Fish Sampling

A total of 60 fish were collected and wrapped in clean polyethylene bags, preserved in an ice chest, and taken to the laboratory for identification of species and analysis of metal levels. Fish length was measured from the tip of the snout to the distal end of the caudal fin ray, using a clean transparent meter rule. Additionally, the weight of fresh fish was measured by using a measuring scale. The coefficient of condition (K) of fish species was determined using Equation (1).

$$K = \frac{100W}{L^3} \quad (1)$$

where W is the fresh weight of fish in grams; L is the fork length of fish measured in cm [11]. All samples were washed and labeled in the laboratory. Portions of the cleaned fish species were taken using a sterile knife, placed in separate sterile containers, and labeled. Collected fish muscle tissues were then kept at 4 °C until needed for further analyses. The content was homogenized and a portion (1 g) of the edible tissue was taken for digestion.

2.3. Digestion of Fish Samples

For each homogenized fish tissue, 1 g (wet weight) was weighed into digestive tubes. Volumes of 2 mL each of nitric acid (HNO_3; 65% purity) and perchloric acid ($HClO_4$; 70% purity) were added and left to stand overnight at room temperature. Afterward, 1 mL of 30% hydrogen peroxide (H_2O_2) was added; the mixture was then digested on a heating block at 100 °C for 4 h. The digested fish samples were then allowed to cool and filtered by using ash-free quantitative filtered through Whatman No. 42 filter paper into a 50 mL volumetric flask and topped with double distilled water to reach the mark. In the same process, the procedural blanks were also digested and made ready. Subsequently, dilute solutions were transferred into a sterile screw-capped plastic container marked and stored at 4 °C until analysis was needed (Table 1).

Table 1. The measurement of ecological characteristics and morphometric (biometrics) of selected fish species (n = 60).

Fish Species	Common Name	Number of Samples	Range of Body Length (cm)	Range of Body Weight (g)	Coefficient of Condition/gcm^{-3} (K)
Hypophthalmichthys molitrix	Silver crap	10	26.10–28.60 (27.33 ± 0.79)	270–330 (304.40 ± 23.98)	1.18–1.59 (1.49 ± 0.12)
Ctenopharyngodon idellus	Grass crap	10	24.30–31.20 (26.09 ± 2.22)	250–320 (273.80 ± 25.66)	1.05–1.76 (1.57 ± 0.23)
Blicca bjoerkna	White bream	10	21.50–24.30 (22.65 ± 0.91)	124–186 (149.70 ± 19.60)	1.15–1.35 (1.28 ± 0.63)
Mylopharyngodon piceus	Black crap	10	20.50–23.70 (22.18 ± 0.86)	173–216 (196.60 ± 14.30)	1.60–2.01 (1.80 ± 0.11)
Carassius carassius	Crucian crap	10	22.80–24.50 (23.52 ± 0.67)	178–206 (193 ± 9.75)	1.27–1.63 (1.49 ± 0.10)
Pelteobagrus fulvidraco	Yellow catfish	10	18.40–21.20 (20.29 ± 0.92)	75–108 (95.00 ± 11.05)	1.04–1.22 (1.13 ± 0.06)

Note: Values in brackets represent the mean value.

2.4. Digestion of Water Samples

Field observation and water sample collection were conducted at the Zhenjiang section of the Yangtze River. Forty water samples for heavy metal analysis were collected in 250 mL acid-washed polyethylene bottles from three different sites of the Yangtze River using a speedboat. To estimate the bioaccumulation factor, a 50 mL aliquot of a well-mixed acidified water sample was transferred into 100 mL Pyrex beakers for total metal analysis. The water samples were heated on a hot plate for it to evaporate until the volume reduced to approximately 30 mL. The digested water samples were allowed to cool at room temperature and then filtrated through an acid-washed glass fiber filter (0.45 mm, Whatman GF/C) using a vacuum, to remove suspended particles, into 50 mL volumetric flasks. The solution was topped up with double distilled water to the mark. For total metal analysis, the digested samples were labeled and kept at 4 °C.

2.5. Heavy Metal Analysis

Atomic Absorption Spectroscophotometer (ZEEnit 700 P Zeeman) coupled with a graphite furnace was used to detect and measure the concentrations of Mn, Co, Cr, Cu, Cd, Pb, Zn, and Ni in the samples of the examined fish species. Blank and drift standards were run after every 20 determinations to maintain instrument calibration.

Quality Control and Quality Assurance Procedures

All chemicals used in this study were of analytical reagent grade or higher purity (Sigma–Aldrich Co. LLC, Shanghai, China). A five-level calibration curve was prepared after adequate dilutions of standard stock solutions. Double distilled water was used to prepare/dilute all reagents, standards, and samples. To preserve analytical reliability, ready-made stock standard solutions, i.e., 10,000 mg/L in 5% HNO_3, were used to prepare the working standards for the calibration curves to estimate various heavy metals. Furthermore, taking into consideration the quality assurance/quality control of atomic absorption determinations at regular intervals, double distilled water was prepared to run through the instrument for a few minutes to clean the instrument to prevent any analyte from being trapped in the instrument [25]. For heavy metals, replicate blanks and reference materials, DORM-3 (fish protein) was used in order to monitor the accuracy and precision of the analytical performance. The accuracy of replicate analyses of reference material was in good agreement with the certified values (relative standard deviation, RSD $\leq$ 4%), with recovery rates ranging from 85 to 115% (Table 2).

Table 2. Elemental values of DORM-3 reference material (mg/kg fresh wt) for the analytical performance.

Element	Certified Value	Measured Value (n = 5)	Recovery (%)	Detection Limits (mg/L)
Cu	380.01 ± 3.10	379.13 ± 5.19	100.22	0.002
Pb	330.00 ± 48.00	329.07 ± 3.92	100.28	0.001
Cd	16.02 ± 2.31	15.63 ± 4.12	102.50	0.004
Mn	500.23 ± 87.11	498.02 ± 87.05	100.44	0.001
Zn	670.23 ± 98.32	656.00 ± 75.00	97.87	0.046
Ni	70.21 ± 13.02	72.31 ± 12.01	97.09	0.001
Cr	70.02 ± 12.61	82.01 ± 11.00	85.31	0.007
Co	50.23 ± 10.53	54.35 ± 12.08	101.61	0.003

2.6. The Bioaccumulation Factor (BAF) of Fish Species

The bioaccumulation factor is the ratio of metal concentration in an organism, relative to that in the ambient environment during steady state or equilibrium [26,27] The uptake of metals could differ in each organism as they follow a passive diffusion mechanism analogous to that of oxygen uptake. The *BAF* was calculated using the formula:

$$BAF = \frac{CmF}{CmW} \tag{2}$$

CmF is the concentration of metal in fish muscle; *CmW* is the total metal concentration in water.

2.7. Health Risk Assessment of Fish Species

Estimated daily intake (*EDI*), the target hazard quotients (THQ), the total target hazard quotient (TTHQ), and the target cancer risk (TR) of detected metals were used to estimate the possible health implications upon consuming the examined fish species [28]. Adult and child health risks were considered separately, based on the hypothesis that children are more prone to the effects of contaminants than adults [29]. Estimated daily intake (EDI) (μg/g/day) [30] was calculated by using Equation (2).

$$EDI = \frac{Mc \times FIR}{BW \times 10^{-3}} \tag{3}$$

where *Mc* is the metal concentration in the fish muscle measured in wet weight (ww), *FIR* is the fish ingestion rate (g/day), and *BW* is the average body weight. It is worth noting the body weight used in this study was that considered for a Chinese adult and child [31].

2.8. Target Hazard Quotient

To assess the hazard of exposure to metals in fish consumed by humans, several recent studies have adopted the concept of the target hazard quotient (*THQ*) as widely reported [32–36], which is the proportion of a single substance exposure level over a definite period (e.g., sub-chronic) to a reference dose (*RfD*) for that substance, derived from a similar exposure period (Equation (4)). The *THQ* adopts an exposure level (i.e., *RfD*) beneath which adverse health effects are unlikely to occur even in vulnerable populations. If the level of exposure is above this threshold (i.e., if *THQ* = *E/RfD* exceeds 1), then potential non-carcinogenic effects may be of concern. Higher values of the *THQ* mean a higher chance of non-carcinogenic long-term effects. The *THQ* for non-carcinogenic risk factors was measured as the ratio of exposure to the reference dose.

$$THQ = \frac{Mc \times FIR \times EF \times ED}{RfD \times BW \times ATn} \times 10^{-3} \quad (4)$$

where *THQ* is the target hazard quotient, M_C is the metal concentration in the tissue of the examined fish species, expressed as wet weight (μg/g ww), *FIR* is the Ingestion Rate of fish (g/day), which indicates the average intake of fish by the local population living in Zhenjiang city (44.9 g $person^{-1}$ day^{-1} [37]). *Ef* is the exposure frequency (365 day $year^{-1}$), *ED* is the exposure duration, (an averaging time of 365 day $year^{-1}$ for 70 years), *RfD* is the oral reference dose (μg/g/day) which is 0.04, 0.003, 0.1, 0.003, 0.0005, 0.0003, 0.14, 0.02 μg/g/day for Cu, Pb, Zn, Cr, Cd, Co, Mn, Ni, respectively [36,38], *BW* is the mean body weight (55.9 kg for adults and 32.7 kg for children [29,39]) and *AT* is the average exposure time for non-carcinogens, which is (365 days $year^{-1}$ × number of exposure years). Cooking was also believed to not affect the toxicity of trace elements in seafood [29,40,41]. For *THQs* < 1, the exposed population is unlikely to experience obvious adverse effects [41]. A $THQ \geq 1$ indicates a potential long-term non-carcinogenic effect or health hazard and therefore the intake of certain species should be restricted, and related interventions and protective measures should be taken [29].

2.9. The Target Cancer Risk

The target cancer risk (*TR*) is used to indicate the carcinogenic risk. *TR* values for carcinogenic contaminants were calculated using Equation (5)

$$TR = \frac{Ef \times ED \times FIR \times Mc \times CSFo}{BW \times ATc} \times 10^{-3} \quad (5)$$

where *FIR, Mc, Ef, ED*, and *BW* have already been explained in equation 4 above, *CSFo* is the oral carcinogenic slope factor from the Integrated Risk Information System database, which is 0.5, 1.7, and 8.5×10^{-3} (μg/g/day) for Cr, Ni, and Pb, respectively [12,40,42] and *ATc* is the averaging time for carcinogens (365 days/year×70 years) [36]. The cancer slope factor (*CSF*) used in our study was according to the USEPA [27] reported values, as shown in Table 3. Chien et al. [41] method, was used to assess the overall potential health risk posed by more than one metal. The sum of the target hazard quotients of each metal can calculate the total target hazard quotient (TTHQ). This is estimated due to different metals exposure or the potential of being subjected to more than one pollutant and suffering combined effects [43]. It was expressed as the arithmetic sum of the individual metal THQ values. TTHQ values exceeding unity indicate an alarm for public health concerns.

$$Total\ THQ(TTHQ) = THQ_{Toxicant\ 1} + THQ_{Toxicant\ 2} + \cdots + THQ_{Toxicant\ n} \quad (6)$$

Table 3. Oral reference dose and cancer slop factor for different metals [27].

Heavy Metal	Reference Dose (mg/kg-day)	Cancer Slope Factor	Reference
Pb	3.0×10^{-3}	8.5×10^{-3}	[27]
Ni	2.0×10^{-2}	1.7	[27]
Cd	5.0×10^{-4}	NA	[27]
Co	3.0×10^{-4}	NA	[27]
Zn	3.0×10^{-1}	NA	[27]
Mn	1.4×10^{-1}	NA	[27]
Cr	3.0×10^{-3}	0.5	[27]
Cu	4.0×10^{-2}	NA	[27]

Note: NA: not available at the time of study.

2.10. Statistical Analysis

Data analyses were achieved using the SPSS statistical package on version 25, Microsoft Office Excel 2019, and Origin 2018. The concentration of metals in examined fish was expressed in micrograms per gram (μg/g) of wet weight (ww). Values are the means ± standard deviation (SD). One-way ANOVA and Tukey's HSD post hoc test were adopted for multiple comparisons of mean heavy metals concentrations among the different fish species where the level of significance was set at $p \leq 0.05$. OriginPro 2018 was used to determine Principal Component Analysis (PCA). The PCA was used to identify the possible sources of pollution of heavy metals and also to know the grouping of metals detected in the examined fish species used in the study area. Loading plots and score plots explained by PCA could be used to infer the plausible sources of heavy metal pollution. The biplot shows the association of the samples (examined fish species) with the variables (heavy metals).

3. Results and Discussion

3.1. Fish Species as Bioindicators of Trace Metal Contamination

The levels of metals in the aquatic ecosystem at any given time represent the current contamination while that found in the aquatic species show an effect of bioaccumulation resulting from a relatively exposure period [44]. Heavy metal accumulation in fish primarily results from life events such as breathing and predation via the food chain. The degree of heavy metals in commercial fish species has received significant attention and growing interest in their use as bioindicators in assessing the integrity of the aquatic environmental systems [29,37,45,46].

Fish absorb and build up heavy metals in their tissues from the ambient water, as they constitute the principal route for the ingestion of waterborne contaminants. Fish have long been used to assess water pollution and therefore are known to be an outstanding biological indicator of aquatic ecosystems [47]. In this study, average daily intake (EDI), the target cancer risk (TR), and the target hazard quotient (THQ) were calculated in order to assess the health risk associated with heavy metal contamination of various fish species living in the Yangtze River. These risk assessment parameters used in this study were those proposed by the U.S. EPA, to quantify the potential health hazard posed by any chemical contaminant over prolonged exposure.

3.2. Coefficient of Condition of the Selected Fish Species

The fish species' mean weight ($n = 10$) ranged from 95 g for *P. fulvidraco* to 304 g for *H. molitrix*, while the mean length ranged from 20.29 cm to 27.33 cm for *P. fulvidraco* and *H. molitrix* (Table 1). The condition factor (*K*) has widely been used for estimating the physiological conditions and health of fish in an aquatic habitat which could reflect the pollution status of the aquatic environment [48,49].

The result of the coefficient of condition "K" obtained for all fish species indicates that all the fish samples significantly differed and were healthy (K > 1). The mean coefficients "K" in *H. molitrix* (1.49) and *C. carassius* (1.49) were similar to *B. bjoerkna* (1.28) and *P. fulvidraco* (1.13). *C. idellus* (1.57) and *M. piceus* (1.80) was found to be greater than

1 (Table 1). Similar results obtained from river Ravi showed a similar mean range of 1.16 and 1.24 in *L. rohita* and *C. catla*, respectively, [50]. Fluctuation in the coefficient of condition "K" of fish species may be due to their age, feeding habit, and environmental and climatic conditions [11,51].

3.3. The Concentration of Metals in Various Fish Species

Trace metals in water and sediment pose a greater threat to aquatic organisms. Fish species mainly accumulate trace metals primarily from surface contact with the water, by breathing, and via the food chain, through adsorption of water and ingestion of particulates in sediment [52]. The mean values for Pb, Ni, Co, Cd, Cr, Cu, Mn, and Zn concentrations, as well as their minimum and maximum values, and standard deviation, recorded in the muscles of the six fish species under investigation, in this study, are presented in Table 4 below. The order of the trace elements in all the fish tissue (not in a particular specie) was as follows: Zn (8.1695) > Mn (6.037) > Cu (1.1792) > Pb (0.7654) > Cr (0.3506) > Co (0.1487) > Ni (0.1138) > Cd (0.1011) all in µg/g ww. The heavy metal contents of the various fish species showed major variations in the present study. In addition, the recorded values were compared with known International reported (USEPA/WHO/FAO) threshold values of trace metals in the aquatic environment. Most of the fish species studied recorded metal concentrations that were within Chinese and foreign organizations' legislative thresholds (Table 4).

Table 4. Summary of Heavy metals concentration in muscle tissue of studied fish species (µg/g wet wt) in the Yangtze River with (minimum and maximum; mean values ± standard deviation in parentheses). (n = 60).

Fish Species	Trace Metals Muscle Tissue								
	Statistics	Pb	Ni	Cd	Co	Zn	Mn	Cr	Cu
Hypophthalmichthys molitrix	Range	0.52–1.44	0.01–0.05	0.03–0.1	0.01–0.81	4.74–8.45	2.25–7.43	0.45–0.86	0.79–1.48
	Mean ± STD	**(1.04 ± 0.39)**	(0.02 ± 0.10)	(0.07 ± 0.02)	**(0.46 ± 0.29)**	(6.13 ± 1.32)	(4.25 ± 1.59)	**(0.63 ± 0.16)**	(0.97 ± 0.19)
Ctenopharyngodon idellus	Range	0.17–0.39	0.08–0.19	0.01–0.07	0.01–0.06	4.24–11.87	3.45–9.25	0.01–0.1	0.62–1.22
	Mean ± STD	(0.29 ± 0.09)	(0.14 ± 0.04)	(0.04 ± 0.02)	(0.03 ± 0.02)	(6.38 ± 2.55)	(6.12 ± 2.01)	(0.05 ± 0.03)	(0.93 ± 0.23)
Blicca bjoerkna	Range	0.62–0.81	0.04–0.19	0.13–0.53	0.07–0.12	7.55–10.85	4.35–7.82	0.27–0.31	0.58–0.89
	Mean ± STD	(0.72 ± 0.08)	(0.12 ± 0.04)	**(0.19 ± 0.12)**	(0.09 ± 0.01)	(8.93 ± 0.96)	(5.91 ± 1.15)	(0.29 ± 0.01)	(0.77 ± 0.09)
Mylopharyngodon piceus	Range	0.32–1.51	0.08–0.17	0.01–0.08	0.09–0.15	5.33–9.87	2.08–6.24	0.12–0.18	0.58–3.22
	Mean ± STD	(0.89 ± 0.40)	(0.13 ± 0.03)	(0.03 ± 0.02)	(0.12 ± 0.01)	(8.20 ± 1.56)	(4.20 ± 1.19)	(0.15 ± 0.02)	**(2.07 ± 1.01)**
Carassius carassius	Range	0.64–0.92	0.09–0.14	0.06–0.21	0.04–0.08	9.05–10.31	4.95–9.47	0.28–0.46	1.01–1.14
	Mean ± STD	(0.86 ± 0.08)	(0.11 ± 0.02)	(0.15 ± 0.04)	(0.06 ± 0.01)	**(9.87 ± 0.39)**	(7.77 ± 1.34)	(0.37 ± 0.07)	(1.07 ± 0.04)
Pelteobagrus fulvidraco	Range	0.23–1.27	0.12–0.18	0.03–0.18	0.1–0.15	8.85–10.03	5.73–9.59	0.18–0.92	1.19–1.34
	Mean ± STD	(0.79 ± 0.29)	**(0.16 ± 0.02)**	(0.12 ± 0.05)	(0.13 ± 0.01)	(9.49 ± 0.42)	**(7.97 ± 1.44)**	(0.61 ± 0.20)	(1.27 ± 0.05)
All samples	Range	0.17—1.51	0.01–0.19	0.01–0.53	0.01–0.81	4.24–11.87	2.08–9.59	0.18–0.92	0.58–3.22
	Mean ± STD	(0.77 ± 0.34)	(0.11 ± 0.01)	(0.10 ± 0.08)	(0.15 ± 0.18)	8.17 ± 1.99	(6.04 ± 2.06)	(0.35 ± 0.24)	1.18 ± 0.59

Note: Bold types indicate the highest mean values ± standard deviation of each heavy metals.

Lead (Pb) in elevated doses can lead to a decrease in survival, rates of growth, metabolism, development, and increased mucus formation [53]. The authors of [45] clearly define Pb as a neurotoxin that leads to behavioral deficits in vertebrates. The maximum mean Pb concentration in the examined fish species was 1.04 µg/g ww recorded in *H. molitrix*, while the minimum mean concentration of 0.289 µg/g ww was in *C. idellus*. The mean concentration of Pb in the examined fish species from the Yangtze River was in the order *H molitrix* > *M. piceus*> *C. carassius* > *P. fulvidraco*> *B. bjoerkna* > *C. idellus* (Table 4). The finding of this study was similar to the results reported by [54] in their studies. The authors of [54] reported a Pb mean concentration of 1.4 µg/g dry w for *H. molitrix* and 1.5 µg/g dry w for silver carp fish, respectively, in the Zarivar Wetland, Western Iran. The mean concentration of Pb differs significantly among examined fish species between *H. molitrix*/*C. idellus*; *C. idellus*/*B. bjoerkna*; *C. idellus*/*M. piceus*; *C. idellus*/*C. carassius*; *C. idellus*/*P. fulvidraco* species at $p \leq 0.05$. The Joint Expert Committee on food and Additives (JECFA) recommends a provisional tolerable weekly intake for Pb to be 25 µg/g/day or below [55]. The WHO also recommends that dietary Pb should not exceed 0.3 µg/g (wet weight basis). Dietary exposure to Pb has been implicated with human disorders for which children neonatal infants, and the fetus are the most vulnerable populations to Pb poisoning [56].

The maximum Mn concentration (9.59 µg/g ww) was recorded in *P. fulvidraco* followed by *C. carassius* (9.47 µg/g ww), *C. idellus* (9.25 µg/g ww), *B. bjoerkna* (7.82 µg/g ww),

H. molitrix (7.43 μg/g ww) and *M. piceus* (6.24 μg/g ww) (Table 4). This study also revealed that Mn levels in muscle of the examined fish samples had a mean concentration ranging from 4.20 μg/g ww to 7.97 μg/g ww. Statistical significant difference existed between the mean concentration of Mn recorded for *H. molitrix/C. carassius; H. molitrix/P. fulvidraco; C. carassius/C. idellus; C. idellus/P. fulvidraco; M. piceus/C. carassius;* and *P. fulvidraco/ M. piceus* species at $p \leq 0.05$.

The highest average value of Cd in the muscle of the examined fish species was detected in *B. bjoerkna* (0.19 μg/g ww). The remaining fish species recorded a mean concentration of 0.03 μg/g ww, 0.04 μg/g ww, 0.07 μg/g ww, 0.12 μg/g ww, 0.15 μg/g ww for *M. piceus, C. idellus, H. molitrix, P. fulvidraco,* and *C. carassius,* respectively (Table 4). The average Cd concentration in the examined fish species was far above those reported by [57] (0.0097 μg/g dry w) in the Banan section of the Three Gorges Reservoir but was within the range of that reported by [58] (0.33 μg/g dry w.) in fish muscle tissue of some selected fish species.

Furthermore, almost all the fish samples except *C. idellus* and *M. piceus* recorded levels of Cd exceeding the Chinese national criterion collection GB2736-94 for a healthy standard for freshwater fish (Table 4). Using the Tukey HSD, it was realized that there was a significant difference in Cd concentration recorded in *H. molitrix/B. bjoerkna; C. idellus/B. bjoerkna; C. idellus/C. carassius; C. idellus/P. fulvidraco; B. bjoerkna/M. piceus; B. bjoerkna/P. fulvidraco; M. piceus/C. carassius* and *M. piceus/P. fulvidraco* at $p \leq 0.05$.

The mean concentration of Cr in muscles of the selected fish species ranged from 0.05 μg/g ww to 0.63 μg/g ww in *C. idellus* and *H. molitrix*, respectively and were in accordance with the range reported by [53] (0.94–2.11 μg/g), which is also far less than that in fish muscle studied in the West Lake (Vietnam) by [58] (2.6 μg/g dry w) in *M. piceus*. The findings in [58] indicate lesser contamination of Cr in fish species from the West lake, Vietnam. The lowest Cr mean concentrations in edible muscles were 0.15 μg/g, 0.29 μg/g, 0.37 μg/g, 0.61 μg/g recorded for *M. piceus, B. bjoerkna, C. carassius,* and *P. fulvidraco*, respectively, all measured in wet weight (ww). The European Union Commission, and the WHO suggested the daily tolerable Cr concentration to be 1 μg/g ww, and 0.15 μg/g ww, respectively [25]. The source of Cr in the Yangtze River could be attributed to paints used in boats, agricultural runoff, and leaching from rocks. The difference in the mean concentration of Cr was statistically significant between the species: *H. molitrix/C. idellus; H. molitrix/B. bjoerkna; H. molitrix/M. piceus; H. molitrix/C. carassius; C. idellus/B. bjoerkna; C. idellus/C. carassius; C. idellus/P. fulvidraco; B. bjoerkna/P. fulvidraco; M. piceus/C. carassius; M. piceus/P. fulvidraco* and *P. fulvidraco/C. carassius* at $p \leq 0.05$.

The mean concentration of Cu in the muscle of fish species varied from 0.77 μg/g ww (*B. bjoerkna*) to 2.07 μg/g ww (*M. piceus*). The highest metal concentration of Cu in muscle recorded in *B. bjoerkna* (0.89 μg/g ww) was much lower than that recorded in *M. piceus* (3.22 μg/g ww) (Table 4). The lowest Cu concentrations were found to be 0.58 μg/g ww recorded in *B. bjoerkna*. The mean concentration of Cu recorded for *M. piceus* was above the mean concentration in the same fish species reported by [58] which is 1.2 μg/g ww (Table 4). Similarly, an average of 1.02 μg/g concentration of Cu was reported by [39] in fish species from the Yangtze River.

Increased boating activities, repeated use of anti-fouling paints, and commercial fishing activities along the Yangtze River could be a plausible cause of Cu contamination in the examined fish species. However, the recorded Cu concentration in the examined fish muscle was far below the Chinese national criterion collection GB 2736-94 [59], which is 10 μg/g. The present study revealed that the mean concentration of Cd differs statistically in the muscle of fish between *M. piceus/H. molitrix, M. piceus-/C. idellus, M. piceus/B. bjoerkna, M. piceus/C. carassius* and *M. piceus/P. fulvidraco* at $p \leq 0.05$. This could provide a baseline for the choice of fish species as a bioindicator for monitoring Cu contamination in the Yangtze River.

The acceptable dietary dose for zinc is 11 mg/day for adults and 5 to 9 mg/day for children [60]. Excess Zn in the body reduces both immune function and the levels of high-density lipoprotein (HDL) [60]. Zn was recorded as the highest metal concentration among all metals analyzed in the examined fish species in the present study. *C. idellus* (11.87 µg/g ww), *B. bjoerkna* (10.85 µg/g ww), and *C. carassius* (10.31 µg/g ww) were the fish species with the highest Zn levels. *P. fulvidraco* (10.03 µg/g ww), *M. piceus* (9.87 µg/g ww), and *H molitrix* (8.45 µg/g ww) also displayed levels worth mentioning. Furthermore, the authors of [39] recorded a mean Zn (12.193 µg/g ww) (Table 5) level above what was recorded in the same study area in selected fish tissue. Furthermore, the authors of [53] recorded the highest Zn value (410 µg/g dry w) in *A. d. dispar* fish muscle, which is in many folds higher than the highest value recorded in our study area (11.87 µg/g ww), *C. idellus*. Excessive oral uptake of zinc may lead to abdominal pain, nausea, vomiting, anemia, and dizziness [61,62]. The amount of Zn recorded in all the fish sample were below the Chinese national criterion collection GB 2736-94 [59] as shown in Table 5 above. Of the selected metals, the difference in the mean concentration of Zn was statistically significant between most of the fish species except *C. idellus/H molitrix; C. idellus/M. piceus; M. piceus/B. bjoerkna; B. bjoerkna/C. carassius; B. bjoerkna/P. fulvidraco, M. piceus/C. carassius; M. piceus/P. fulvidraco* and *C. carassius/P. fulvidraco* at $p \leq 0.05$.

Table 5. Comparison of heavy mean metals concentrations in fish muscle tissue observed by different authors and recognized Standard values (µg/g).

Area	Heavy Metals								Reference
	Pb	Ni	Cd	Co	Zn	Mn	Cr	Cu	
Yangtze River(China) [a]	1.04	0.02	0.07	0.46	6.13	4.25	0.63	0.97	Present study
Yangtze River, (China) [a]	0.117	NA	0.062	NA	12.193	NA	0.420	1.02	[39]
Chenab River, (Pakistan) [a]	0.1	0.9	NA	0.1	16	1.98	1.3	1.1	[63]
Wadi Hanifah, (Saudi Arabia) [b]	5.12	5.77	2.48	NA	301.42	43.61	2.11	NA	[53]
Taihu Lake, (China) [a]	0.61	NA	0.12	NA	NA	NA	0.34	0.21	[64]
ZarivarWetland (Western Iran) [b]	1.4	0.3	0.1	NA	NA	NA	NA	2.1	[54]
Chaohu Lake, (China) [b]	0.13	NA	0.007	NA	42.54	NA	0.92	1.65	[65]
Criterion	0.5 [d]	10 [d]	0.1 [c]	0.02	50 [c]	0.1	0.5 [c]	40 [d]	

Note: [a] mean value expressed as wet wt. [b] Mean values expressed as dry wt. [c] Chinese national criterion collection GB 2736-94: [59] [d] European Commission (2006). NA: not available at time of study.

Fish are known to accumulate Ni in various tissues when exposed to their environment [66]. It was found that Ni was present in the muscle of all the examined fish species. The maximum mean concentration of Ni was recorded in *P. fulvidraco* (0.16 µg/g ww), while the minimum Level of Ni was recorded in *H. molitrix* (0.02 µg/g ww). Similarly, 0.14 µg/g ww in *Cirrhinus reba* and 0.34 µg/g ww in *C. catla* have been detected in fish samples from river Chenab [12]. The mean concentrations of Ni detected in other fish species were: 0.14 µg/g (*C. idellus*), 0.12 µg/g (*B. bjoerkna*), 0.11 µg/g (*C. carassius*), and 0.13 µg/g (*M. piceus*) all in wet weight (ww). The results of our study ranging from (0.02 to 0.16 µg/g ww) were far lower than Ni values reported by [53] ranging from 2.92 µg/g to 5.77 µg/g dry w from Wadi Hanifah in Saudi Arabia (Table 4). Furthermore, the mean concentration of Ni in the studied fish species was statistically significant between *H. molitrix/C. idellus; H. molitrix/B. bjoerkna; H. molitrix/M. piceus; H. molitrix/C. carassius; H. molitrix/P. fulvidraco; B. bjoerkna/P. fulvidraco* and *C. carassius/P. fulvidraco* at ($p < 0.5$).

The International Agency for Research on Cancer (IARC) has determined that Co is a possible carcinogenic to humans upon acute and chronic exposure [66]. The mean concentration of Co in fish muscles ranges from (0.06 to 0.81 µg/g ww) in the examined fish species, with the highest mean Co concentration of 0.46 µg/g ww recorded in *H. molitrix* followed by *P. fulvidraco* (0.13 µg/g ww), *M. piceus* (0.12 µg/g ww), *B. bjoerkna* (0.09 µg/g ww), *C. carassius* (0.06 µg/g ww) and *C. idellus* (0.03 µg/g ww). Our findings were comparable to those concentrations reported by [12] in fish muscle ranging from (0.01 µg/g to 0.16 µg/g ww). However, the findings of our study were in many folds lower than those values reported by [63], where 0.01 to 3.2 (µg/g ww) were recorded for different fish species from river Chenab in Pakistan. The mean concentration of Co for *H. molitrix* was found to be statistically significant for all the studied fish species at $p \leq 0.05$.

3.4. Biological Accumulation Factor (BAF) of Selected Fish Species

The mean concentration of trace elements in water samples and that recorded in the fish muscle tissues were used to calculate the biological accumulation factor of the various fish species (Figure 2). The bioaccumulation factor of the various metals, including *Pb, Ni, Cd, Co, Zn, Cr, Mn,* and *Cu* for individual species were estimated and presented in Figure 2. The metals are ranked in the following according to the measured BAF values: *Cr* > *Mn* > *Pb* > *Cu* > *Co* > *Zn* > *Cd* > *Ni*. The BAF of different fish species with respect to metal accumulation could help predict which fish species is most suitable in monitoring metal contamination in the Yangtze River. Regarding the BAF of *Cr, H. molitrix* recorded the highest value of 2134.37, whereas *C. idellus* recorded the lowest BAF with a value of 152.94. *Mn* was found to be highly accumulated in *P. fulvidraco* species of the present study, and it is attributed to the high amounts of industrial and domestic wastes from the surroundings. Mn recorded a BAF range value of 1047.42 to 1984.92. This observation indicates that the pollution level of Cr in the Yangtze River could regularly be monitored using *H. molitrix* as the preferred bioindicator.

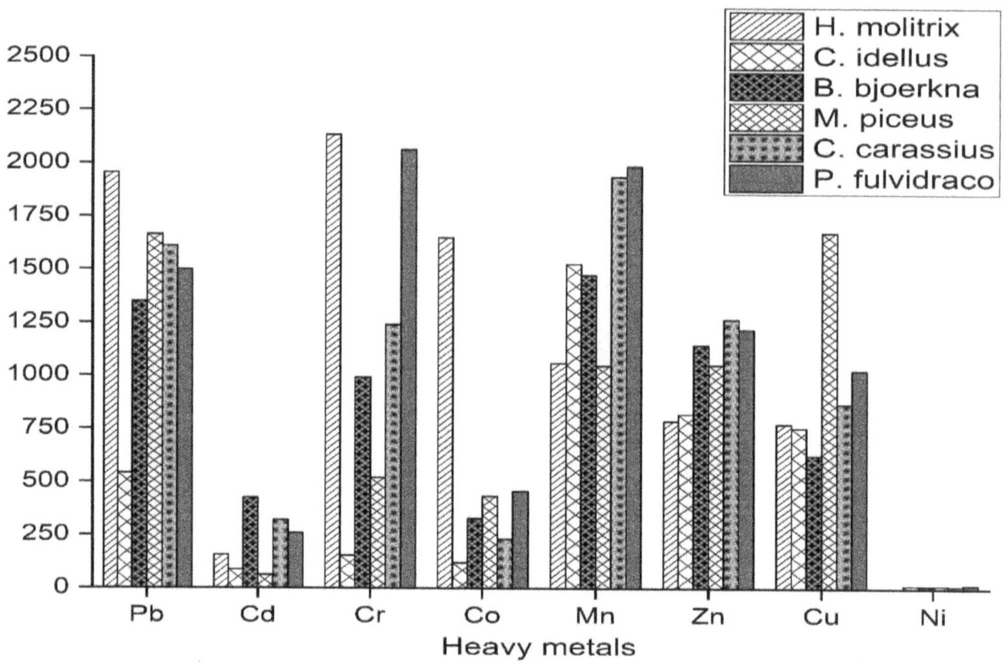

Figure 2. The BAF of heavy metals in selected fish species.

3.5. Non-Carcinogenic Risk Assessment of Heavy Metals in Fish Species

3.5.1. Estimated Daily Intake (EDI)

Trace metals in food could pose danger to consumers, and this effect is dependent on the metal concentration/dose ingested. Fish dietary exposure is a reliable tool for researching a population's diet in terms of nutrient intake, biologically active compounds, and pollutants, offering important information on the possible nutritional deficiencies and exposure to food contaminants [67]. Figure 3 indicate the average EDI values of trace metals through the consumption of the studied fish species.

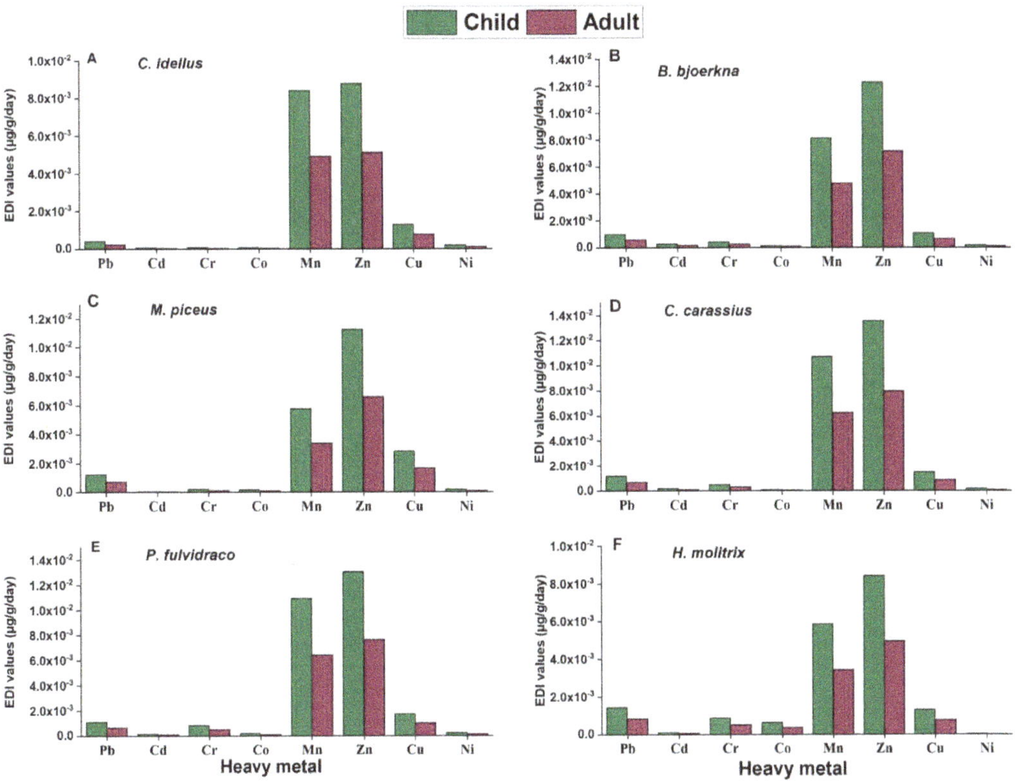

Figure 3. EDI (µg/g/day) of trace metals through the consumption of studied fish species.

The *EDI* values for this study were calculated and presented in Figure 4. As depicted in Figure 4, the calculated *EDI's* of *Pb*, *Ni*, *Cd*, *Co*, *Zn*, *Cr*, *Mn*, and *Cu* through the ingestion of selected fish species, from the Yangtze River, by both children and adults was subsequently compared with the oral reference dose (*RfD*) to estimate the potential health risk since the bodyweights of a child and adult are not the same.

Except for *Co*, *H. molitrix* for both child and adult, the average EDIs of metals in all examined fish species were less than the RfD for child and adult exposures (Figure 3), signifying that consumption of the analyzed metals in the examined fish species from the Yangtze River may not result in non-carcinogenic health implication to human. Dietary intakes of *Co* surpassed the *RfD* through consumption of *H. molitrix* by adults and children, which was mainly attributed to the fact that high concentrations of *Co* were accumulated in the fish species. Since *Co* can also elicit toxic effects such as liver and kidney damage and even death at a high level, it is necessary to avoid excessive consumption of *H. molitrix* in the Yangtze River to prevent the detrimental effects caused by *Co* accumulation.

The average *EDI* value of *Pb* was 0.00105 µg/g/day for a child and 0.00061 µg/g/day for an adult please, for which fish species. For a child, *M. piceus* recorded the highest *EDI* value of 0.00121 µg/g/day, whereas *C. idellus* recorded the least Pb value of 0.00039 µg/g/day. Furthermore, maximum *EDI* values for *Mn* and *Zn* were identified in *P. fulvidraco* (0.01094 µg/g/day) and *C. carassius* (0.01355 µg/g/day), respectively upon child's consumption of such fish species.

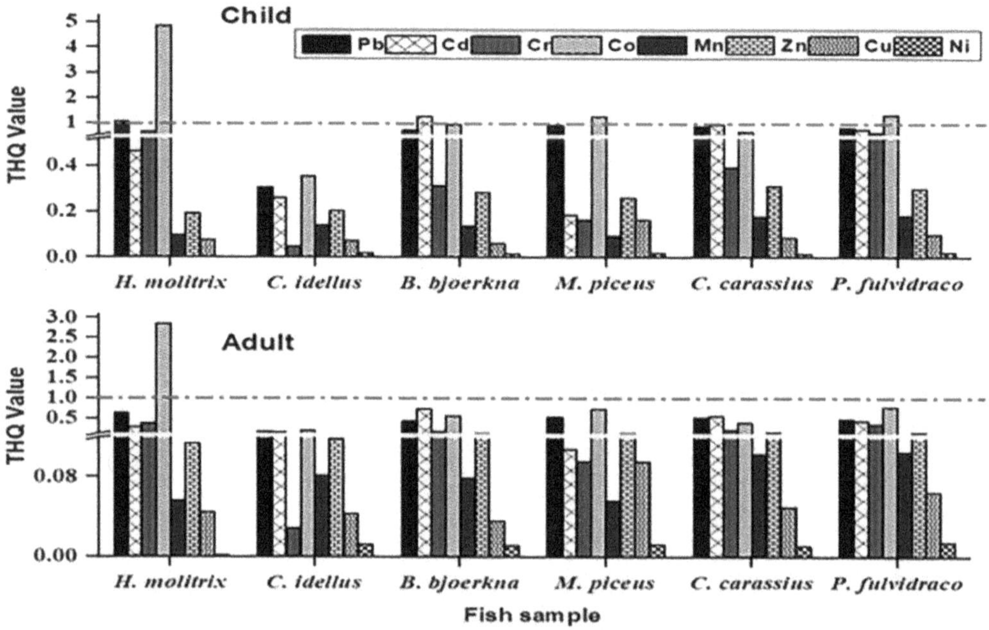

Figure 4. Target hazard quotient (THQ) of eight heavy metals in fish species from the Yangtze River for child and adult exposure.

3.5.2. The Target Hazard Quotient (THQ)

The *THQ* is a dimensionless quantity. It is a proportion of the heavy metal concentration in the food product to its *RfD*, weighted by consumption duration, frequency of exposure, intake quantity, and body weight. In the present study, *Pb*, *Cd*, and *Co* showed *THQ* values >1. The *THQ* values for all heavy metals in children are comparatively higher than in adults. Figure 4 shows that *Co* had the highest *THQ* value, and thus, it could pose a non-carcinogenic risk to the population for which children are more susceptible. The mean *THQ* value of each metal from the consumption of fish was generally less than 1 for both child and adult, which suggests that individuals would not significantly suffer adverse health risks from their intake of fish. While the *THQ* deals with individual metals, the *TTHQ* is the sum of all the *THQ's* calculated values for the individual metals. The *TTHQ* values for both children and adults with respect to the examined fish species exceeded 1 except for *C. idellus*, which recorded 0.826 for Adults. Moreover, children are at a relatively higher non-carcinogenic risk compared to Adults upon consuming the examined fish species from the Yangtze River with respect to the *TTHQ*. *H. molitrix* had the highest *TTHQ* value of 7.506 for children, whereas *P. fulvidraco* had the highest value of 2.479 for Adults. This result implies a potential non-carcinogenic risk of metals to consumers upon fish dietary from the Yangtze River.

3.5.3. Carcinogenic Risk Assessment of Heavy Metals in Fish Species from the Yangtze River

Target carcinogenic (*TR*) risks resulting from *Pb*, *Cr* and *Ni* intake have been determined as they may cause non-carcinogenic as well as carcinogenic effects depending upon exposure dose. The mean value of *TR* for *Pb*, *Cr*, and *Ni* for a child intake ranged between *C. carassius* (1.00×10^{-5}) and *P. fulvidraco* (9.32×10^{-6}), *M. piceus* (1.06×10^{-4}) and *C. idellus* (3.11×10^{-5}), and *C. carassius* (2.65×10^{-4}) and *H. molitrix* (4.37×10^{-5}), respectively. For adults, the mean concentration ranged between *C. idellus* (1.97×10^{-06}) and *H. molitrix* (7.11×10^{-6}); *B. bjoerkna* (1.18×10^{-4}) and *M. piceus* (6.18×10^{-5}), and *C. carassius* (1.55×10^{-4}) and *H. molitrix* (2.55×10^{-5}) for *Pb*, *Cr*, and *Ni*, respectively. Moreover, according to New York State Department of Health [NYSDOH (New York State Department of Health) [68,69], the *TR* categories are described as: if $TR \leq 10^{-6}$ = Low; 10^{-4} to 10^{-3} = moderate; 10^{-3} to 10^{-1} = high; $\geq 10^{-1}$ = very high.

This result indicates that *TR* values for *Pb* for the studied fish species were below 10^{-4} and can, therefore, be termed negligible as shown in Table 6. *Cr* and *Ni* showed low or moderate risk to the exposed population. Therefore, comparing the target cancer risk values with guideline values, it can be inferred that fish from the Zhenjiang section of the Yangtze River is free from carcinogenic effects, but we recommend further research should be conducted to ensure the local inhabitants' health.

Table 6. Lifetime carcinogenic risks ($1 \times 10^{-4} \times 10^{-6}$) of selected trace metals due to the consumption of sampled fish species collected from the Zhenjiang part of the Yangtze River, China.

Heavy Metals	CSF	Child (Bw = 32.7 kg)		Fish Species				Adult (Bw = 55.9 kg)		Fish Species			
		H. molitrix	*C. idellus*	*B. bjoerkna*	*M. piceus*	*C. carassius*	*P. fulvidraco*	*H. molitrix*	*C. idellus*	*B. bjoerkna*	*M. piceus*	*C. carassius*	*P. fulvidraco*
Pb	0.0085	1.22×10^{-5}	3.37×10^{-6}	8.39×10^{-6}	1.03×10^{-5}	1.00×10^{-5}	9.32×10^{-6}	7.11×10^{-6}	1.97×10^{-6}	4.91×10^{-6}	6.05×10^{-6}	5.86×10^{-6}	5.45×10^{-6}
Cr	0.5	4.34×10^{-4}	3.11×10^{-5}	2.02×10^{-4}	1.06×10^{-4}	2.52×10^{-4}	4.19×10^{-4}	2.54×10^{-4}	1.82×10^{-5}	1.18×10^{-4}	6.18×10^{-5}	1.48×10^{-4}	2.45×10^{-4}
Ni	1.7	4.37×10^{-5}	3.16×10^{-4}	2.85×10^{-4}	3.06×10^{-4}	2.65×10^{-4}	3.77×10^{-4}	2.55×10^{-5}	1.85×10^{-4}	1.67×10^{-4}	1.79×10^{-4}	1.55×10^{-4}	2.21×10^{-4}

Note: Bw: body weight.

3.6. *Multivariate Analysis of Heavy Metals with Fish Species*

To investigate the associations among eight heavy metals in the six examined fish species, PCA was conducted by reducing the dataset to several determining factors [70]. PCA has proven to be an effective tool for detecting sources of pollution in environmental media [71]. Four principal components, applied on a standardized dataset through Z-scale transformation, with eigenvalues greater than 1.0 (Kaiser Criterion), with a total contribution of 52.37% were extracted to identify the possible sources of metals in the Yangtze River of Zhenjiang city (Figure 5). The relationship between heavy metals could theoretically suggest their common origins, similar paths, and provide a basis to determine the homogeneity of the investigated sources of heavy metals [72]. The correlation matrix results depicted in Table 7 revealed that *Cr* and *Pb* had a positive significant correlation, indicating the possibility of their common contamination. Furthermore, the correlation between the metals *Ni*/*Mn* and *Zn*/*Cd* were all strongly positive; also, *Co* had a significantly positive correlation only with *Pb* and *Cr*.

Moreover, the results of the loading plot shown in Figure 5B revealed that *Cr*, *Pb*, and *Co* could have a similar sources of pollution in the river, whereas Cd, Mn, and Zn could also have similar sources of pollution in the Yangtze River. This observation suggests that human activities such as agricultural practices, industrial activities, and metal processing around the Yangtze River could contribute to the observed levels of *Co*, *Zn*, *Pb*, *Mn*, *Cu*, *Cr*, *Cd*, and *Ni* in the river. Furthermore, traffic sources from Zhenjiang city and speed boat activities for recreation on the Yangtze River may be the major source of *Pb*, *Cr*, and *Zn* pollution in the river. The high *Cd*, *Mn*, and *Zn* concentrations may be due to agricultural management practices such as fertilizer and pesticide application that is close to the river [73].

Ni concentration may be a result of the combustion of diesel and lubricant oil and brake abrasion, which also coincided with other studies in rapidly developing Asian countries and tertiary industries [74,75]. The PCA scores plot in Figure 5C helps to deduce the close association of heavy metals with the examined fish species in the river. Fish species *P. fulvidraco* and *C. carassius* were closely related, indicating a similar source of feeding with regard to the metal. The PCA biplot (Figure 5D) reveals that *Zn*, *Mn*, and *Cd* showed association with three examined fish species (*B. bjoerkna*, *P. fulvidraco*, and *C. carassius*) used in this study and confirms that these metals are available in the river for uptake by these three examined fish species. Ni was closely associated with *C. idellus*, *M. piceus*, and *P. fulvidraco*. Deduction made from these observations is the similarity in pollution sources of *Co*, *Pb*, and *Cr* in the river. From Figure 5D, *Co* is more closely related to *Pb* than *Cr*. *Pb* and *Ni* were in separate clusters, indicating their dissimilarity in sources of contamination. *Ni* pollution in the river is likely to be associated with farming activities within the reservoir's catchment, and further, provides a basis for predicting the predominant use of *Ni* fertilizers for agricultural purposes within the reservoir's catchment.

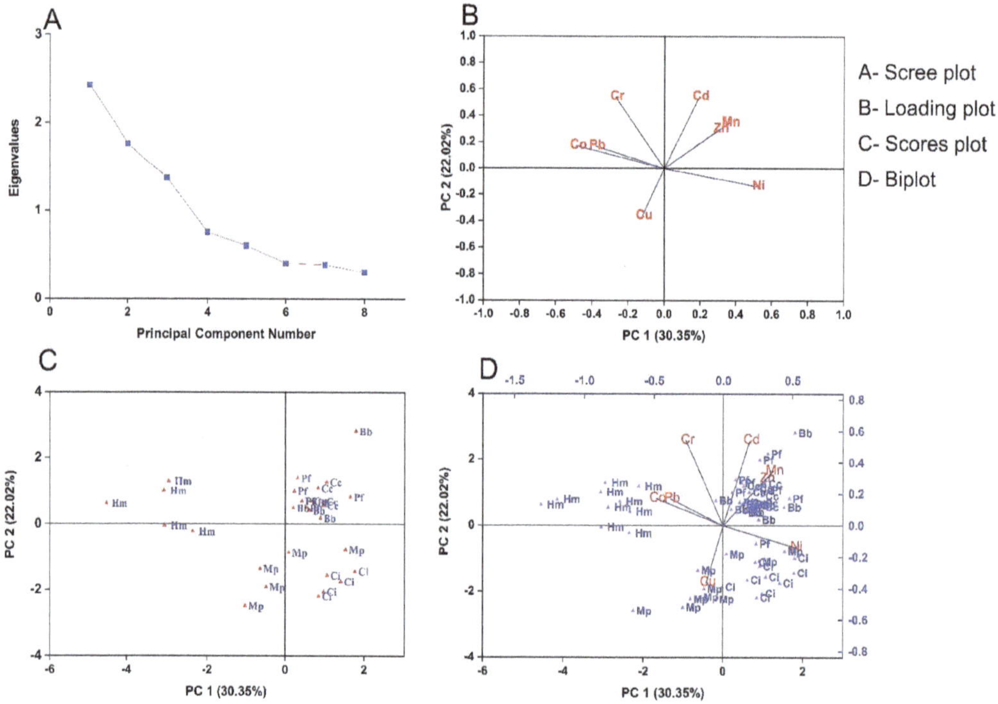

Figure 5. (**A**). Scree plot of PCA. (**B**). Loading plot of PCA. (**C**). Score plot of PCA. (**D**). PCA Biplot of heavy metals and examined fish species.

Table 7. The multiple correlation matrix analysis between heavy metal concentrations.

	Pb	Cd	Cr	Co	Mn	Zn	Cu	Ni
Pb	1	0.04029	0.39321	0.31756	−0.22217	0.02883	0.38017	−0.33448
Cd		1	0.22772	−0.09784	0.25401	0.36339	−0.3314	0.12108
Cr			1	0.36396	0.13868	0.04868	−0.13303	−0.35649
Co				1	−0.2851	−0.18781	−0.02192	−0.57229
Mn					1	0.3008	−0.24814	0.30827
Zn						1	0.05698	0.38415
Cu							1	0.12149
Ni								1

4. Conclusions

The present study assessed heavy metal concentrations in the muscle of six edible fish species from the Yangtze River, as one of the most important rivers in China. Multivariate statistical analysis including principal component analysis was employed to identify the pattern of heavy metal concentrations to a common pollution source and its relation to the distribution of fish species. Aquatic organisms take up contaminants from water and accumulate them in their tissues at higher levels than that of their surrounding environment.

The metal concentrations in the examined fish species were in the order of *Zn* > *Mn* > *Cu* > *Pb* > *Cr* > *Co* > *Cd* > *Ni*. Estimation of daily intake, the target hazard quotient, the total target hazard quotient, and the target cancer risk indicated that consuming the examined fish species, notably *H. molitrix* from the Yangtze River, could pose non-carcinogenic and carcinogenic effects on humans, especially children. Although heavy metal concentrations of the Yangtze River are largely below the prescribed level, the impact of its bioaccumulating may, in the future, be of considerable concern and thus management of the river should be enforced to minimize anthropogenic discharges in the river and constantly track it; otherwise, increased levels of heavy metal pollution will pose major problems to human health and aquatic lives. This study also recommends periodic monitoring and assessment of the river and its as to help mitigate potential health risks and protect aquatic life more effectively.

Author Contributions: Methodology, P.K. and E.G.; Software, M.H.; Validation, A.A.; Formal analysis, P.K.; Investigation, P.K., E.G. and M.H.; Resources, A.A.; Data curation, M.H.; Writing—original draft, P.K.; Writing—review & editing, S.S., E.G. and A.A. All authors have read and agreed to the published version of the manuscript.

Funding: This work was financially supported by the Zhenjiang Social Development Foundation Project of China (SH2020004, SH2020018), the 5TH phase "169 Project" training fund of Zhenjiang and Jiangsu Collaborative Innovation Center of Technology and Material of Water Treatment.

Data Availability Statement: The datasets generated during and/or analysed during the current study are available from the corresponding author on reasonable request.

Conflicts of Interest: The authors declare no conflict of interest.

References

1. Li, L.; Huang, J.; Almutairi, A.W.; Lan, X.; Zheng, L.; Lin, Y.; Chen, L.; Fu, N.; Lin, Z.; Abomohra, A.E.-F. Integrated approach for enhanced bio-oil recovery from disposed face masks through co-hydrothermal liquefaction with Spirulina platensis grown in wastewater. *Biomass-Convers. Biorefinery* **2021**, 1–12. [CrossRef] [PubMed]
2. Abomohra, A.E.-F.; El-Hefnawy, M.E.; Wang, Q.; Huang, J.; Li, L.; Tang, J.; Mohammed, S. Sequential bioethanol and biogas production coupled with heavy metal removal using dry seaweeds: Towards enhanced economic feasibility. *J. Clean. Prod.* **2021**, *316*, 128341. [CrossRef]
3. Rahman, M.M.; Islam, M.A.; Khan, M.B. Status of heavy metal pollution of water and fishes in Balu and Brahmaputra rivers. *Progress. Agric.* **2016**, *27*, 444–452. [CrossRef]
4. Lokhande, R.S.; Singare, P.U.; Pimple, D.S. Toxicity study of heavy metals pollutants in waste water effluent samples collected from Taloja Industrial Estate of Mumbai, India. *Resour. Environ.* **2011**, *1*, 13–19.
5. Cáceres-Saez, I.; Haro, D.; Blank, O.; Aguayo Lobo, A.; Dougnac, C.; Arredondo, C.; Cappozzo, H.L.; Guevara, S.R. High status of mercury and selenium in false killer whales (Pseudorca crassidens, Owen 1846) stranded on Southern South America: A possible toxicological concern? *Chemosphere* **2018**, *199*, 637–646. [CrossRef]
6. Watanabe, T.; Kiron, V.; Satoh, S. Trace minerals in fish nutrition. *Aquaculture* **1997**, *151*, 185–207. [CrossRef]
7. Kumar, B.; Mukherjee, D.; Kumar, S.; Mishra, M.; Prakash, D.; Singh, S.; Sharma, C. Bioaccumulation of heavy metals in muscle tissue of fishes from selected aquaculture ponds in east Kolkata wetlands. *Ann. Biol. Res.* **2011**, *2*, 125–134.
8. Hohle, T.H.; O'Brian, M.R. Manganese is required for oxidative metabolism in unstressed *Bradyrhizobium japonicum* cells. *Mol. Microbiol.* **2012**, *84*, 766–777. [CrossRef]
9. FAO. *The State of World Fisheries and Aquaculture 2018-Meeting the Sustainable Development Goals*; Food and Agriculture Organization of the United Nations: Rome, Italy, 2018.
10. Ali, S.S.; Shaaban, M.T.; Abomohra, A.E.-F.; El-Safity, K. Macroalgal activity against multiple drug resistant Aeromonas hydrophila: A novel treatment study towards enhancement of fish growth performance. *Microb. Pathog.* **2016**, *101*, 89–95. [CrossRef]
11. Gyimah, E.; Akoto, O.; Mensah, J.K.; Bortey-Sam, N. Bioaccumulation factors and multivariate analysis of heavy metals of three edible fish species from the Barekese reservoir in Kumasi, Ghana. *Environ. Monit. Assess.* **2018**, *190*, 553. [CrossRef]
12. Alamdar, A.; Eqani, S.A.M.A.S.; Hanif, N.; Ali, S.M.; Fasola, M.; Bokhari, H.; Katsoyiannis, I.A.; Shen, H.J.C. Human exposure to trace metals and arsenic via consumption of fish from river Chenab, Pakistan and associated health risks. *Chemosphere* **2017**, *168*, 1004–1012. [CrossRef]
13. Renieri, E.A.; Safenkova, I.V.; Alegakis, A.K.; Slutskaya, E.S.; Kokaraki, V.; Kentouri, M.; Dzantiev, B.B.; Tsatsakis, A.M. Cadmium, lead and mercury in muscle tissue of gilthead seabream and seabass: Risk evaluation for consumers. *Food Chem. Toxicol.* **2019**, *124*, 439–449. [CrossRef]

14. Ramos-Miras, J.J.; Sanchez-Muros, M.J.; Morote, E.; Torrijos, M.; Gil, C.; Zamani-Ahmadmahmoodi, R.; Rodríguez Martin, J.A. Potentially toxic elements in commonly consumed fish species from the western Mediterranean Sea (Almería Bay): Bioaccumulation in liver and muscle tissues in relation to biometric parameters. *Sci. Total Environ.* **2019**, *671*, 280–287. [CrossRef]
15. Copat, C.; Bella, F.; Castaing, M.; Fallico, R.; Sciacca, S.; Ferrante, M. Heavy metals concentrations in fish from Sicily (Mediterranean Sea) and evaluation of possible health risks to consumers. *Bull. Environ. Contam. Toxicol.* **2012**, *88*, 78–83. [CrossRef]
16. El-Moselhy, K.M.; Othman, A.; Abd El-Azem, H.; El-Metwally, M. Bioaccumulation of heavy metals in some tissues of fish in the Red Sea, Egypt. *Egypt. J. Basic Appl. Sci.* **2014**, *1*, 97–105. [CrossRef]
17. Murtala, B.A.; Abdul, W.O.; Akinyemi, A.A. Bioaccumulation of heavy metals in fish (Hydrocynus forskahlii, Hyperopisus bebe occidentalis and Clarias gariepinus) organs in downstream Ogun coastal water, Nigeria. *J. Agric. Sci.* **2012**, *4*, 51. [CrossRef]
18. Malakootian, M.; Tahergorabi, M.; Daneshpajooh, M.; Amirtaheri, K. Determination of Pb, Cd, Ni, and Zn concentrations in canned fish in Southern Iran. *Sacha J. Environ. Stud.* **2011**, *1*, 94–100.
19. Sun, J.; Pan, L.; Zhan, Y.; Lu, H.; Tsang, D.C.; Liu, W.; Wang, X.; Li, X.; Zhu, L. Contamination of phthalate esters, organochlorine pesticides and polybrominated diphenyl ethers in agricultural soils from the Yangtze River Delta of China. *Sci. Total. Environ.* **2016**, *544*, 670–676. [CrossRef]
20. Yi, Y.; Yang, Z.; Zhang, S. Ecological risk assessment of heavy metals in sediment and human health risk assessment of heavy metals in fishes in the middle and lower reaches of the Yangtze River basin. *Environ. Pollut.* **2011**, *159*, 2575–2585. [CrossRef]
21. Wang, H.; Wu, Q.; Hu, W.; Huang, B.; Dong, L.; Liu, G. Using multi-medium factors analysis to assess heavy metal health risks along the Yangtze River in Nanjing, Southeast China. *Environ. Pollut.* **2018**, *243*, 1047–1056. [CrossRef]
22. Xiong, X.; Qian, Z.; Mei, Z.; Wu, J.; Hao, Y.; Wang, K.; Wu, C.; Wang, D. Trace elements accumulation in the Yangtze finless porpoise (Neophocaena asiaeorientalis asiaeorientalis)—A threat to the endangered freshwater cetacean. *Sci. Total Environ.* **2019**, *686*, 797–804. [CrossRef] [PubMed]
23. Zhao, L.; Gong, D.; Zhao, W.; Lin, L.; Yang, W.; Guo, W.; Tang, X.; Li, Q. Spatial-temporal distribution characteristics and health risk assessment of heavy metals in surface water of the Three Gorges Reservoir, China. *Sci. Total Environ.* **2020**, *704*, 134883. [CrossRef] [PubMed]
24. Zhenjiang People's Government. Available online: http://szb.zhenjiang.gov.cn/zjgk/201905/t20190528_2142453.htm (accessed on 28 May 2019).
25. Li, L.; Jiang, M.; Liu, Y.; Shen, X. Heavy metals inter-annual variability and distribution in the Yangtze River estuary sediment, China. *Mar. Pollut. Bull.* **2019**, *141*, 514–520. [CrossRef] [PubMed]
26. Maurya, P.K.; Malik, D.S.; Yadav, K.K.; Kumar, A.; Kumar, S.; Kamyab, H. Bioaccumulation and potential sources of heavy metal contamination in fish species in River Ganga basin: Possible human health risks evaluation. *Toxicol. Rep.* **2019**, *6*, 472–481. [CrossRef] [PubMed]
27. USEPA Human Health Risk Assessment; Regional Screening Level (RSL) Fish ingestion Table. Washington, DC. March 2019. Available online: http://www.epa.gov/reg3hwmd/risk/human/pdf/NOV2012FISH.pdf (accessed on 28 May 2019).
28. Borgå, K. *Ecotoxicology: Bioaccumulation*; Elsevier: Amsterdam, The Netherlands, 2013.
29. Bhupander, K.; Mukherjee, D.P. Assessment of human health risk for arsenic, copper, nickel, mercury and zinc in fish collected from tropical wetlands in India. *Adv. Life Sci. Technol.* **2011**, *2*, 13–24.
30. Wang, X.; Sato, T.; Xing, B.; Tao, S. Health risks of heavy metals to the general public in Tianjin, China via consumption of vegetables and fish. *Sci. Total Environ.* **2005**, *350*, 28–37. [CrossRef]
31. Bo, S.O.N.G.; Mei, L.E.I.; Tongbin, C.H.E.N.; Zheng, Y.; Yunfeng, X.I.E.; Xiaoyan, L.I.; Ding, G.A.O. Assessing the health risk of heavy metals in vegetables to the general population in Beijing, China. *J. Environ. Sci.* **2009**, *21*, 1702–1709.
32. Li, Z.Y.; Ma, Z.W.; van der Kuijp, T.J.; Yuan, Z.W.; Huang, L. A review of soil heavy metal pollution from mines in China: Pollution and health risk assessment. *Sci. Total Environ.* **2014**, *468*, 843–853. [CrossRef]
33. Gutiérrez, A.J.; Rubio, C.; Moreno, I.M.; González, A.G.; Gonzalez-Weller, D.; Bencharki, N.; Hardisson, A.; Revert, C. Estimation of dietary intake and target hazard quotients for metals by consumption of wines from the Canary Islands. *Food Chem. Toxicol.* **2017**, *108*, 10–18. [CrossRef]
34. Proshad, R.; Rahman, M.M.; Kumar, T.; Islam, M.S.; Mursheed, N.; Howladar, R. Human health hazard implications of heavy metals in agricultural foodstuff grown around brick kiln areas in Bangladesh. *J. Environ. Sci. Comput. Sci. Eng. Technol.* **2017**, *6*, 138–156.
35. Kundu, G.K.; Alauddin, M.; Akter, M.S.; Khan, M.S.; Islam, M.M.; Mondal, G.; Islam, D.; Mohanta, L.C.; Huque, A. Metal contamination of commercial fish feed and quality aspects of farmed tilapia (*Oreochromis niloticus*) in Bangladesh. *Bioresearch Commun.-BRC* **2017**, *3*, 345–353.
36. Yi, Y.; Tang, C.; Yi, T.; Yang, Z.; Zhang, S. Health risk assessment of heavy metals in fish and accumulation patterns in food web in the upper Yangtze River, China. *Ecotoxicol. Environ. Saf.* **2017**, *145*, 295–302. [CrossRef] [PubMed]
37. USEPA. *USEPA Regional Screening Level (RSL) Summary Table: November 2011*; United States Environmental Protection Agency: Washington, DC, USA, 2011.
38. Du, Z.-Y.; Zhang, J.; Wang, C.; Li, L.; Man, Q.; Lundebye, A.-K.; Frøyland, L. Risk–benefit evaluation of fish from Chinese markets: Nutrients and contaminants in 24 fish species from five big cities and related assessment for human health. *Sci. Total. Environ.* **2012**, *416*, 187–199. [CrossRef] [PubMed]

39. GB2762. National Food Safety Standards Contaminant Limit in Food; 2017. Available online: https://food.chemlinked.com/database/view/803 (accessed on 14 March 2019).
40. Islam, M.S.; Ahmed, M.K.; Habibullah-Al-Mamun, M. Determination of heavy metals in fish and vegetables in Bangladesh and health implications. *Hum. Ecol. Risk Assess. Int. J.* **2015**, *21*, 986–1006. [CrossRef]
41. Chien, L.-C.; Hung, T.-C.; Choang, K.-Y.; Yeh, C.-Y.; Meng, P.-J.; Shieh, M.-J.; Han, B.-C. Daily intake of TBT, Cu, Zn, Cd and As for fishermen in Taiwan. *Sci. Total Environ.* **2002**, *285*, 177–185. [CrossRef]
42. Council, N.R. *Review of EPA's Integrated Risk Information System (IRIS) Process*; National Academies Press: Washington, DC, USA, 2014.
43. Madden, E.F. The role of combined metal interactions in metal carcinogenesis: A review. *Rev. Environ. Health* **2003**, *18*, 91–110. [CrossRef]
44. Castaldo, G.; Pillet, M.; Slootmaekers, B.; Bervoets, L.; Town, R.; Blust, R.; De Boeck, G. Investigating the effects of a sub-lethal metal mixture of Cu, Zn and Cd on bioaccumulation and ionoregulation in common carp, Cyprinus carpio. *Aquat. Toxicol.* **2020**, *218*, 105363. [CrossRef]
45. Chi, Q.-Q.; Zhu, G.-W.; Langdon, A. Bioaccumulation of heavy metals in fishes from Taihu Lake, China. *J. Environ. Sci.* **2007**, *19*, 1500–1504. [CrossRef]
46. Jiang, Q.; Lee, T.; Chen, K.; Wong, H.; Zheng, J.; Giesy, J.; Lo, K.; Yamashita, N.; Lam, P. Human health risk assessment of organochlorines associated with fish consumption in a coastal city in China. *Environ. Pollut.* **2005**, *136*, 155–165. [CrossRef]
47. Zaidi, J.; Pal, A. Chapter-6 Research in India on Aquatic Plants and Their Role in Remediation of Metal Pollution. *Recent Trends* **2019**, 87.
48. Sauliutė, G.; Svecevičius, G. Environmental Protection Engineering. Siesarties ir Vilnios upių ekotoksikologinės būklės įvertinimas pagal atlantinės lašišos (*Salmo salar* L.) jauniklių morfologinius rodiklius. *Moksl.–Liet. Ateitis/Sci.–Future Lith. Environ. Prot. Eng.* **2015**, *7*, 424–429.
49. Anene, A. Condition factor of four Cichlid species of a man-made lake in Imo State, Southeastern Nigeria. *Turk. J. Fish. Aquat. Sci.* **2005**, *5*, 43–47.
50. Shakir, H.A.; Qazi, J.I. Impact of industrial and municipal discharges on growth coefficient and condition factor of major carps from Lahore stretch of River Ravi. *J. Anim. Plant Sci.* **2013**, *23*, 167–173.
51. Lizama, M.D.L.A.P.; Ambrósio, A.M. Condition factor in nine species of fish of the Characidae family in the upper Paraná river floodplain, Brazil. *Braz. J. Biol.* **2002**, *62*, 113–124. [CrossRef]
52. Casado-Martinez, M.C.; Smith, B.D.; Luoma, S.N.; Rainbow, P.S. Bioaccumulation of arsenic from water and sediment by a deposit-feeding polychaete (*Arenicola marina*): A biodynamic modelling approach. *Aquat. Toxicol.* **2010**, *98*, 34–43. [CrossRef]
53. Mahboob, S.; Al-Balawi, H.F.A.; Al-Misned, F.; Al-Quraishy, S.; Ahmad, Z. Tissue metal distribution and risk assessment for important fish species from Saudi Arabia. *Bull. Environ. Contam. Toxicol.* **2014**, *92*, 61–66. [CrossRef]
54. Majnoni, F.; Rezaei, M.; Mansouri, B.; Hamidian, A.H. Metal concentrations in tissues of common carp, cyprinus carpio, and silver carp, hypophthalmichthys molitrix from the Zarivar Wetland in Western Iran. *Arch. Pol. Fish.* **2013**, *21*, 11–18. [CrossRef]
55. Hellberg, R.S.; DeWitt, C.A.M.; Morrissey, M.T. Risk-benefit analysis of seafood consumption: A review. *Compr. Rev. Food Sci. Food Saf.* **2012**, *11*, 490–517. [CrossRef]
56. Li, Y.; Wu, S.; Xiang, Y.; Liang, X. An investigation of outpatient children's blood lead level in Wuhan China. *PLoS ONE* **2014**, *9*, e95284. [CrossRef]
57. Zhang, Z.; He, L.; Li, J.; Wu, Z.-B. Analysis of heavy metals of muscle and intestine tissue in fish-in Banan section of Chongqing from three gorges reservoir, China. *Pol. J. Environ. Stud.* **2007**, *16*, 949.
58. Pham, N.T.T.; Pulkownik, A.; Buckney, R.T. Assessment of heavy metals in sediments and aquatic organisms in West Lake (Ho Tay), Hanoi, Vietnam. *Lakes Reserv. Res. Manag.* **2007**, *12*, 285–294. [CrossRef]
59. *GB 2736-94*; Healthy Standard for Fresh Water Fish. Chinese National Criterion Press: Beijing, China, 1994; Volume 32.
60. Porea, T.J.; Belmont, J.W.; Mahoney, D.H.J. Zinc-induced anemia and neutropenia in an adolescent. *J. Pediatr.* **2000**, *136*, 688–690. [CrossRef] [PubMed]
61. Plum, L.M.; Rink, L.; Haase, H. The essential toxin: Impact of zinc on human health. *Int. J. Environ. Res. Public Health* **2010**, *7*, 1342–1365. [CrossRef] [PubMed]
62. Tchounwou, P.B.; Yedjou, C.G.; Patlolla, A.K.; Sutton, D.J. Heavy metal toxicity and the environment. *Exp. Suppl.* **2012**, *101*, 133–164. [CrossRef] [PubMed]
63. Waheed, S.; Kamal, A.; Malik, R.N. Human health risk from organ-specific accumulation of toxic metals and response of antioxidants in edible fish species from Chenab River, Pakistan. *Environ. Sci. Pollut. Res.* **2014**, *21*, 4409–4417. [CrossRef]
64. Rajeshkumar, S.; Li, X. Bioaccumulation of heavy metals in fish species from the Meiliang Bay, Taihu Lake, China. *Toxicol. Rep.* **2018**, *5*, 288–295. [CrossRef]
65. Fang, T.; Lu, W.; Cui, K.; Li, J.; Yang, K.; Zhao, X.; Liang, Y.; Li, H. Distribution, bioaccumulation and trophic transfer of trace metals in the food web of Chaohu Lake, Anhui, China. *Chemosphere* **2019**, *218*, 1122–1130. [CrossRef]
66. ATSDR, U. *Toxicological Profile for Polybrominated Biphenyls and Polybrominated Diphenyl Ethers (PBBs and PBDEs)*; ATSDR: Atlanta, GA, USA, 2004.

67. Safiur Rahman, M.; Solaiman Hossain, M.; Ahmed, M.K.; Akther, S.; Jolly, Y.N.; Akhter, S.; Jamiul Kabir, M.; Choudhury, T.R. Assessment of heavy metals contamination in selected tropical marine fish species in Bangladesh and their impact on human health. *Environ. Nanotechnol. Monit. Manag.* **2019**, *11*, 100210. [CrossRef]
68. NYSDOH. *Hopewell Precision Area Contamination: Appendix C-NYS DOH. Procedure for Evaluating Potential Health Risks for Contaminants of Concern*; New York State Department of Health: New York, NY, USA, 2007.
69. Javed, M.; Usmani, N. Accumulation of heavy metals and human health risk assessment via the consumption of freshwater fish Mastacembelus armatus inhabiting, thermal power plant effluent loaded canal. *SpringerPlus* **2016**, *5*, 776. [CrossRef]
70. Guan, Q.; Wang, L.; Wang, F.; Pan, B.; Song, N.; Li, F.; Lu, M. Phosphorus in the catchment of high sediment load river: A case of the Yellow River, China. *Sci. Total Environ.* **2016**, *572*, 660–670. [CrossRef]
71. Singh, U.K.; Kumar, B. Pathways of heavy metals contamination and associated human health risk in Ajay River basin, India. *Chemosphere* **2017**, *174*, 183–199. [CrossRef]
72. Soliman, N.F.; El Zokm, G.M.; Okbah, M.A. Risk assessment and chemical fractionation of selected elements in surface sediments from Lake Qarun, Egypt using modified BCR technique. *Chemosphere* **2018**, *191*, 262–271. [CrossRef]
73. Cai, L.; Xu, Z.; Bao, P.; He, M.; Dou, L.; Chen, L.; Zhou, Y.; Zhu, Y.-G. Multivariate and geostatistical analyses of the spatial distribution and source of arsenic and heavy metals in the agricultural soils in Shunde, Southeast China. *J. Geochem. Explor.* **2015**, *148*, 189–195. [CrossRef]
74. Pulles, T.; van der Gon, H.D.; Appelman, W.; Verheul, M. Emission factors for heavy metals from diesel and petrol used in European vehicles. *Atmos. Environ.* **2012**, *61*, 641–651. [CrossRef]
75. Kumar, V.; Sharma, A.; Kumar, R.; Bhardwaj, R.; Kumar Thukral, A.; Rodrigo-Comino, J. Assessment of heavy-metal pollution in three different Indian water bodies by combination of multivariate analysis and water pollution indices. *Hum. Ecol. Risk Assess. Int. J.* **2018**, *26*, 1–16. [CrossRef]

MDPI AG
Grosspeteranlage 5
4052 Basel
Switzerland
Tel.: +41 61 683 77 34

Water Editorial Office
E-mail: water@mdpi.com
www.mdpi.com/journal/water

www.ingramcontent.com/pod-product-compliance
Lightning Source LLC
Chambersburg PA
CBHW040742250926
51850CB00007BB/1388
9783725848799